Chemistry at the Beginning of the Third Millennium

Springer-Verlag Berlin Heidelberg GmbH

Luigi Fabbrizzi, Antonio Poggi (Eds.)

Chemistry at the Beginning of the Third Millennium

Molecular Design, Supramolecules, Nanotechnology and Beyond

Proceedings of the German-Italian Meeting of Coimbra Group Universities Pavia, 7-10 October, 1999

With 239 Figures, 27 Tables and 23 Schemes

Springer

Editors:

Prof. Luigi Fabbrizzi
Prof. Antonio Poggi

Department of General Chemistry
University of Pavia
Via Taramelli 12
27100 Pavia, Italy

DOI 10.1007/978-3-662-04154-3

CIP-Data aplied for

Die Deutsche Bibliothek - CIP-Einheitsaufnahme
The chemistry at the beginning of the third millennium : molecular design, supramolecules, nanotechnology and beyond ; proceedings of the German Italian meeting of Coimbra Group Universities Pavia, 7 - 10 October, 1999 / Luigi Fabbrizzi ; Antonio Poggi (ed.). -
Berlin ; Heidelberg ; New York ; Barcelona ; Hong Kong ; London ; Milan ; Paris ; Singapore ; Tokyo : Springer, 2000

Originally published by Springer-Verlag Berlin Heidelberg New York in 2000.

MyCopy version of the original edition 2000

Typesetting: Camera-ready copy by authors
Cover-design: E. Kirchner, Heidelberg
Printed on acid-free paper SPIN: 10737968 2 / 3020 hu - 5 4 3 2 1 0
www.springer.com/mycopy

Preface

This Volume is based on the Lectures presented at the Meeting "Chemistry at the Beginning of the Third Millennium", which was held in Pavia, Italy, during the period 7-10 October, 1999. The Meeting involved the participation of scientists from German and Italian Universities of the 'Coimbra Group'. The 'Coimbra Group', which was founded in 1987, gathers the most ancient and prestigious European Universities, with the aim to promote initiatives in both research and teaching and to provide guidelines for the progress and development of the University system. German and Italian Universities within the Coimbra Group propose every year a theme for scientific discussion, which originates a Meeting to be held in a German or Italian University. The Meeting in Pavia was the fifth of the series and followed those of Bologna (1995), Jena (1996), Siena (1997), Heidelberg (1998). Each Meeting is centred on a topic from either humanistic or natural sciences and consists in a series of lectures presented by distinguished scientists from the six participanting Universities. For the Pavia Meeting, the Steering Committee chose Chemistry as the topic and gathered researchers with experience in almost all fields of chemistry. In particular, during the Meeting, lectures were presented on many up-to-date subjects of chemistry, including: materials science, superconductors, supramolecular chemistry, bioinorganic chemistry, fullerenes, liquid crystals, photoinduced electron transfer, etc. The different topics were covered by distinguished and renown researchers of the various fields.

The date of the Meeting, less than three months to the turn of the Millennium, and the variety and fascination of the discussed themes account for the title of both the Meeting and this book. One could argue on the aptness of the title, as Chemistry is a rather young field of investigation, whose lifetime is measured in centuries (two or a little more), and not in millennia. Moreover, chemists, probably due to the strongly experimental character of their work, are rather centred on the present and, being deeply interested in their current activity, are in general reluctant to look too far ahead. This attitude may be due to the fact that chemists know that the development of their research plans can suddenly change due to unexpected or serendipitous findings, which eventually lead to an outstanding discovery, and, on the other hand, they are afraid that, at any moment, an unpredictable practical obstacle can prevent the attainment of their ambitious designs. Thus, we agree that a less fashionable, but more credible title should better refer to the next decade, rather than to the next millennium. What is true is that chemistry, as an expression of the human tendency to manipulate matter, will continue to progress for decades, centuries and perhaps millennia, thanks to the enthusiast contributions of millions of researchers. The ultimate reason of this enthusiasm is that each researcher knows that his own work can be more or less significant, but in any case it contributes to its own extent to the progress of chemistry. This can be the message which, at the turn of the Millennium, is passed to the new generations of researchers in Chemistry. In this connection, it is to be mentioned that a number of young scientists (Ph. D. students and post-doctoral fellows) from the six participating Universities attended the Meeting, thus profiting from the opportunity to approach frontline topics of modern chemistry and to interact with leading scientists of the field in an informal and pleasant atmosphere. We could invite these young researchers to the Meeting thanks to the liberality of the Rector of the University of Pavia and, in particular, of *Fondazione CARIPLO*, whose support is warmly acknowledged.

Pavia, February 2000

Luigi Fabbrizzi
Antonio Poggi

Table of Contents

Bottom-up Approach to Nanotechnology: Molecular-Level Devices

Vincenzo Balzani, Paola Ceroni

Dipartimento di Chimica "G. Ciamician", Università di Bologna, via Selmi 2, I-40126 Bologna, Italy.
E-mail: vbalzani@ciam.unibo.it

Introduction

In the same way as combination of atoms leads to molecules, combination of molecular components leads to supramolecular species (supermolecules).[1-3] Because of the extraordinary progress made by synthetic methods, it is now possible to say that almost any desired supramolecular species can be prepared, provided that sufficient resources are applied to the problem. Several supramolecular species are very appealing from an aesthetic viewpoint.[4] More important, by assembling in an appropriate way suitably chosen molecular components it is now possible to design and construct supramolecular species capable of performing useful functions (*molecular-level devices*).[2,3,5,6]
To better understand what we mean for molecular-level devices, let us make a simple example.[5] In everyday life we make extensive use of *macroscopic devices*. A macroscopic device is an assembly of components designed to achieve a specific function. Each component of the device performs a simple act, while the entire device performs a more complex function, characteristic of the assembly.

For example (Figure 1a), the function performed by a hairdryer (production of hot wind) is the result of acts performed by a switch, a heater, and a fan, suitably connected by electric wires and assembled in an appropriate framework.

The concept of device can be extended to the molecular level.[2,5] A *molecular-level device* can be defined as an assembly of a discrete number of structurally organized and functionally integrated molecular components (that is, a *supramolecular* structure, Figure 1b) designed to achieve a specific function. Each molecular component performs a single act (e.g., absorption of a photon, transfer of an electron), while the entire supramolecular structure performs a more complex function (e.g., photoinduced charge-separation), which results from the ensemble of the acts performed by its molecular components.

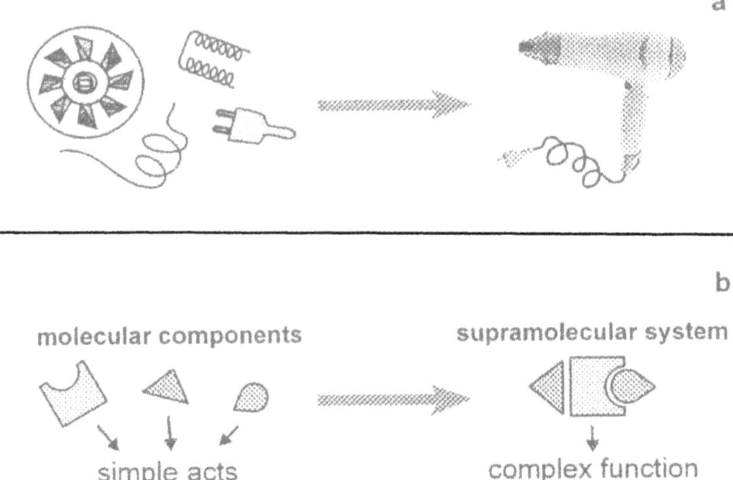

Fig. 1. Macroscopic (a) and molecular-level (b) devices.

The extension of the concept of device to the molecular level is of interest not only for basic research, but also for the growth of nanoscience and the development of nanotechnology.[7] The "top down" approach to nanostructures has intrinsic limitations, and it seems likely that, for further miniaturization, the "bottom up" approach will result more convenient (Figure 2).

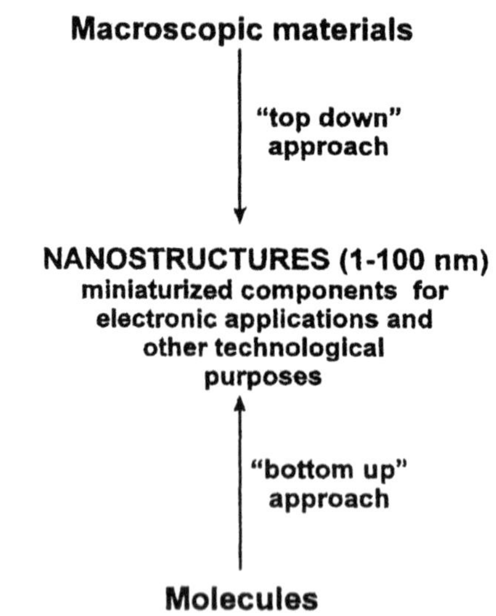

Fig. 2. "Top down" and "bottom up" approaches to nanostructures

Molecular-level devices operate via electronic and/or nuclear rearrangements, and like macroscopic devices they need energy to operate and signals to communicate with the operator. Energy can be supplied by photons, electrodic redox processes, or various kinds of chemical reactions. The most commonly used signals are the spectroscopic ones, but other techniques (in particular, electrochemistry) can prove very useful in several instances.

In the last few years several research groups have investigated the possibility to construct two kinds of molecular-level devices, namely (i) molecular-level electronic components (i.e., molecular-level systems that play the same functions played by macroscopic components in macroscopic electronic devices), and (ii) molecular-level machines (i.e., molecular-level systems that perform mechanical movements analogous to those observed in macroscopic machines). The development of molecular-level electronic components and molecular-level machines is of the greatest importance in the fields of information storage, display, and processing.[8-10] An extremely interesting goal is the construction of molecular-based (chemical) computers.

For space reasons, we will only illustrate a few of the systems investigated by our research group in collaboration with other laboratories. A substantial part of the work performed by other groups is described in refs. 11-26. In most cases, the molecular-level devices described in this paper are driven by light excitation and rely on the excited state properties of transition-metal complexes.

Wires

Light absorption by a molecule generates electronic energy. Since light is going to play a major role in signal generation, processing, and storage, an important function is represented by the possibility of transmitting electronic energy, over a more or less long distance, between molecular components in a supramolecular species. Another possibility to process light inputs is to take advantage of the increased oxidation and reduction power of an excited state to transfer an electron, again over a more or less long distance, between a photoexcited molecular component and another molecular component of a supramolecular species. It should also be noticed that photoinduced energy- and electron-transfer processes, at the molecular level, can be discussed on the basis of related theoretical treatments.[5]

Rod-like molecular-level wires capable of performing such functions can be obtained by assembling suitable molecular components in an appropriate sequence, organized in the dimensions of space, energy, and time [5] to achieve the desired sequence of energy- or electron-transfer. In collaboration with the groups of P. Belser and A. von Zelewsky (Fribourg, CH), F. Vögtle (Bonn, D), J.-P. Sauvage (Strasbourg, F), and F. Barigelletti (CNR, Bologna, I) we have investigated energy- and electron-transfer processes in a number of hetero-dinuclear Ru-Os complexes of a variety of polypyridine bridging ligands (Figure 3), with metal-to-metal distance up to 4.2 nm.[27,28]

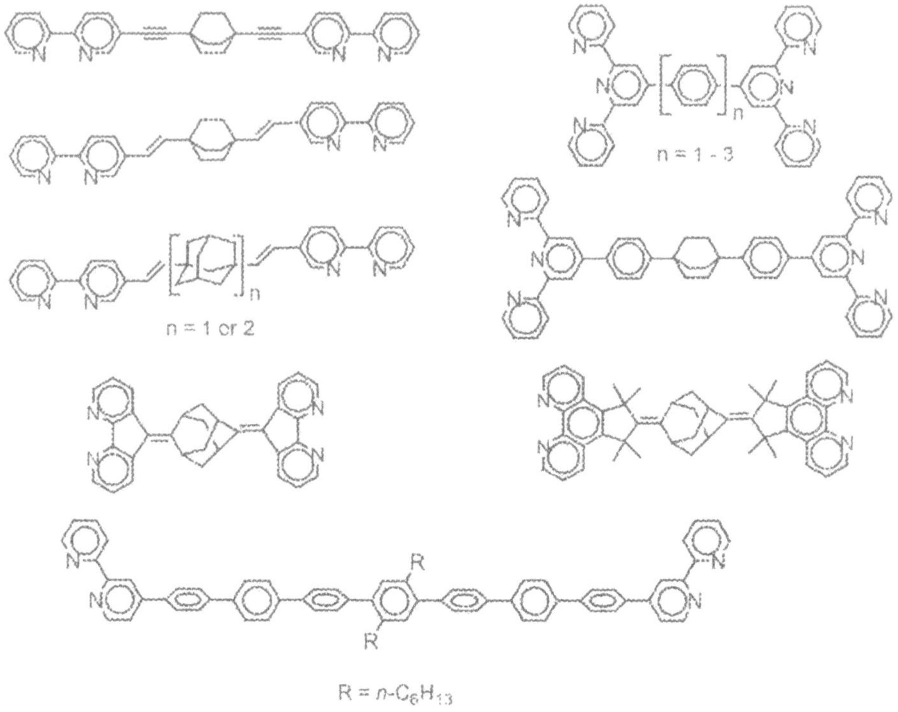

Fig. 3. (a) Schematic representation of rod-like bridging ligands.

In the Ru(II)-Os(II) species presented in Figure 4, the Ru-based luminescence is quenched and the Os-based luminescence is sensitized as a consequence of the energy transfer. The lifetimes of the excited states and the rate of the energy-transfer processes can be measured by pulsed laser excitation and single photon counting techniques. The observed energy-transfer rate constants are in the range 10^6-10^{12} s^{-1}, depending on the length and chemical nature of the spacer. In the complex shown in Figure 4a, the rate constant for energy transfer from the Ru-based to the Os-based moiety is 4.4×10^6 s^{-1}.[28d] When the bicyclo[2.2.2]octane module is removed (Figure 4b), the rate of energy transfer increases by more than four orders of magnitude [28c]. In the homogeneous series shown in Figure 4c, the energy-transfer rate constant decreases with increasing length of the oligophenylene spacer (in MeCN solution at 293 K, $k_{en}=6.7 \times 10^8$ s^{-1} for n=1, 1.0×10^7 s^{-1} for n=2, and 1.3×10^6 s^{-1} for n=3).[28h] In the Ru(II)-Os(III) species, which can be obtained by selective chemical oxidation of the Os(II) moiety, excitation of the Ru(II) component causes the transfer of an electron to the Os(III) one with rate constants 3.9×10^8 s^{-1} and 3.4×10^7 s^{-1} for n=1 and 2, respectively. The Ru(III)-Os(II) species then undergo back electron transfer to the ground state Ru(II)-Os(III) species with rate constants 1.2×10^7 s^{-1} and 2.7×10^5 s^{-1} for n=1 and 2, respectively.[28i]

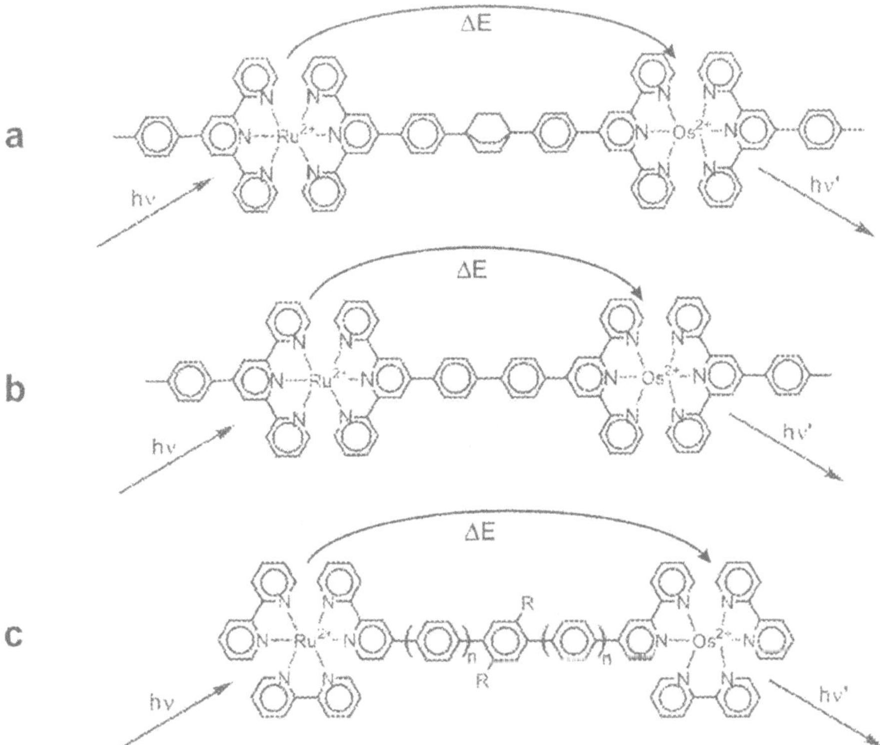

Fig. 4. Energy-transfer processes in rod-like dinuclear complexes.

Switches

In the above discussed systems the energy levels of the spacers lie at higher energies than the donor and acceptor energy levels involved in the energy (Figure 5a) or electron transfer and the process takes place by the so-called superexchange mechanism.[5] Spacers with energy levels in between those of the donor and acceptor may help energy (Figure 5b) or electron transfer. Spacers whose redox levels can be manipulated by an external stimulus can play the role of switches (Figure 5c).

An example of the situation schematized in Figure 5c is given by the supramolecular species shown in Figure 6, where a Ru(II) and an Os(II) components are separated by an anthracene-based spacer. In deaerated solution, excitation of the Ru(II) component is followed by fast and complete energy transfer to anthracene which, in its turn, transfers energy to the Os(II) component, as expected because of the relative energy of the lowest excited states in the three components. In aerated solution, however, the energy flow stops because the

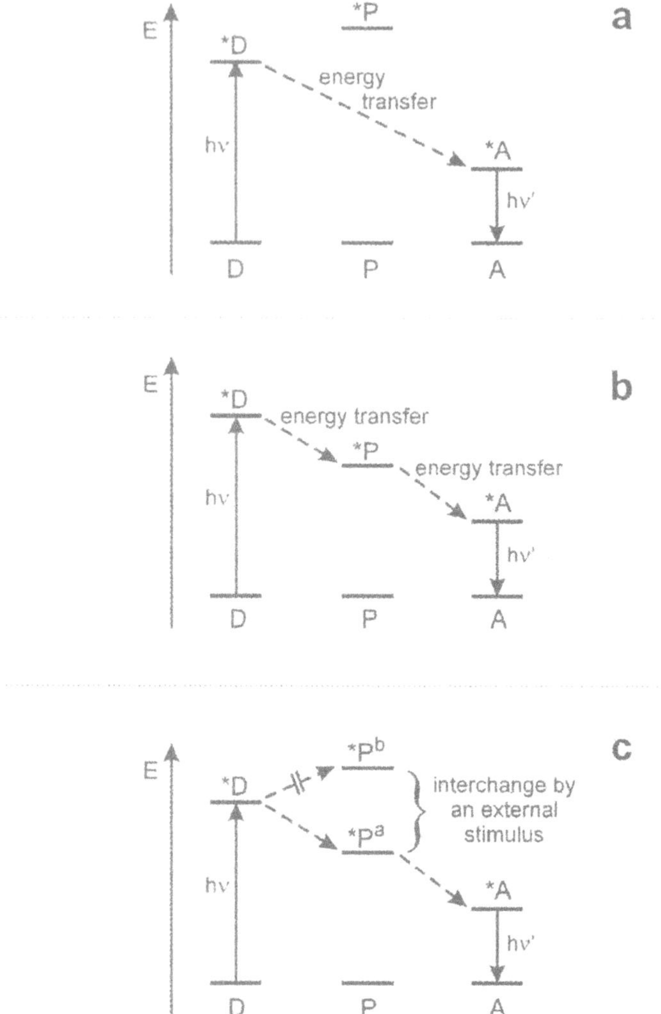

Fig. 5. Energy transfer processes: (a) through a spacer by a superexchange mechanism; (b) involving an energy level of the spacer (hopping mechanism). (c) On/off switching of energy transfer by manipulation of the lowest energy level of the spacer by means of external inputs.

anthracene component reacts with singlet oxygen produced by the oxygen quenching of the excited state of the Os(II) component, giving rise to an anthracene endoperoxide and other anthracene derivatives whose lowest excited state lies higher in energy than the Ru-based excited state.[29] Unfortunately this system does not show a fully reversible behavior.

Other supramolecular systems where energy or electron transfer can be switched on/off have been described [30] and the idea of an electrochemically controlled three-pole supramolecular switch has recently been discussed.[31]

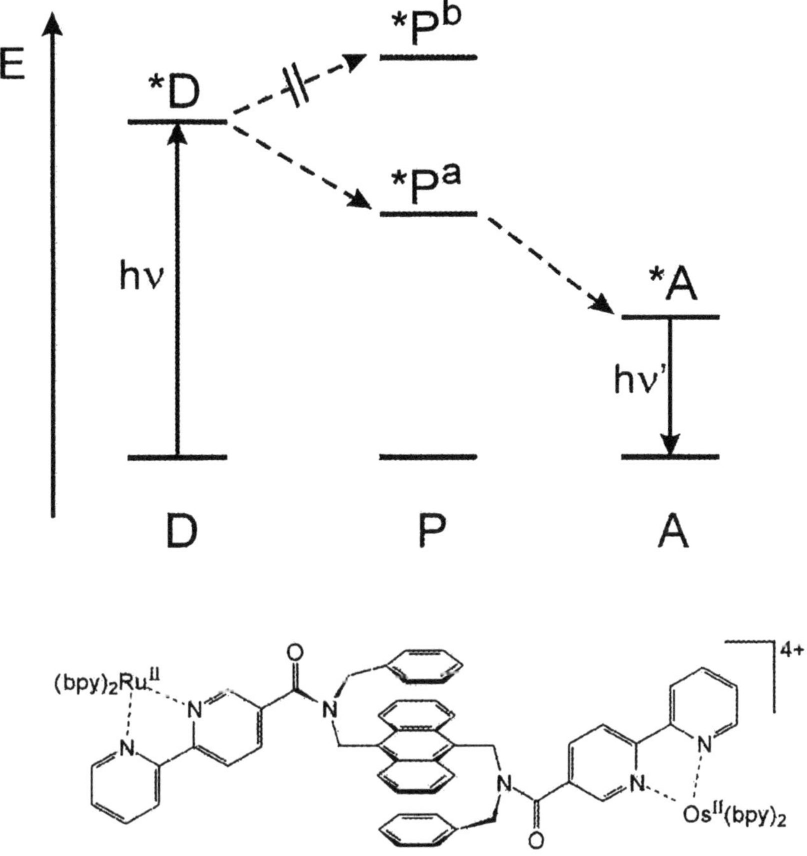

Fig. 6. Energy transfer from a Ru(II) to an Os(II) component linked by an anthracene-type spacer. $*P^a$ is the lowest triplet excited state of anthracene. Oxidation of anthracene to anthracene endoperoxide moves the lowest energy level of the spacer to $*P^b$. For more details, see ref 29.

Antennas

In collaboration with the groups of S. Campagna and S. Serroni (University of Messina) and G. Denti (University of Pisa) we have synthesized a number of tree-like (dendritic) multicenter transition-metal complexes based on Ru and Os as metals, 2,3-bis(2-pyridyl)pyrazine) (2,3-dpp) and 2,5-bis(2-pyridyl)pyrazine) (2,5-dpp) as bridging ligands, and 2,2'-bipyridine (bpy) and 2,2'-biquinoline (biq) as terminal ligands. The largest compounds so far prepared contain 22 metal atoms, 21 bridging ligands (2,3-dpp), and 24 terminal ligands (bpy). They comprise1090 atoms, with a molecular weight of 10890 daltons (for the docosanuclear Ru complex), and an estimated size of 5 nm. Since the properties of the modular components are known and different modules can be located in the desired positions of the dendrimer array, synthetic control of the various properties can be obtained. It is therefore possible to construct arrays where the electronic energy migration pattern can be predetermined, so as to channel the energy created by light absorption on the various components towards a selected module (antenna effect). Since this work has been extensively reviewed,[32] it will not be further discussed here. We would only like to recall that such complexes have recently been shown to exhibit outstanding electrochemical properties.

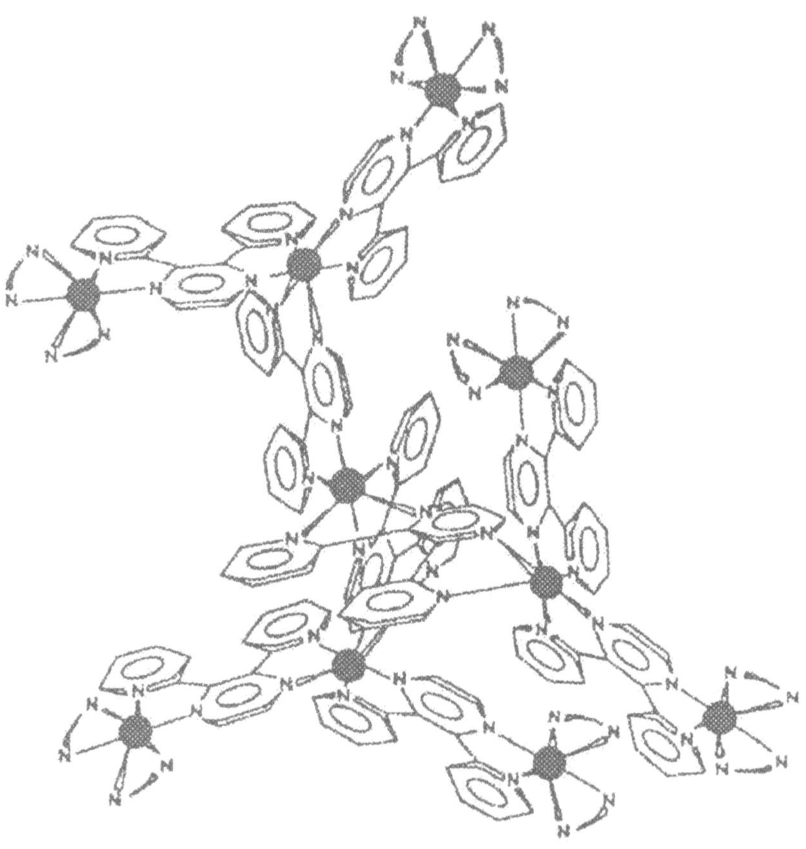

Fig. 7. Schematic representation of a Ru^{II} decanuclear complex, $[Ru\{(\mu\text{-}2,3\text{-dpp})Ru(\mu\text{-}2,3\text{-dpp})Ru(bpy)_2]_2\}_3](PF_6)_{20}$, containing 2,3-bis(2-pyridyl)pyrazine as bridging ligands and 2,2'-bipyridine as terminal ligands.

For example, in liquid SO_2 the decanuclear $[Ru\{(\mu\text{-}2,3\text{-dpp})Ru(\mu\text{-}2,3\text{-dpp})Ru(bpy)_2]_2\}_3](PF_6)_{20}$ complex (Figure 7) undergoes as many as ten metal based oxidation processes involving, as the potential becomes more positive, three sets of electrochemically equivalent Ru ions: (i) the six outer-shell metal centers, (ii) the central one and (iii) the three intermediate ones, respectively.[33] On the other hand, in dimethylformamide the hexanucler $\{[bpy_2Ru(\mu\text{-}2,3\text{-}dpp)]2Ru(\mu\text{-}2,3\text{-}dpp)Ru[(\mu\text{-}2,3\text{-}dpp)Ru(bpy)_2]_2]\}(PF_6)_{12}$ complex exhibits a ligand centered redox series comprising up to twentysix reversible reduction processes.[34]

Plug/socket and extension systems

A macroscopic plug/socket system is characterized by two features: (i) possibility to connect/disconnect the two components in a reversible way; (ii) electron flow from the socket to the plug when the two components are connected. In collaboration with the group of L. Mandolini (Rome, I), we have recently designed and constructed supramolecular systems that may be considered as molecular-level plug/socket devices (see, e.g., Figure 8): "plug in/plug out" is reversibly controlled by acid/base reactions, and photoinduced flow of electronic energy takes place in the "plug in" state, as shown by the quenching of the binaphthyl–type fluorescence which is accompanied by the sensitization of the fluorescence of the anthracenyl unit of the ammonium ion.[35]

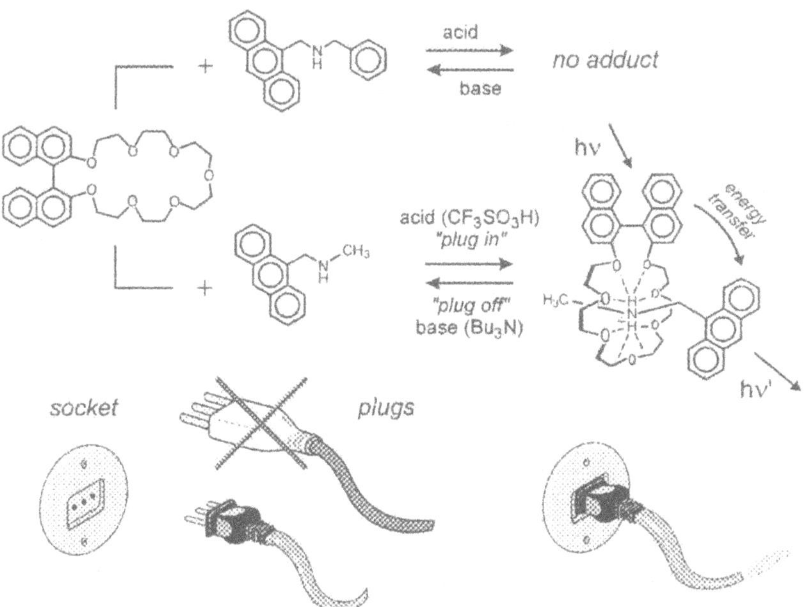

Fig. 8. A molecular-level plug/socket system.

The plug/socket molecular-level concept can be straightforwardly extended to the construction of molecular-scale extensions [36] and to the design of systems where (a) light excitation induces an electron flow instead of an energy flow, and (b) the plug in/plug out function is stereoselective (the enantiomeric recognition of chiral ammonium ions by chiral crown ethers is well known).[37]

Logic gates

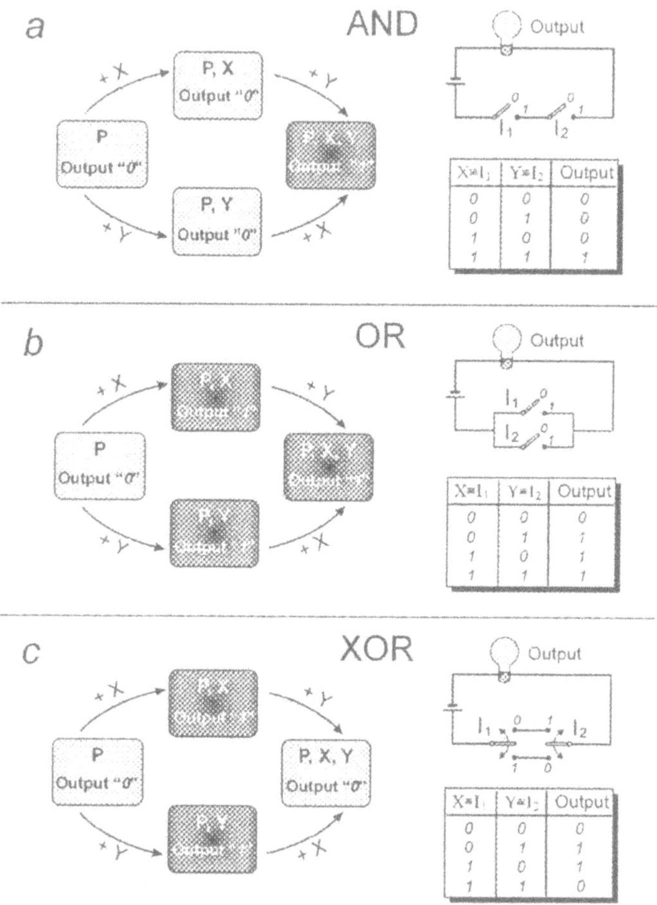

Fig. 9. Schematic representation of a chemical system (P) which performs the AND (a), OR (b), and XOR (c) logic operations under the action of two chemical inputs (X and Y). The truth tables of such operations are also shown, along with their representations based on electric circuit schemes.

Computers are based on semiconductor logic gates, which perform binary arithmetic and logical operations. Logic gates are switches whose output state (*0* or *1*) depends on the input conditions (*0* or *1*). Several examples of supramolecular systems capable to behave as logic gates have recently been reported, particularly by the group of A. P. de Silva.[38] In general, they are constituted by a fluorophore connected with one or more receptors; the excited state behavior of the fluorophore depends on whether or not the receptors are interacting with external species.

YES and NOT single-input gates are the simplest logic devices. Molecular systems that can perform such simple logic operations are very common and fluorescence is a particularly useful signal to monitor such operations. A molecule that fluoresces (output) only in acidic media (i.e., in the presence of a proton input) acts as a YES gate. Conversely, a fluorescent molecule whose emission (output) disappears under the action of an input of protons can be regarded as a NOT logic device. In order to perform more complex logic operations, however, carefully designed multi-component chemical systems are needed.

Figure 9 shows schematically the changes which have to occur in a chemical system in order to perform the AND, OR, and XOR (eXclusive OR) logic operations under the action of two chemical inputs (X and Y). For illustration purposes, the equivalent (from a logic viewpoint) electric circuits are also shown. It should be pointed out, however, that the comparison between the characteristics of logic gates as different as chemical systems and electric circuits is only formal and should be made with great care. Under some aspects, such a comparison could also be inappropriate or misleading.

Since examples of supramolecular systems capable of performing as AND, OR, and XOR logic gates have recently been reviewed and discussed,[6,17,37,39] this topic will not be further dealt with in this paper.

Memories

Bistable molecular or supramolecular species presenting two forms whose interconversion can be modulated by an external stimulus can be used as molecular-level memories.[2,3,5,13] Typical bistable species are the so-called photochromic compounds, molecules that can be interconverted between two forms (**X** and **Y**) exhibiting different colors.[40] Most photochromic compounds change their color by photoexcitation and revert more or less slowly to their initial state when kept in the dark (Figure 10a). Compounds exhibiting this behavior are useless for information storage since the written information is spontaneously erased (back converted) after a relatively short time. Other photochromic compounds do not return to the initial state thermally, but can undergo reversible photoisomerization (Figure 10b). Such compounds can be used for optoelectronic devices. However they present a severe problem. The light used for reading the written data causes the back-conversion of the sampled molecules and therefore the gradual loss of information. A general approach to avoid destructive reading is to

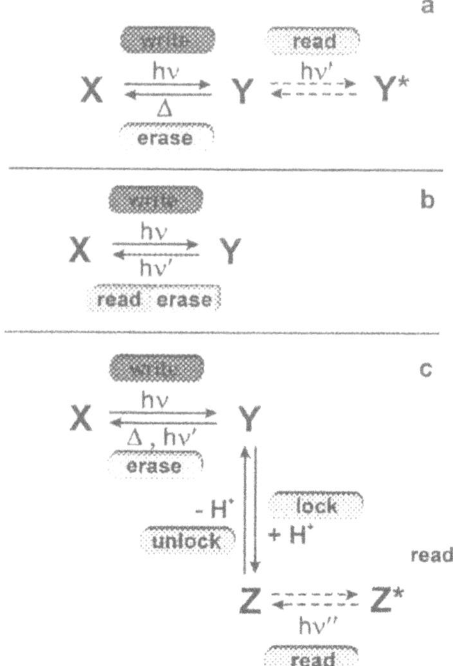

Fig. 10. Schematic representation of the behavior of three types of photochromic systems. (a) The photochemical reaction of the form **X** is thermally reverted in the dark. (b) The photochemical reaction of the form **X** can be reverted only by light excitation of the form **Y**. (c) The form **Y**, which goes back to **X** under light excitation, can be transformed by a second stimulus (e.g., an acid-base reaction) into another form **Z** which is stable toward light excitation and, when necessary, can be reconverted to **Y**. For more details, see text.

combine two reversible processes that can be addressed by means of two different stimuli (dual-mode systems).[41] In such systems (Figure 10c), light is used to convert **X** to **Y** (*write*), then a second stimulus (e.g., a proton as in Figure 10c) is employed to transform **Y** (which would be reconverted back to **X** by a direct photon reading process) into **Z**, another stable state of the system (*lock*), that can be optically detected without being destroyed (*read*). Through this process, the change caused by the writing photon is safeguarded. When the written information has to be erased, **Z** is reconverted back to **Y** (*unlock*; e.g., by addition of base in the example of Figure 10c) and **Y** is then reconverted back to **X** (*erase*). The second stimulus can also be another photon, an electron, heat or even something subtler such as formation of a hydrogen bond. A *write-lock-read-unlock-erase* cycle can constitute the basis for an optical memory system with multiple storage and nondestructive readout capacity.

In collaboration with the group of F. Pina (Universidade Nova de Lisboa, P) we have investigated several photochromic systems, based on synthetic flavylium salts in aqueous solution, which can undergo such a *write-lock-read-unlock-erase* cycle.[42] The behavior of the system has been characterized under a variety of experimental conditions, including continuous and pulsed excitation. Other interesting aspects of the chemistry of photochromic flavylium salts have been recently reviewed.[43]

Light-fueled piston-cylinder machines

In collaboration with the group of J.F. Stoddart (UCLA, USA), we have investigated the possibility of controlling the dethreading/rethreading of the wire and ring components of a pseudorotaxane, a movement that reminds that of a piston in a cylinder. We have shown that this is indeed possible by using chemical energy, electrical energy, or light.[44]

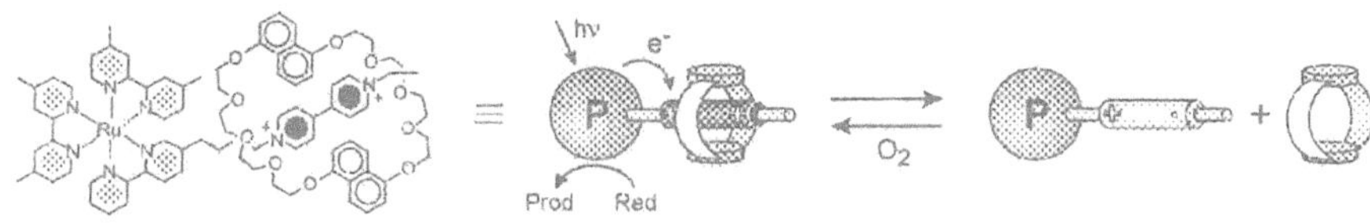

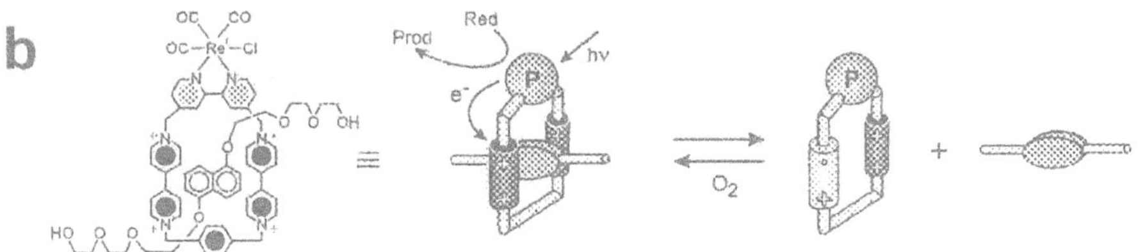

Fig. 11. Light-driven dethreading of pseudorotaxanes incorporating a photosensitiser (a) as a stopper in the wire-type component, and (b) in the ring component.

Two light-fueled piston-cylinder machines designed and constructed by our research groups are schematically shown in Figure 11. An electron-transfer photosensitizer (namely, a suitable metal complex) has been incorporated in the wire (Figure 11a) [45] or in the macrocyclic ring (Figure 11b).[46] As a result of a CT interactions, the wire and ring component self-assemble in aqueous solution to give the pseudorotaxane structure. Light excitation of the photosensitizer causes reduction of a bipyridinium-type unit of the wire or of the ring, respectively, while the back electron-transfer reaction is prevented by the presence of a sacrificial reductant like triethanolamine. As a consequence, the donor-acceptor interaction responsible for the self-assembly is destroyed and the wire dethreads from the ring. If oxygen is allowed to enter the solution, oxidation of the reduced bipyridinium-type unit restores the donor-acceptor interaction and causes rethreading. The threading, dethreading, and rethreading processes can be followed by absorption and fluorescence spectroscopy.

Molecular abacus

In collaboration with the group of J.F. Stoddart we have designed and constructed several types of molecular-level abacus.[44] They are based on rotaxanes, interlocked structures made of dumbbell-shaped and ring components which exhibit some kind of interaction originating from complementary chemical properties. In rotaxanes containing two different recognition sites in the dumbbell-shaped component, it is possible to switch the position of the ring between the two "stations" by an external stimulus.[44,47-49] A system of this kind [47] is shown in Figure 12. It is made of a DB24C8 crown ether ring and a dumbbell-shaped

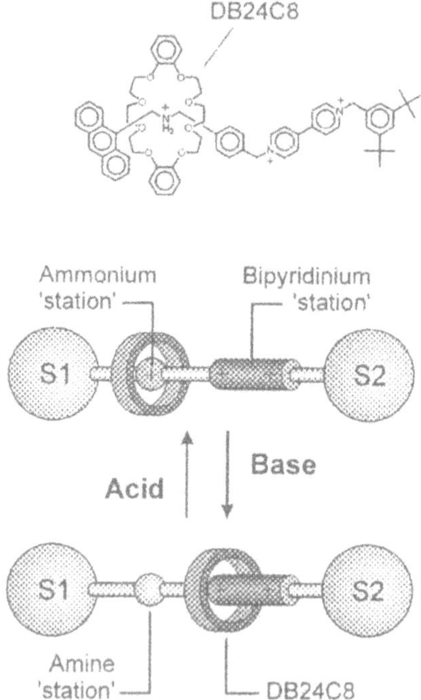

Fig. 12. Acid/base driven movement of the DB24C8 crown ether along a dumbbell-shaped component containing two different recognition sites.

component containing a secondary dialkylammonium center and a 4,4'-bipyridinium unit. An anthracene moiety is used as a stopper because its absorption, luminescence, and redox properties are useful to monitor the state of the system. The DB24C8 ring exhibits 100% selectivity for the ammonium recognition site because of formation of strong hydrogen bonds and therefore the [2]rotaxane exists as only one of the two possible translational isomers, as evidenced by X-ray crystallography. Deprotonation of the ammonium center, however, causes displacement of the ring component to the bipyridinium unit, a configuration that allows a moderately strong charge-transfer interaction between the ring and the dumbbell components. Reprotonation of the amine dumbbell site leads back the crown ring on the ammonium center. Such a switching process can be investigated by ^1H NMR spectroscopy and by electrochemical and photophysical measurements. It is fully reversible and 100% efficient.

Photochemically-induced molecular motions in a different kind of rotaxane have been investigated in collaboration with the group of J.-P. Sauvage.[50]

Rotation of a ring in catenanes

Catenanes are supramolecular species made of interlocked rings. When one of the two rings contains two non-equivalent units, structural changes caused by rotation of one ring with respect to the other can be evidenced . We have investigated the problem of inducing ring rotation in catenanes by means of chemical, electrochemical and photochemical stimulations. Much work has been performed in collaboration with the group of J.F. Stoddart on catenanes featuring donor-acceptor interactions.[44,51-53] The study that will be briefly mentioned here is the photoinduced motion in a disymmetrical copper [2]catenane carried out in collaboration with the group of J.P. Sauvage.[54] The investigated system is schematically represented in Figure 13.[55]

The starting Cu(I) complex $Cu(I)N_4$ is a stable, tetracoordinated species where two phenanthroline ligands surround the metal ion. Upon oxidation of the metal ion, the coordination environment is no longer appropriate since Cu(II) prefers pentacoordination. Therefore, the $Cu(II)N_4$ species produced by oxidation rearranges to the $Cu(II)N_5$ species where the metal ion is surrounded by a bidentate phenanthroline and a tridendate terpyridyl ligand. At this stage, reduction of the metal ion leads to a $Cu(I)N_5$ species that spontaneously rearranges to the stable $Cu(I)N_4$ configuration. We have shown that the oxidation process that causes the first step of the motion can be induced photochemically. Excitation of the $Cu(I)N_4$ species by visible light leads to a relatively long-lived excited state, $^*Cu(I)N_4$, which in the presence of a *p*-nitrobenzylbromide, *p*-$NO_2C_6H_4CH_2Br$, undergoes an excited state oxidation process with formation of $Cu(II)N_5$.

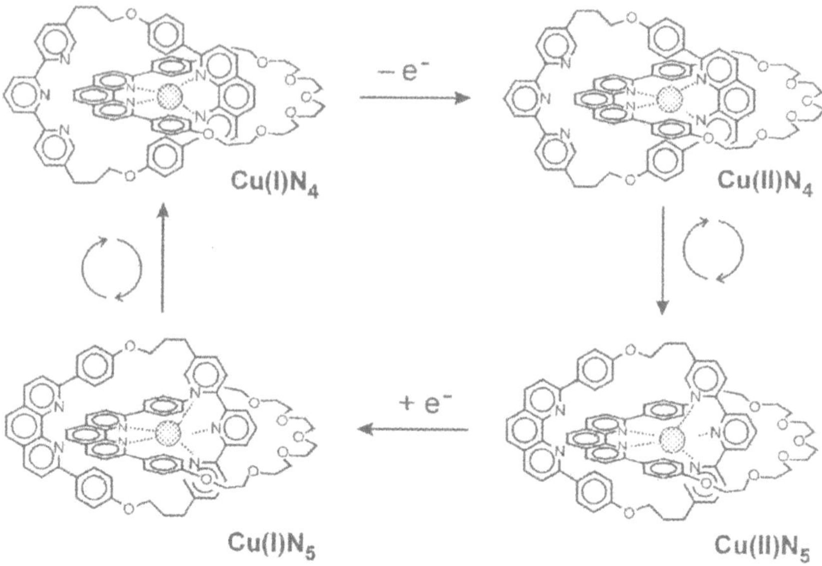

Fig. 13. Molecular square scheme illustrating the response of the catenate Cu(I)N4 to oxidation and successive reduction. The red and green circles stand respectively for Cu(I) and Cu(II). Each complex is distinguished through its number of coordinated nitrogen atoms (CuN_x, x=4 or 5). Oxidation generates a metastable Cu(II)N4 complex that rearranges to adopt the Cu(II)N5 coordination mode which best fits the new oxidation state of the metal center. Back reduction leads to a metastable Cu(I)N5 complex that rearranges to the stable Cu(I)N4 configuration.

Photoswitchable dendritic box

In collaboration with the group of F. Vögtle, we are investigating the properties of functional dendrimers.[56] One aim is that of constructing photoswitchable dendritic host which might be useful in practical application (e.g., drug delivery). With this purpose, we have studied the photochemical behavior of dendrimers bearing up to 32 photoisomerizable azobenzene groups.[57,58] Azobenzene derivatives have in fact been used to construct photoswitchable devices for many years [11] since photoisomerization of azobenzene is one of the clearest and best known photoreactions.[40] We have found that all-E azobenzene dendrimers can be reversibly switched to the Z form by light excitation (Figure 14).[57]

In the presence of eosin Y, the E→Z and Z→E photoreaction are sensitized by eosin and the eosin fluorescence is quenched. Quantitative examination of the obtained results showed that eosin is hosted by the dendrimers and that the Z form is a more efficient host than the E form.[58]

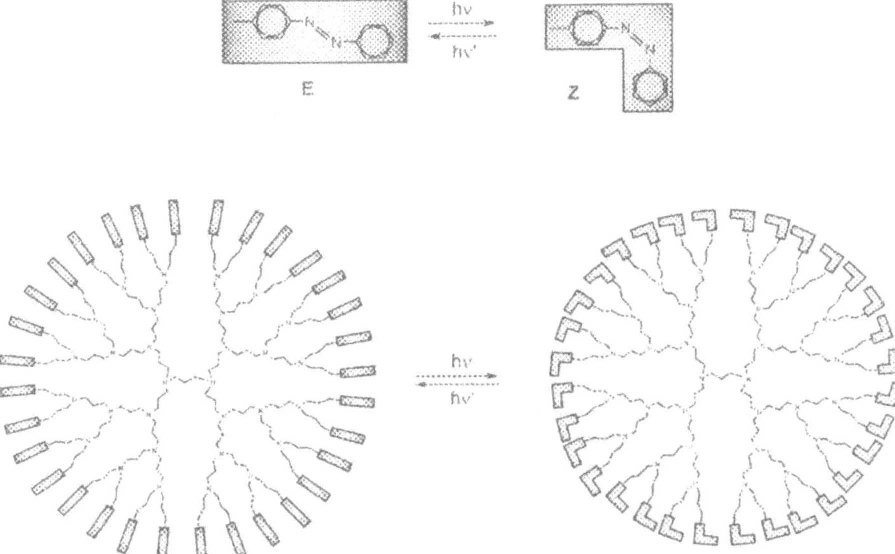

Fig. 14. Photoisomerization of azobenzene and of the fourth generation dendrimers bearing 32 photoisomerizable azobenzene groups in the periphery.

Conclusions

We would like to point out that the construction of the described systems is the result of (i) deep knowledge of the properties of the molecular components, (ii) accurate design of their assembly into a supramolecular structure, and (iii) integration of the acts performed by the various molecular components into more complex functions performed by the supramolecular species.

We are well aware that the systems described in this paper relate to investigations performed in the solution phase where incoherence remains a major impediment to designing and realizing devices that work. Of course, much progress has still to be made to arrive at the construction of molecular-based computers or other useful molecular-based devices. Nevertheless, we believe that the bottom up approach to nanotechnology is more promising. Last but not least, we would also like to emphasize that looking at supramolecular chemistry from the viewpoint of functions with reference to devices of the macroscopic world is a very interesting exercise that helps the development of chemistry by introducing new concepts.

Acknowledgements

This work has been supported by the Università di Bologna (Funds for selected research topics) and by the EC TMR Contract FMRX-CT96-0031.

References and Notes

[1] According to most authors (see, e.g., ref. 2) the term *supramolecular* should only be applied with reference to species obtained by association of molecular components via non-covalent bonds. Such a limitation, however, does not seem useful in the fields of photochemistry and electrochemistry and, more generally, for the discussion of the concept of molecular-level device. With the terms supramolecular species and supermolecules we refer here to systems made by a small number of weakly interacting molecular components held together by any kind of bonding. For the distinction between supramolecular species and large molecules, see ref. 3.

[2] J.-M. Lehn, Supramolecular Chemistry, VCH, Weinheim, 1995.

[3] (a) V. Balzani, *Tetrahedron*, **1992**, *48*, 10443. (b) V. Balzani, F. Scandola *Comprehensive Supramolecular Chemistry* (Eds.: J. L. Atwood, J. E. D. Davies, D. D. MacNicol, F. Vögtle), Pergamon, Oxford, UK, 1996, Vol. 10, p. 1.

[4] For a recent example, see: A. Soi, A. Hirsh, *New J. Chem.* **1998**, 1337.

[5] V. Balzani, V. Scandola, *Supramolecular Photochemistry*, Horwood, New York, USA, 1991.

[6] V. Balzani, A. Credi, M. Venturi, *Supramolecular Science: Where It Is and Where It Is Going* (Eds.: R. Ungaro, E. Dalcanale), Kluwer, Dordrecht, The Netherlands, 1999, p. 1.

[7] (a) K. E. Drexler, *Nanosystems: Molecular Machinery, Manufacturing, and Computation*, Wiley, New York, 1992. (b) W. Göpel, Ch. Ziegler, (Eds.) *Nanostructures Based on Molecular Materials*, VCH, Weinheim, 1992. (c) A.J. Bard, *Integrated Chemical Systems*, Wiley, New York, 1994.

[8] J. Jortner, M. Ratner (Eds.) *Molecular Electronics*, Wiley–VCH, Weinheim, Germany, 1998.

[9] C.P. Collier, E.W. Wong, M. Belohradsky, F.M. Raymo, J.F. Stoddart, P.J. Kuekes, R.S. Williams, J.R. Heath, *Science*, **1999**, *285*, 391.

[10] D. Rouvray, *Chem. Br.* **1998**, *34*(2), 26–29.

[11] S. Shinkai, O. Manabe, *Top. Curr. Chem.* **1984**, *121*, 67.

[12] G.L. Closs, J.R. Miller, *Science*, **1988**, *240*, 440.

[13] B. L. Feringa, W. F. Jager, B. de Lange, *Tetrahedron* **1993**, *49*, 8267.

[14] L. Fabbrizzi, A. Poggi, *Chem. Soc. Rev.* **1995**, *24*, 197.

[15] M. Verdaguer, *Science*, **1996**, *272*, 698.

[16] A. Harriman, R. Ziessel, *Chem. Commun.* **1996**, 1707–1716.

[17] A. P. de Silva, H. Q. N. Gunaratne, T. Gunnlaugsson, A. J. M. Huxley, C. P. McCoy, J. T. Rademacher, T. E. Rice, *Chem. Rev.* **1997**, *97*, 1515.

[18] D. Astruc, *Acc. Chem. Res.*, **1997**, *30*, 383.

[19] J. M.Seminario, J.M. Tour, *Molecular Electronics-Science and Technology* (Eds.: A. Aviran, M.A. Ratner, New York Academy of Science, New York, 1998, p. 69.

[20] J.-P. Sauvage, *Acc. Chem. Res.*, **1998**, *31*, 611.

[21] P.L. Boulas, M. Gomez–Kaifer, L. Echegoyen, *Angew. Chem. Int. Ed.*, **1998**, *37*, 216.

[22] G. Steinberg–Yfrach,J.-L. Rigaud, E.N. Durantini, A.L. Moore, D. Gust, T.A. Moore, *Nature*, **1998**, *392*, 479.

[23] W.B. Davis, W.A. Svec, M.A. Ratner, M.R. Wasielewski, *Nature*, **1998**, *396*, 60.

[24] F. Li, S.Y. Yang, Y. Ciringh, J. Seth, C.H. Martin III, D.L. Singh, D. Kim, R.R. Birge, D.F. Bocian, D. Holten, J.S. Lindsey, *J. Am. Chem. Soc.* **1998**, *120*, 10001.

[25] A. Niemz, V. Rotello, *Acc. Chem. Res.*, **1999**, *32*, 405.

[26] A. Kaifer, *Acc. Chem. Res.*, **1999**, *32*, 62.

[27] For reviews, see: (a) V. Balzani, A. Juris, M. Venturi, S. Campagna, S Serroni, *Chem. Rev.*,**1996**, *96*, 759. (b) L. De Cola, P. Belser, *Coord. Chem. Rev.*, **1998**, *177*, 301.

[28] For representative papers, see: (a) L. De Cola, V. Balzani, F. Barigelletti, L. Flamigni, P. Belser, A. von Zelewsky, M. Frank, F. Vögtle, *Inorg. Chem.*, **1993**, *32*, 5228. (b) F. Vögtle, M. Frank, M. Nieger, P. Belser, A. von Zelewsky, V. Balzani, F. Barigelletti, L. De Cola, L. Flamigni, *Angew. Chem. Int. Ed. Engl.*, **1993**, *32*, 1643. (c) F. Barigelletti, L. Flamigni, V. Balzani, J.-P. Collin, J.-P. Sauvage, A. Sour, E. C. Constable, A.M.W. Cargill Thompson, *J. Am. Chem. Soc.*, **1994**, *116*, 7692. (d) F. Barigelletti, L. Flamigni, V. Balzani, J.-P. Collin, J.-P. Sauvage, A. Sour, *New. J. Chem.*, **1995**, *19*, 793. (e) L. De Cola, V. Balzani, F. Barigelletti, L. Flamigni, P. Belser, S. Bernhard, *Rec. Trav. Chim. Pays Bas*, **1995**, *114*, 534. (f) M. Frank, M. Nieger, F. Vögtle, P. Belser, L. De Cola, V. Balzani, F. Barigelletti, L. Flamigni, *Inorg. Chim. Acta*, **1996**, *242*, 281. (g) V. Balzani, F. Barigelletti, P. Belser, S. Bernhard, L. De Cola, L. Flamigni, *J. Phys. Chem.*, **1996**, *100*, 9635. (h) B. Schlicke, P. Belser, L. De Cola, E. Sabbioni, V. Balzani, *J. Am. Chem. Soc.* **1999**, *121*, 4207. (i) L.De Cola, V. Balzani, unpublished results.

[29] (a) P. Belser, R. Dux, M. Baak, L. De Cola, V. Balzani, *Angew. Chem. Int. Ed. Engl.*, **34**, 595 (1995); (b) L. De Cola, V. Balzani, P. Belser, R. Dux, M. Baak, *Supramol. Chem.*, **5**, 297 (1995).

[30] (a) J. Walz, K. Ulrich, H. Port, H.C. Wolf, J. Wonner, F. Effenberger, *Chem. Phys. Lett.*, **213**, 321 (1993); (b) S.L. Gilat, S.H. Kawai, J.-M. Lehn, *J. Chem. Soc., Chem. Commun.*, 1993, 1439. (c) R.W. Wagner, J.S. Lindsey, J. Seth, V. Palianappan, D.F. Bocian, *J. Am. Chem. Soc.* **1996**, *118*, 3996.

[31] P.R. Ashton, V. Balzani, J. Becher, A. Credi, M.C.T. Fyfe, G. Mattersteig, S. Menzer, M. Nielsen, F.M. Raymo, J.F. Stoddart, M. Venturi, D.J. Williams, *J. Am. Chem. Soc.***1999**, *121*, 3951.

[32] (a) V. Balzani, S. Campagna, G. Denti, A. Juris, S. Serroni, M. Venturi, *Accounts Chem. Res.*, **1998**, *31*, 26. (b) M. Venturi, S. Serroni, A. Juris, S. Campagna, V. Balzani, *Topics Curr. Chem.*, **1998**, *197*, 193. (c) V. Balzani, A.Juris, M. Pink, M. Venturi, S. Campagna, S. Serroni, *Conjugated polymers, oligomers, and dendrimers: from polyacetylene to DNA* (Ed.: J.-L. Bredas), DeBoeck-Université Publisher, Bruxelles, 1999, p. 291.

[33] P. Ceroni, F. Paolucci, C. Paradisi, A. Juris, S. Roffia, S. Serroni, S. Campagna, A. J. Bard, *J. Am. Chem. Soc.* **1998**, *120*, 5480.

[34] M. Marcaccio, F. Paolucci, C. Paradisi, S.Roffia, C. Fontanesi, L.J. Yellowlees, S. Serroni, S. Campagna, G. Denti, V. Balzani, *J. Am. Chem. Soc.*, in press.

[35] E. Ishow, A. Credi, V Balzani, F Spadola, L Mandolini, *Chem. Eur. J.*, **1999**, *5*, 984.

[36] Work in progress in collaboration with the group of J.F. Stoddart.

[37] (a) E. B. Kyba, K. Koga, L. R. Sousa, M. G. Siegel, D. J. Cram, *J. Am. Chem. Soc.* **1973**, *95*, 2692. (b) X. X. Zhang, J. S. Bradshow, R. M. Izatt, *Chem. Rev.* **1997**, *97*, 3313.

[38] (a) A. P. de Silva, H. Q. N. Gunaratne, C. P. McCoy, *Nature* **1993**, *364*, 42. (b) A.P. de Silva, C.P. McCoy, *Chem. Ind.*, **1994**, December 19, 992. (c) A. P. de Silva, I.M. Dixon, H. Q. N. Gunaratne, C. P. McCoy, *J. Am. Chem. Soc*,. **1997**, *119*, 7891. (d) A. P. de Silva, H. Q. N. Gunaratne, T. Gunnlaugsson, P.R.S. MAxwell, T. E. Rice, *J. Am. Chem. Soc*,. **1999**, *121*, 1393.

[39] (a) A. Credi, V. Balzani, S. J. Langford, J. F. Stoddart, *J. Am. Chem. Soc.* **1997**, *119*, 2679. (b) F. Pina, A. Roque, M.J. Melo, M. Maestri, L. Belladelli, V. Balzani, *Chem. Eur. J.* **1998**, *4*, 1184.

[40] H.Dürr, H.Bouas-Laurent (Eds.) *Photochromism - Molecules and Systems*, Elsevier: Amsterdam, **1990**.

[41] (a) Kawai, S.H.; Gilat, S.L.; Posinet, R.; Lehn, J.-M., *Chem. Eur. J.*, **1995**, *1*, 285. (b) Uchida, K.; Irie, M., *J. Am. Chem. Soc.* **1993**, *115* , 6442.

[42] (a) F. Pina, M. J. Melo, M. Maestri, R. Ballardini, V. Balzani, *J. Am. Chem. Soc.*, **119**, 5556 (1997). (b) F. Pina, M.J. Melo, P. Passaniti, M. Maestri, N. Camaioni, V. Balzani, *Eur. J. Org. Chem.*, in press. (c) A. Roque, F. Pina, S. Alves, R. Ballardini, M. Maestri, V. Balzani, *J. Mater. Chem.*, in press.

[43] V. Balzani, M. Maestri, F. Pina, *Chem. Commun.*, **1999**, 107.

[44] For a review on "Molecular machines", see: V. Balzani, M. Gomez-Lopez, J.F. Stoddart, *Acc. Chem. Res.*, **1998**, *31*, 405.

[46] P.R. Ashton, R. Ballardini, V. Balzani, E.C. Constable, A. Credi, O. Kocian, S.J. Langford, J.A. Preece, L. Prodi, E.R. Schofield, N. Spencer, J.F. Stoddart, S. Wenger, *Chem. Eur. J.*, **1998**, *4*, 2411.

[47] P. R. Ashton, V. Balzani, O. Kocian, L. Prodi, N. Spencer, J.F. Stoddart, *J. Am. Chem. Soc.*,**1998**, *120*, 11190.

[48] P.R. Ashton, R. Ballardini, V. Balzani, I. Baxter, A. Credi, M.C.T. Fyfe, M.T. Gandolfi, M. Gomez-Lopez, M. V. Martínez-Díaz, A. Piersanti, N. Spencer, J.F. Stoddart, M. Venturi, A.J.P. White, D.J. Williams, *J. Am. Chem. Soc.*, **1998**, *120*, 11932.

[49] P.R. Ashton, R. Ballardini, V. Balzani, M.C.T. Fyfe, M.T. Gandolfi, M.V. Martínez-Díaz, M. Morosini, C. Schiavo, K. Shibata, J.F. Stoddart, A.J.P. White, D.J. Williams:, *Chem. Eur. J.*, **1998**, *4*, 2332.

[50] A light-driven molecular abacus is currently under investigation in our laboratories.

[51] N. Armaroli, V. Balzani, J.-P. Collin, P. Gavina, J.-P. Sauvage, B. Ventura, *J. Am. Chem. Soc.*, **1998**, *121*, 4397.

[52] P.R. Ashton, R. Ballardini, V. Balzani, A. Credi, M.T. Gandolfi, D.J.-F. Marquis, S. Menzer, L. Pérez-Garcia, L. Prodi, J..F. Stoddart, M. Venturi, A.J.P. White, and D.J. Williams, *J. Am. Chem. Soc.*, **1995**, *117*, 11171.

[53] M. Asakawa, P. Ashton, V. Balzani, A. Credi, C. Hamers, G. Mattersteig, M. Montalti, A.N. Shipway, N. Spencer, J. F. Stoddart, M.S. Tolley, M. Venturi, A. J. P. White and D. J. Williams, *Angew. Chem. Int. Ed. Engl.*, **1998**, *37*, 333.

[54] V. Balzani, A. Credi, G. Mattersteig, O.A. Mattews, F.M. Raymo, J.F. Stoddart, M. Venturi, A.J.P. White, D.J. Williams, submitted for publication.

[55] For a review of the molecular machines investigated by Sauvage and co-workers, see ref.**20**.

[56] A. Livoreil, J.-P. Sauvage, N. Armaroli, V. Balzani, L. Flamigni, B. Ventura, *J. Am. Chem. Soc.*, **1997**, *119*, 12114.

[57] For some recent papers, see: (a) M. Plevoets, F. Vögtle, L. De Cola, V. Balzani, *New. J. Chem.*, **23,** 63-69 (1999). (b) F. Vögtle, M. Plevoets, M. Nieger, G.C. Azzellini, A. Credi, L. De Cola, V. De Marchis, M. Venturi, V. Balzani, *J. Am. Chem. Soc.*,**1999,**

121, 6290-98. (c) F. Vögtle, S. Gestermann, C. Kauffmann, P. Ceroni, V. Vicinelli, L. De Cola, V. Balzani, *J. Am. Chem. Soc.*, in press.
[58] A. Archut, F. Vögtle, L. De Cola, G.C. Azzellini, V. Balzani, P.S. Ramanujam, R.H. Berg, *Chem. Eur. J.*, **1998**, *4*, 699-706.
[59] A. Archut, G. C. Azzellini, V. Balzani, L. De Cola, F. Vögtle, *J. Am. Chem. Soc.*, **1998**, *120*, 12187.

From Theory and Organometallic Model Chemistry to Catalysis: A New Class of Extremely Efficient, Single Site Homogeneous Ruthenium ROMP Catalysts

Martin A. O. Volland, S. Michael Hansen, Peter Hofmann*

Organisch-Chemisches Institut, Ruprecht-Karls-Universität Heidelberg,
Im Neuenheimer Feld 270, D-69120 Heidelberg (Germany)

Abstract. The chemistry of efficient, well-defined single component homogeneous catalysts is an area of active organometallic research. The facile synthesis of a new class of neutral ruthenium carbene complexes, containing bis(di-*t*-butylphosphino)methane (dtbpm) as a chelating ligand, is described. These compounds display moderate reactivity in metathesis but can be conveniently transformed into novel cationic ruthenium carbenes with unprecedented catalytic activity in ROMP reactions. For both types of systems their molecular (X-ray) and electronic (EH, DFT) structures are outlined. Mechanistic aspects of olefin metathesis with ruthenium carbenes, as derived from experimental and theoretical work, are reported.

1 Introduction

The search for efficient, well-defined, single component homogeneous olefin metathesis catalysts is an area of active research [1]. The discovery of stable, catalytically active transition metal carbene complexes, introduced predominantly by Schrock [2] and Grubbs [3], has opened unprecedented perspectives and opportunities for organic synthesis (Fig. 1).

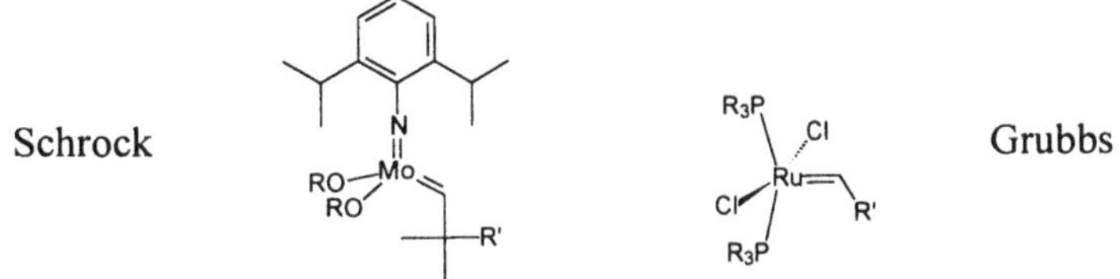

Fig. 1. Carbene complexes as developed by Schrock and by Grubbs

Both types of complexes have found wide applications in ring closing metathesis (RCM), ring opening metathesis polymerization (ROMP), acyclic diene metathesis (ADMET) and intermolecular olefin cross metathesis (Scheme 1).

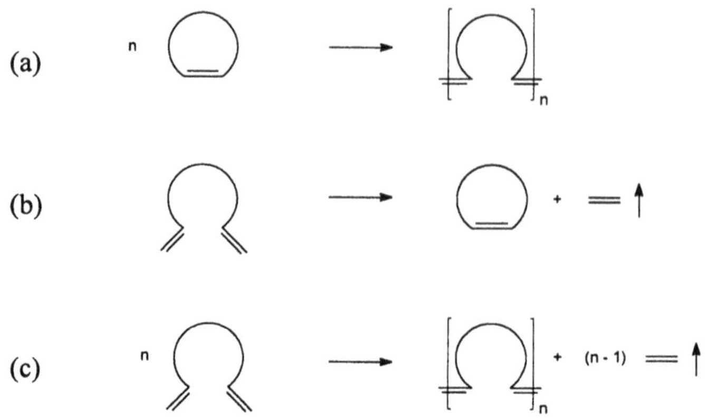

(a) n

(b) +

(c) n + (n - 1)

Scheme 1. (a) ROMP- , (b) RCM- and (c) ADMET-reactions

Due to their stability and functional group tolerance in metathesis reactions, especially Grubbs type square pyramidal ruthenium(II)carbenes were intensively investigated in order to improve their catalytic performance. Variations of these ruthenium systems (Fig. 2) have focused upon carbene [4], phosphine [5] or anionic ligand [6] exchange, phosphine and chloro ligand replacement by chelating Schiff-bases [7] and upon Lewis acid addition [8].

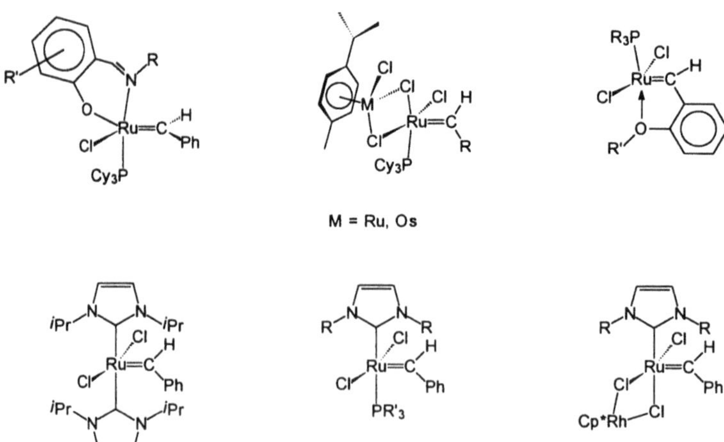

Fig. 2. Recent developments in the area of Grubbs type carbene complexes

Quite recently Grubbs type ruthenium carbenes with one or two unsaturated or saturated Wanzlick-Arduengo type carbene ligands [9,10] replacing one or both of the *trans* phosphines have been reported. Based upon available mechanistic and theoretical studies of Grubbs type *trans*-$(PR_3)_2Cl_2Ru{=}CHR'$ compounds [6b,11], the investigation of complexes with chelating bisphosphine ligands, enforcing *cis* coordination of two phosphorus atoms at the metal center, seemed interesting for various reasons, as will be discussed in detail below. Previous experience with bis(di-*t*-butylphosphino)methane, $^tBu_2PCH_2P^tBu_2$ (dtbpm) as a ligand, enabled us to prepare neutral *cis*-(κ^2-dtbpm)$Cl_2Ru{=}CHR$ [12] (**1**) as well as a novel class of cationic Ru^{II} carbene complexes, $[(\kappa^2\text{-dtbpm})(\mu\text{-Cl})Ru{=}CHR]_2(O_3SCF_3)_2$ [13] (**2**), which are most efficient catalysts in ROMP reactions (Fig. 3).

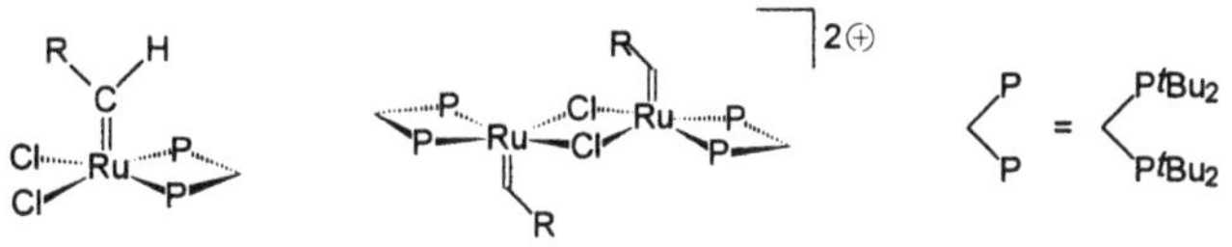

1 **2**

Fig. 3. Novel metathesis catalysts with *cis* phosphine stereochemistry

In this article the synthesis and structures as well as theoretical studies related to the mechanistic behavior of these compounds will be reviewed.

2 Mechanism of Olefin Metathesis with Grubbs Type Carbene Complexes

2.1 Experimental and Theoretical Investigations

Experimental studies [6b,11,14] and theoretical investigations [15] of the reaction mechanism and of catalyst activity for typical Grubbs type *trans*-$(PR_3)_2Cl_2Ru{=}CHR'$ systems have been carried out. Two competing processes were postulated for Ru-catalyzed olefin metathesis: an associative pathway with direct olefin attack and subsequent metallacycle formation at the five-coordinate ruthenium(II) center, and a parallel, dominant dissociative path, in which loss of one of the phosphines precedes olefin coordination and metallacycle formation at the four-coordinate 14 electron fragment $(PR_3)Cl_2Ru{=}CHR'$ (Scheme 2). Especially bulky, electron-rich phosphines PR_3 (R = alkyl) with a high donor strength lead to complexes with remarkable activity in catalysis.

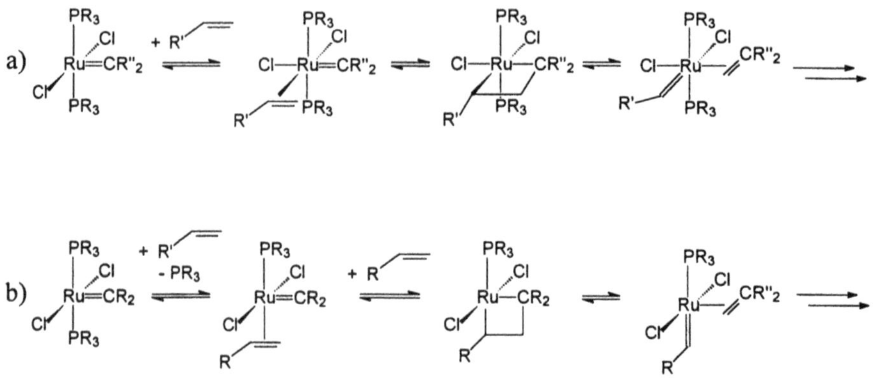

Scheme 2. Simplified mechanistic scheme for Ru-catalyzed olefin metathesis: (a) associative (minor) and (b) dissociative (major) pathway (adapted from ref. 6b)

These mechanistic proposals, based upon kinetic studies by Grubbs et al., are supported by an isolated ruthenium chelate carbene olefin complex, in which the olefin sidearm of the carbene moiety is intramolecularly coordinated to the Ru center and replaces one monodentate phosphine in the *trans* position [17]. Furthermore it should be mentioned that gas phase experiments (ESI-MS/MS) by Chen et al. [14] as well as quantum molecular dynamics calculations (DFT) by Meier et al. [15] also seem to support the proposed mechanism. On the other hand, the existence of ruthenacyclobutane structures as true intermediates in olefin metathesis has been questioned on the basis of most recent gas phase data [16], and their nature as transition states rather than intermediates is still a matter of intensive debate.

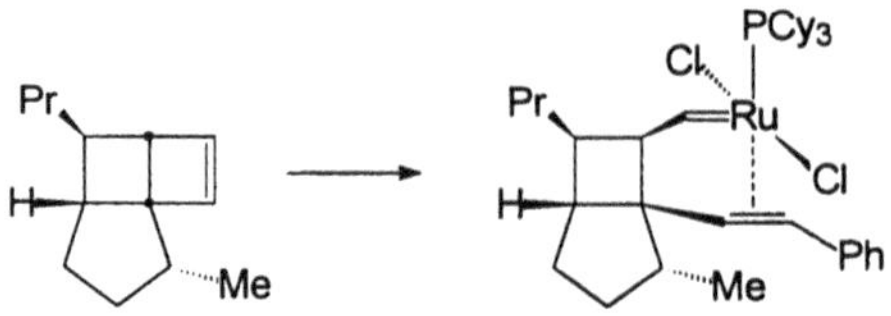

It should be pointed out, however, that Grubbs' mechanistic conclusions with respect to the importance of phosphine dissociation (Scheme 2, path b) as the initiating step in olefin metathesis rely strongly upon kinetic RCM experiments, in which the presence of additional phosphine (e.g. PCy_3) was shown to slow down the reaction considerably. It is important to note in this context, that ligand replacement experiments with dtbpm (Scheme 3) in which the replacement of the PPh_3 ligands of the parent Grubbs type complex *trans*-$(PPh_3)_2Cl_2Ru{=}CHPh$ by the chelating phosphine was intended, only lead to the formation of the stable, dinuclear ylid complex **4** [12].

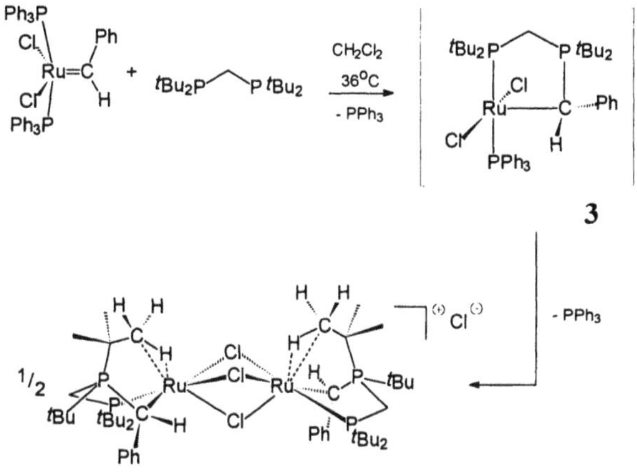

Scheme 3. Ligand replacement experiment leading to an agostic ylid complex

Undoubtedly compound **4**, which has been fully characterized by NMR and single crystal X-ray diffraction, is formed by attack of a dtbpm phosphorus center at the carbene carbon atom, most probably via intermediate **3**, which then dimerizes with further PPh_3 loss to form **4**. This interesting dinuclear species with two formally 16 electron, five-coordinate Ru^{II} centers is stabilized by two strong agostic interactions with one methyl group of a phosphorus ^{t}butyl substituent on each side. Based upon these observations with *trans*-$(PPh_3)_2Cl_2Ru{=}CHPh$ and dtbpm, the formation of ylidic intermediates in the presence of additional strongly donating phosphines in olefin metathesis catalyzed by complexes *trans*-$(PR_3)_2Cl_2Ru{=}CHR'$ might also operate in these cases. Of course, the formation of P-ylid complexes from transition metal carbenes and phosphines has been known for a long time, since E. O. Fischer's work. This alternative pathway should also be considered as a possibility in mechanistic discussions, because ylid formation also would lead to a $1/[PR_3]$ term in the overall kinetic equation for olefin metathesis in the presence of added phosphines PR_3. Then, however, the slowing down of the reaction rate in e.g. RCM kinetic experiments by added phosphine, would not be indicative of phosphine dissociation in the rate determining step.

2.2
Anionic Ligand Dissociation vs. Phosphine Dissociation

Our investigation towards Grubbs type carbene complexes with chelating phophines, which enforce *cis* coordination of phosphines at the metal center, was initiated primarily for two reasons. First of all, we noted the observation of Grubbs et al. in 1993 [18] that according to their NMR spectra the complex *trans*-$[(PR_3)_2Cl_2Ru{=}CH\text{-}CH{=}CPh_2]$ (R = Cy, ^{i}Pr) contained about 16-20% of the *cis* isomer both in solution and in the isolated product. It seemed highly surprising that sterically demanding phophines with large cone angles such as PCy_3 and P^iPr_3 should allow the formation of *cis* complexes in observable quantities, given the well known nonrigidity of d^6-ML_5 complexes. The use of the chelating ligand dtbpm seemed promising for the isolation of stable *cis* coordinated carbene complexes of the Grubbs type. More importantly however, the *cis* phosphine coordination and consequently the enforced mutual *trans* arrangement of phosphines and chloride ligands at the ruthenium center presumably would lead to systems with rather different electronic properties compared to the typical *trans* compounds. The *trans* influence of highly electron-rich phosphorus centers seemed ideal for facilitating a possible ***chloride anion dissociation*** as an initiating step in the catalyzed olefin metathesis, leading to ***cationic*** 14 electron species. With these objectives in mind, the employment of dtbpm seemed ideal for the preparation of these complexes due to the well established steric bulk and the high donor capability of this bisphosphine and the preference of this ligand for chelation (*gem.*-dialkyl or Thorpe-Ingold effect) rather than metal-metal bridging.

It should be also mentioned, that all our own attempts to synthesize stable or at least isolatable cationic Grubbs type carbene complexes $[(PR_3)_2(Cl)Ru{=}CHR']^+$ with non coordinating anions from normal *trans* carbene complex precursors $[(PR_3)_2Cl_2Ru{=}CHR']$ by halide abstraction, have been unsuccessful. Halide abstraction experiments, starting from various complexes *trans*-$(PR_3)_2Cl_2Ru{=}CHR'$ only gave unidentified and untractable product mixtures.

Werner et al. [19], in an attempt to create such species by protonating hydrido vinylidene complexes, were only able to isolate unstable, isomeric, octahedral base-coordinated carbyne hydrido complexes as shown in Fig. 4, which are very active metathesis catalysts, but which decompose in solution at ambient temperature within 30 minutes. The catalytically active species most likely is the isomeric carbene compound $[(PCy_3)_2Cl(L)Ru{=}CH\text{-}CH_3]^+$, which however could not be identified spectroscopically.

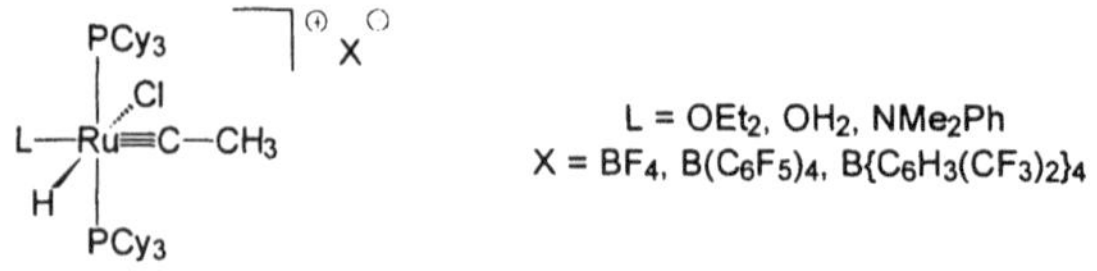

Fig. 4. Cationic ruthenium carbyne complexes of Werner et al.

3 Novel Grubbs-Type Metathesis Catalysts with *cis* Phosphine Stereochemistry

3.1 Synthesis of a Carbene Complex Precursor

Different preparative approaches to carbene complexes with dtbpm as a ligand were investigated. As already outlined in Scheme 3, phosphine exchange reactions starting from *trans*-$(PPh_3)_2Cl_2Ru{=}CHR'$ lead to the formation of ylidic dinuclear ruthenium compounds and do not allow access to the desired carbene complexes. The use of various mononuclear metal hydride complexes with monodentate phosphine ligands for the synthesis of transition metal carbene complexes is well established [20]. Hence, an analogous route starting from appropriate hydride complexes with diphosphinomethane ligands seemed attractive, and it was necessary to find synthetic pathways to such species.

3.2 Survey of Possible Ruthenium Hydride Precursors

Different synthetic routes were probed in order to get access to a suitable hydride precursor. Starting from $Ru(\eta^4\text{-COD})(\eta^3\text{-}C_4H_7)_2$ (**5**), a novel dinuclear hydride complex (**6**) containing terminal and bridging hydride and terminal dihydrogen ligands was isolated and fully characterized structurally and spectroscopically [21].

2 **5** + 2 dtbpm $\xrightarrow[\text{80°C, } -2\ C_8H_{16},\ -4\ C_4H_{10}]{\text{THF, 30 bar } H_2}$ **6**

Unfortunately, the highly fluxional hydride dihydrogen complex **6** - a compound of quite some interest in its own right - turned out not to be a suitable precursor for the desired carbene complexes. On the other hand, treatment of the polymer $[Ru(\eta^4\text{-COD})Cl_2]_x$ (COD = 1,5-cyclooctadiene) with hydrogen and the dtbpm ligand in THF, gave the unusual triply chloro bridged dinuclear hydride dihydrogen complex 7 [12], which could also be characterized by spectroscopic means and by single crystal X-ray diffraction. Detailed NMR (T_{1min} values, spin saturation transfer experiments, NOESY spectra, isotope labelling) studies were conducted to establish an intramolecular H exchange mechanism between the hydride and dihydrogen ligands [21], which occurs independent of the intermolecular exchange of the η^2-coordinated H_2 with H_2 of the gas phase.

$^2/_x$ $[RuCl_2(\eta^4\text{-COD})]_x$ + 2 dtbpm → (THF, 80°C, 10 bar H_2, - HCl, -2 C_8H_{10}) **7**

When the same reaction was carried out in the presence of triethylamine, the dinuclear dihydride complex $[(\kappa^2\text{-dtbpm})RuH(\mu\text{-Cl})]_2$ (**8**) was isolated in high yields [12]. The formation of this product is easily explained by HCl elimination from **7** by triethylamine and is a consequence of the acidity of the metal coordinated η^2-dihydrogen ligand.

$^2/_x$ $[Ru(COD)Cl_2]_x$ → (THF, 80°C, dtbpm, NEt_3, H_2 (10 bar), -2 $NEt_3H^+Cl^-$, -2 C_8H_{16}) **8**

The dinuclear dihydride **8**, which can be regarded as the dimer of the unsaturated 14 electron monohydride fragment $[(\kappa^2\text{-dtbpm})Ru(H)Cl]$, turned out to be an ideal precursor for carbene complex synthesis and could indeed be successfully employed as a suitable, easily accessible staring material for the synthesis of various carbene complexes with dtbpm as a *cis* chelating ligand.

3.3
Synthesis and Structure of *cis* Ruthenium Carbene Complexes

Grubbs et al. have shown that treatment of the dihydrogen complex $(PCy_3)_2ClRu(H)(H_2)$, first prepared by Chaudret et al. and formally a stabilized derivative of the 14 electron monohydride moiety $(PCy_3)_2ClRu(H)$, with propargyl- and vinyl chlorides [20] leads to the formation of carbene complexes *trans*-$(PR_3)_2Cl_2Ru{=}CHR'$. As mentioned above, our new unsaturated dihydride **8** formally is the dimer of a similarly unsaturated, four-coordinate $[(\kappa^2\text{-dtbpm})Ru(H)(Cl)]$ unit with Ru^{II} and also 14 valence electrons, comparable to *trans*-$(PCy_3)_2ClRu(H)$. Therefore it was hoped that reactions of **8** with propargyl- and vinyl chlorides would lead to the formation of the related *cis* carbene complexes. In fact these substrates turned out to be very efficient in the synthesis of the desired complexes **9**. Furthermore chloroallenes were established as another convenient class of organic precursors (Scheme 4), replacing propargyl chlorides.

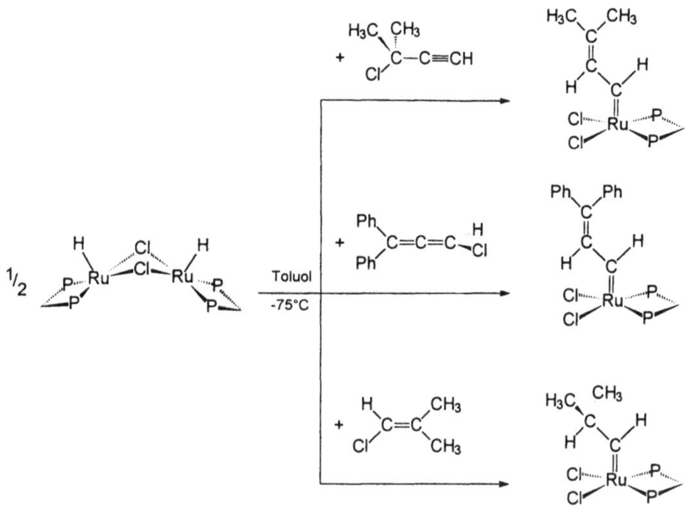

Scheme 4. Synthesis of neutral *cis* carbene complexes

Compounds **9a-c** exhibit the expected spectroscopic properties. For each of them a singlet for the equivalent phosphorus atoms is detected in the ^{31}P NMR; in ^{1}H NMR spectra the resonance for Ru=C*H* is shifted downfield and displays the expected coupling pattern with a doublet of triplets showing up as a pseudo quartet.

The molecular geometry of these compounds in the solid state, determined by single crystal X-ray diffraction, is displayed for **9a** as a typical representative in Fig. 5. Complex **9a** has a distorted square-pyramidal structure, in which the carbene moiety occupies an apical position, the P-Ru-P bite angle in the chelate ring is 73.8°, and the Ru-C distance of 1.858(2) Å is identical to that of *trans*-$[(PCy_3)_2Cl_2Ru{=}CH\text{-}CH{=}CPh_2)]$ [1.851(21) Å]. The orientation of the carbene moiety in the crystal minimizes steric interaction with the ^{t}Bu-groups of the dtbpm ligand. The carbene plane does not bisect the P1-Ru-P2 and Cl1-Ru-Cl2 angles, but is orthogonal to the P1-Ru-Cl1 axis. As a consequence, the acceptor orbital of the carbene carbon atom is parallel to P1-Ru-Cl1, and this decreases the angle P1-Ru-Cl1 to only 154.9° (vs. 165.3° for P2-Ru-Cl2), with one chlorine atom bent out of the basal plane. This is precisely the structural feature which was calculated (vide infra) for the simplified model *cis*-$(PH_3)_2Cl_2Ru{=}CH_2$, in which the minimum energy orientation of the CH_2 plane is identical to that observed in the solid state orientation of **9a**.

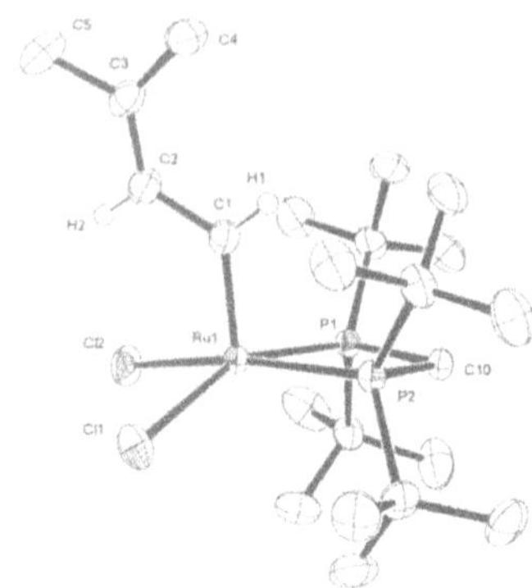

Fig. 5. Molecular structure of **9a** in the crystal (ORTEP, 50 % probability; all CH hydrogen atoms, except for H1 and H2 of the vinylidene group, omitted for clarity)

The complexes (κ^2-dtbpm)Cl_2Ru=CHR (**9a-c**) catalyze the ROMP of the strained olefin norbornene. They are however less active than the standard Grubbs catalyst *trans*-$(PCy_3)_2Cl_2Ru$=CHPh. This decrease in activity may be taken as a hint towards the acceleration of olefin metathesis by phosphine dissociation: for the chelate ligand dtbpm phosphine dissociation either does not play a role at all or is severely hampered, and the (slower) associative pathway **b** of Scheme 2 is the only one that operates.

3.4
Electronic Structures of Ruthenium Carbene Complexes with *cis* and *trans* Phosphine Ligands: Quantum Chemistry Model Calculations

Both Extended Hückel (EH) and density functional (DFT) [22] calculations were carried out on the simplified model complexes $(PH_3)_2Cl_2Ru$=CH_2, in which steric preferences should not overshadow electronic effects, in order to gain a better insight into the electronic structure of these carbene complexes [12]. The expected distorted square-pyramidal minimum energy geometry (EH) was found for *trans*-$(PH_3)_2Cl_2Ru$=CH_2 (**10**), in good agreement with X-ray diffraction data and B3LYP calculations. However we also found a square pyramidal structure of similar energy for *cis*-$(PH_3)_2Cl_2Ru$=CH_2 (**11**). Both structures have a significant HOMO-LUMO gap of approximately 1.7 eV in EHT, in accord with a stable 16-electron system. These results were corroborated by DFT calculations (B3LYP/LANL2DZ). The DFT-optimized geometries for **10** and **11** are shown in Fig. 6. The energy difference is only 10 kcal mol^{-1} in favor of the *trans* isomer **10** vs. the *cis* isomer **11** at the B3LYP/LANL2DZ level, and this decreases to 7 kcal mol^{-1} with a larger basis set [23].

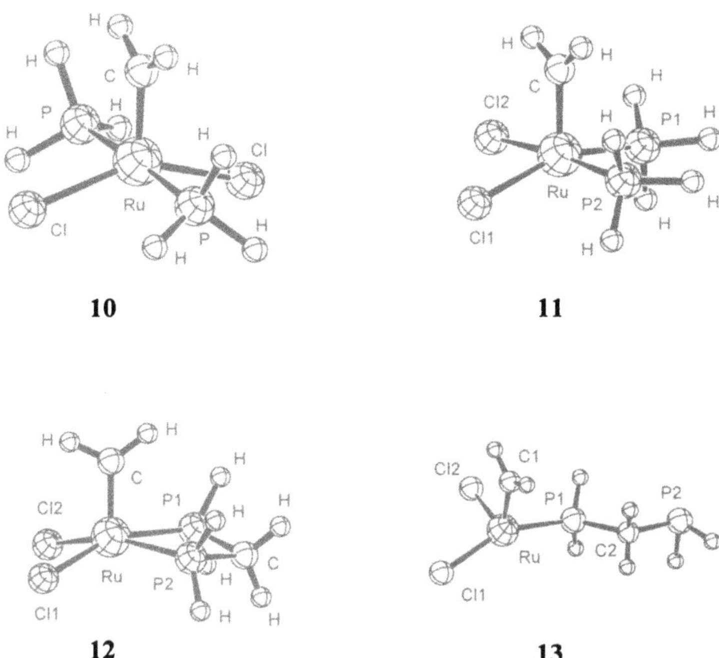

Fig. 6. Geometry-optimized structures of **10**, **11**, **12**, **13** (B3LYP/LANL2DZ)

The orientation of the carbene moiety allows for optimal backbonding from the metal fragment to the carbene acceptor orbital in both structures, but the Ru=CH_2 rotation barriers are very small (<9 kcal mol^{-1}). In **11** the P1-Ru-Cl1 unit that interacts with the π-acceptor orbital of the methylene carbon atom is more strongly bent (157.1°) than the P2-Ru-Cl2 unit (175.0°). The LUMOs of both **10** and **11** are predominantly localized on the methylene ligand both at the EH and DFT level of theory. In the *trans* stereoisomer **10** the LUMO is practically degenerate with the LUMO-1, an empty metal orbital (the equivalent of an empty d^2sp^3 hybrid of an octahedron) that points towards the open coordination site opposite to the carbene ligand, while in the *cis* structure **11** the LUMO lies at distinctly lower energy (DFT: 0.5 eV). This feature would predict a frontier orbital controlled attack of nucleophiles to occur at the carbene carbon atom, precisely as we found it to occur with the electron-rich phosphine dtbpm (vide supra).

Given the fluxionalty of d^6ML_5 systems, the calculated small energy difference between **10** and **11** also implies only a marginal energy barrier for the *cis-trans* isomerization of $(PR_3)_2Cl_2Ru{=}CR^1R^2$ complexes with monodentate phosphines, with only a small preference for the *trans* geometry. Thus the isolation of Grubbs type *cis* complexes should only be possible if the rearrangement to *trans* structures is blocked by chelating diphosphine ligands with small bite angles such as diphosphinomethanes. The optimized geometry (B3LYP/LANL2DZ) of the model system $(\kappa^2\text{-}H_2PCH_2PH_2)Cl_2Ru{=}CH_2$ (**12**) is shown in Fig. 6. This complex is electronically identical to **11**, except that in the geometry optimized structure of lowest energy the carbene plane now bisects the P-Ru-P as well as the Cl-Ru-Cl angle in EH and DFT calculations, leading to C_s symmetry. The orbital structure of **12** closely resembles that of **11**, with the carbene centered p orbital (60%) as the LUMO and a predominantly d type (70%) metal acceptor level *trans* to the methylene unit; the latter is around 0.7 eV (DFT) higher in energy than the LUMO. Molecular orbital contour plots (EH) for these two lowest lying empty levels LUMO and LUMO-1 of structure **12** are displayed in Fig. 7.

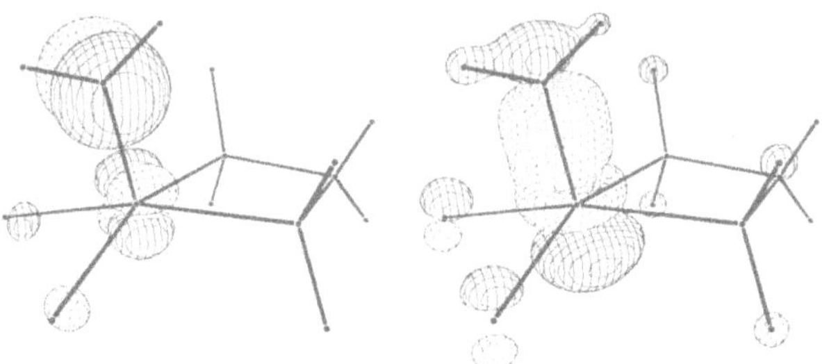

Fig. 7. Three-dimensional MO contour plots (EH MOs, CACAO plots [26]) of the LUMO (left) and LUMO-1 (right) of model complex **12**

With respect to phosphine dissociation as the initiating step in the mechanism of olefin metathesis under the influence of Grubbs type carbene complexes with monodentate phosphines, it was interesting to examine the d^6ML_4 14 electron species (κ^1-$H_2PCH_2PH_2$)Cl_2Ru=CH_2 (**13**) which would be generated if the chelate ring in **12** were to open. The geometry optimized structure (B3LYP/LANL2DZ) of **13** is displayed in Fig. 6. This pseudotetrahedral complex, with one of the phosphine arms pending, lies 9 kcal mol^{-1} higher in energy than **11**. **13** is a true minimum on the energy surface (no imaginary frequencies) and is calculated to have a closed shell ground state structure. It should be mentioned that complexes such as **13** should be increasingly energetically disfavored with a rising steric demand (Thorpe-Ingold effect) and increasing electron donor character of the substituents in the ligand $R_2PCH_2PR_2$. As a consequence, phosphine vs. chloride dissociation was expected to be even more disfavored with dtbpm as a ligand.

4
Cationic Ruthenium Carbene Complexes

4.1
Synthesis and Structure of a Novel Class of Cationic Ru^{II} Carbene Complexes

As already discussed above, the *trans* arrangement of phosphine and chloride ligands in complexes (κ^2-dtbpm)Cl_2Ru=CHR should favor chloride dissociation due to the *trans* influence of the electron-rich phosphorus centers. With the neutral complexes at hand, it was attempted to obtain stable cationic complexes.

Treatment of the neutral precursors **9a-c** with trimethylsilyl triflate (TMSOTf) leads to the desired compounds in high yields, as shown for **9a** in Scheme 5 [13].

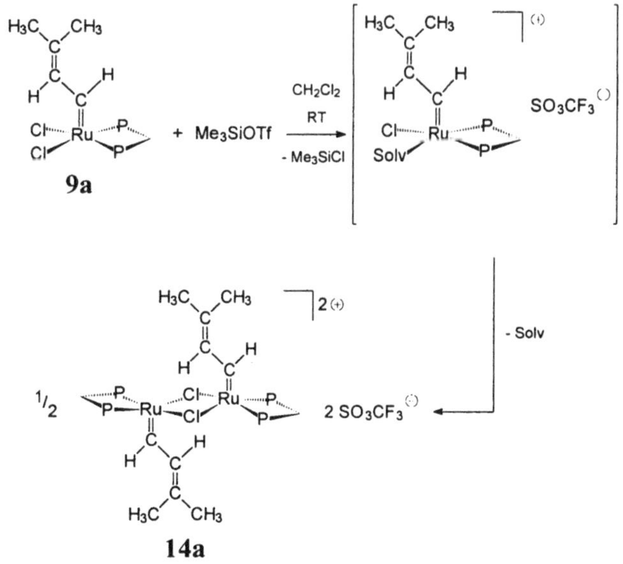

Scheme 5. Synthesis of stable dinuclear dicationic ruthenium carbene complexes

Trimethylsilyl triflate turned out to be a very efficient and convenient Lewis acid for chloride abstraction, because the irreversible formation of the volatile trimethylsilyl chloride enables facile workup of the reaction mixtures. The molecular structures of these air stable binuclear cations **14**, isolated as triflates, were determined by single crystal X-ray diffraction. The solid state structure of compound **14a** is displayed in Fig. 8 as a typical representative.

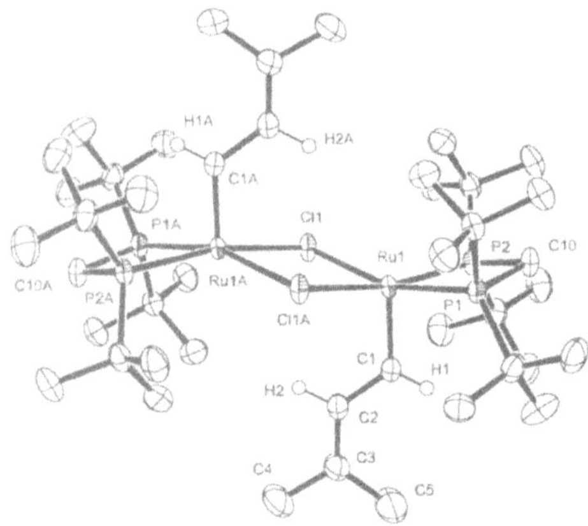

Fig. 8. Molecular structure of dication **14a** in the solid state (ORTEP, 50% probability, CH-hydrogens except H1, H1A, H2 and H2A of the carbene moieties as well as the non coordinating triflate anions are omitted for clarity)

As in the neutral precursors, each ruthenium center of **14a** in the crystal is in a distorted square pyramidal ligand environment with the P atoms and the μ-chloro ligands in the basal plane and the carbene moiety in the apical position, nearly bisecting the P-Ru-P angle. In the solid state the two carbene fragments are arranged in a *trans* fashion to each other. The triflate anions are not coordinating to the ruthenium centers.

In solution however, ^{13}C and ^{31}P NMR spectra show a completely doubled set of signals, indicating the presence of two isomers in solution. Variable temperature (VT) NMR as well as EXSY spectra were recorded in order to determine the nature of these isomers. The VT and EXSY spectra reveal a dynamic equilibrium between two dimeric species with *cis*- and *trans*-arrangement (*cis*-**14a** / *trans*-**14a**) of the carbene moieties (Fig. 9).

trans-**14a** *cis*-**14a**

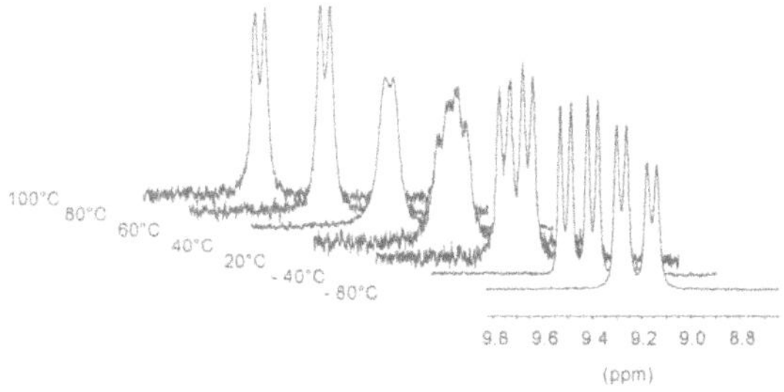

Fig. 9. VT NMR spectra of **14a** in CD_2Cl_2 under 10 bar argon pressure (vinyl proton resonances indicating interconversion between *trans*-**14a** and *cis*-**14a**)

The EXSY spectrum of **14a** shown in Fig. 10 confirms the interpretation of the VT NMR data, showing the interconversion of *cis* and *trans* species, as indicated by the appropriate cross peaks.

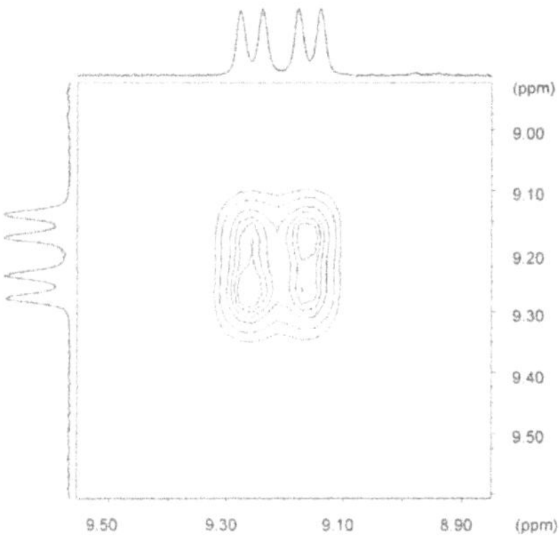

Fig. 10. EXSY spectrum of **14a**: resonances of the vinylic protons (T=233K) in CD_2Cl_2

Surprisingly, complex $[(\kappa^2\text{-dtbpm})(\mu\text{-Cl})Ru{=}CH\text{-}CH{=}CPh_2]_2[OTf]_2$ (**15**) with rather sterically demanding carbene fragments displays a *cis* arrangement of the carbene ligands in the solid state (Fig. 11). It is impossible at this point to decide if crystal packing or attractive, intramolecular arene π-π stacking is the reason for the observed *cis* configuration in this case. A nearly parallel phenyl orientation is seen in the crystal, but π-π interactions between phenyl rings have been shown to be repulsive in the gas phase, while attractive π-π stacking, which stabilizes the packing of molecules, has been noted in porphyrin crystal structures as well.

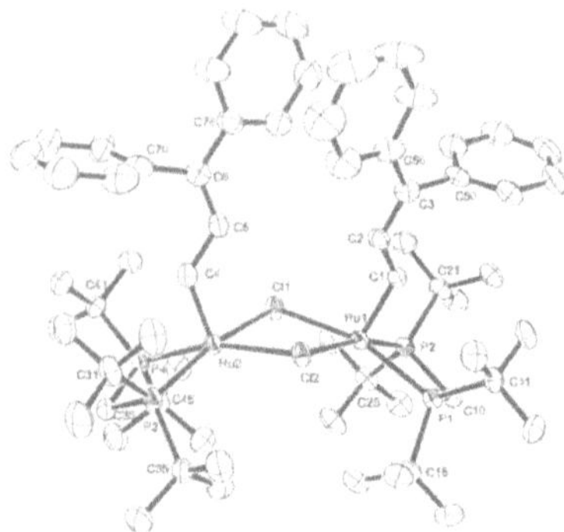

Fig. 11. Molecular structure of dication **15** in the solid state (ORTEP, 50% probability, hydrogens and the non coordinating triflate anions are omitted for clarity)

In solution the *cis*/*trans*-equilibration of these dicationic systems occurs through mononuclear, solvent stabilized, five-coordinate cationic carbene complexes [(κ^2-dtbpm)(μ-Cl)Ru=CHR(solv)]$^+$. We are able to observe such a monomeric species in solution for [(κ^2-dtbpm)(μ-Cl)Ru=CH-CH=C(CH_3)$_2$(CH_3CN)]$^+$ in ^{1}H and ^{31}P spectra of **14a** in CD_2Cl_2 / CH_3CN. Furthermore in a 1:1 mixture of **14a** and [(κ^2-dtbpm)(μ-Cl)Ru=CH^{i}Pr]$_2$[OTf]$_2$ (**14c**) in CD_2Cl_2 the signals of all dinuclear, dicationic cross products are detected. In Fig. 12 the vinyl proton resonances are displayed, unambiguously indicating the formation of the respective crossover products.

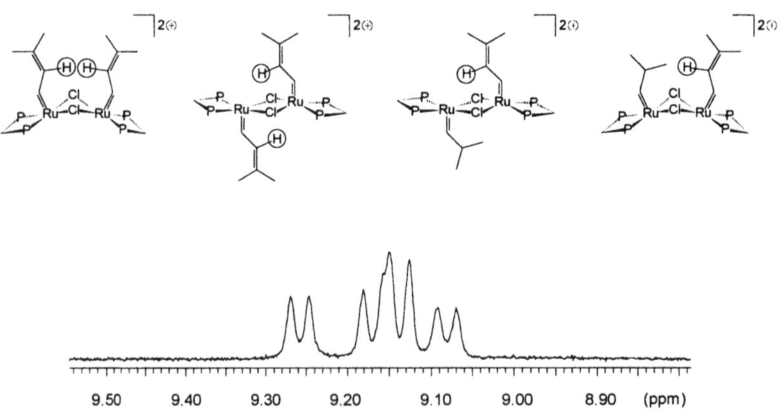

Fig. 12. NMR crossover experiment for compounds **14a** and **14c**, 1:1 mixture in CD_2Cl_2; only vinyl proton resonances (doublets) are displayed for clarity

4.2 Catalytic Activity of Cationic Ruthenium Carbene Complexes

The new cationic carbene complexes are very efficient catalysts for ring opening metathesis polymerization (ROMP) of strained cyclic olefins. Contrasting the cationic carbyne systems of Werner et al. (cf. Fig. 4) they are thermally stable in CD_2Cl_2 solution for at least 3 days at ambient temperature under an atmosphere of argon (also see Fig. 9). To obtain a precise comparison with the standard Grubbs catalyst $(PCy_3)_2Cl_2Ru{=}CHPh$, the ROMP production of polyoctenamer from cyclooctene was studied kinetically with our systems in ^1H NMR experiments. The published activity and polyoctenamer production curve for the Grubbs type catalyst $(PCy_3)_2Cl_2Ru{=}CHPh$, with a cyclooctene:Ru (substrate:catalyst) ratio of 250:1 [10] was reproduced by our measurements [13] and so $(PCy_3)_2Cl_2Ru{=}CHPh$ could be taken as a standard. Due to the much higher activity of our systems in this reaction the cyclooctene:Ru ratio was raised to 12500:1. Under these conditions polycyclooctenamer production as determined by ^1H NMR is shown in Fig. 13.

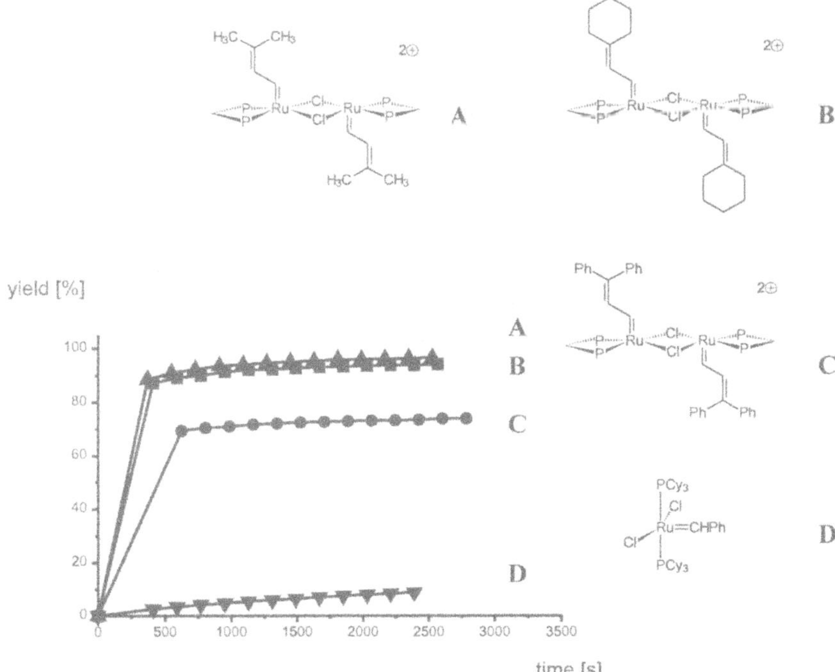

Fig. 13. ROMP of cyclooctene as followed by ^1H NMR; (0.5 mL CD_2Cl_2; T=23°C)
For **A, B, D** : [cyclooctene]:[Ru]=12500:1 (=0.008 mol% Ru)
For **C** : [cyclooctene]:[Ru]=24800:1 (=0.004 mol% Ru)

For compound $[(\kappa^2\text{-dtbpm})(\mu\text{-Cl})Ru{=}CH\text{-}CH{=}CPh_2]_2[OTf]_2$, the cyclooctene:Ru ratio was even raised to 24800:1. In this case, the polymerization stops after approximately 75% of the monomer is converted. The catalyst may either be poisoned by unidentifiable substances at these very low concentrations or the polymerization stops due to slow diffusion in the viscous reaction mixture. Evidently, another alternative is the decomposition of the catalytically active species. It should be noted that these novel cationic systems also catalyze the ring closing metathesis of 1,7-octadiene to cyclohexene. The full potential of these catalysts for other substrates remains to be determined in detail.

4.3 Quantum Chemistry Model Calculations for Cationic Ruthenium Carbene Complexes

The high catalytic activity of the cationic ruthenium compounds for olefin metathesis that are described above, makes them an interesting subject for theoretical investigations using quantum chemical methods. One of the initial perspectives for the synthesis of these Grubbs type compounds was the possibility of a halide dissociation step initiating the process of olefin metathesis. The drastic improvement going from our neutral systems $(\kappa^2\text{-dtbpm})Cl_2Ru{=}CHR$ to the dinuclear dicationic systems $[(\kappa^2\text{-dtbpm})(\mu\text{-Cl})Ru{=}CHR]_2^{2+}$ approved our strategy for the design of these catalysts. The question remains, which species are actually catalytically active. It seems reasonable to assume that the observed mononuclear, highly electrophilic, solvent stabilized mononuclear cations $[(\kappa^2\text{-dtbpm})ClRu{=}CHR(solv)]^+$ function as active catalysts. This assumption is the starting point for theoretical investigations to be described in the following.

Based upon the established Chauvin mechanism [24] for olefin metathesis, DFT (B3LYP/LANL2DZ) calculations [25] of the postulated catalytic cycle shown in Scheme 6 were carried out.

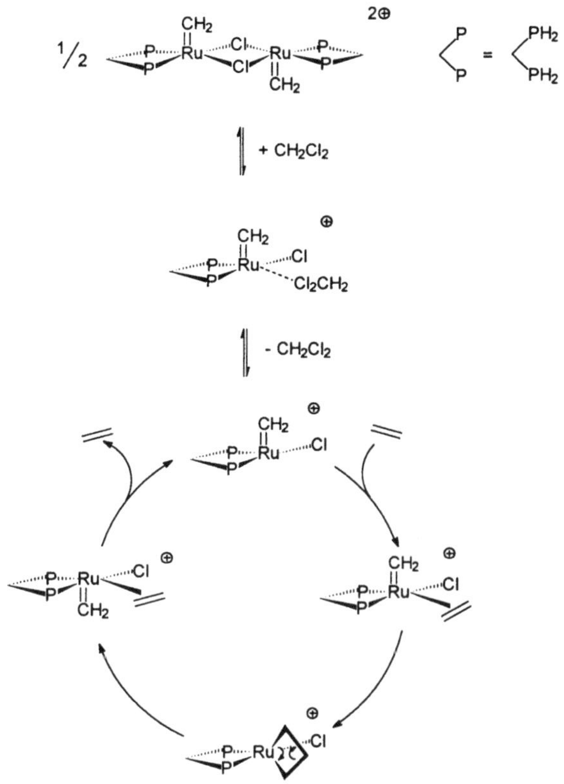

Scheme 6. Calculated (B3LYP/LANL2DZ) catalytic cycle for olefin metathesis for the model system $[(\kappa^2\text{-}H_2PCH_2PH_2)(\mu\text{-}Cl)Ru{=}CH_2]_2^{2+}$

All structures that are displayed in Scheme 6 were fully geometry optimized. They represent true minima on the energy surface and have closed shell ground states. In the computed energy profile for the above catalytic cycle (Fig. 14) the energies of all species are given relative to the dinuclear dication $[(\kappa^2\text{-}H_2PCH_2PH_2)(\mu\text{-}Cl)Ru{=}CH_2]_2^{2+}$. The mere dissociation of the dimeric dication into the mononuclear counterparts (in the gas phase) is found to be slightly endothermic (+2.5 kcal mol^{-1}). Coordination of a solvent molecule (i.e. CH_2Cl_2), however, stabilizes the monomeric cation by 14.1 kcal mol^{-1}. The solvent molecule may be displaced by an olefin - in this model system by ethylene - in an exothermic reaction (7.7 kcal mol^{-1}). At this level of DFT theory the metallacyclobutane intermediate, a closed shell structure with a similarly large HOMO-LUMO gap both in DFT and EHT calculations, is 3.0 kcal mol^{-1} more stable than the olefin complex precursor.

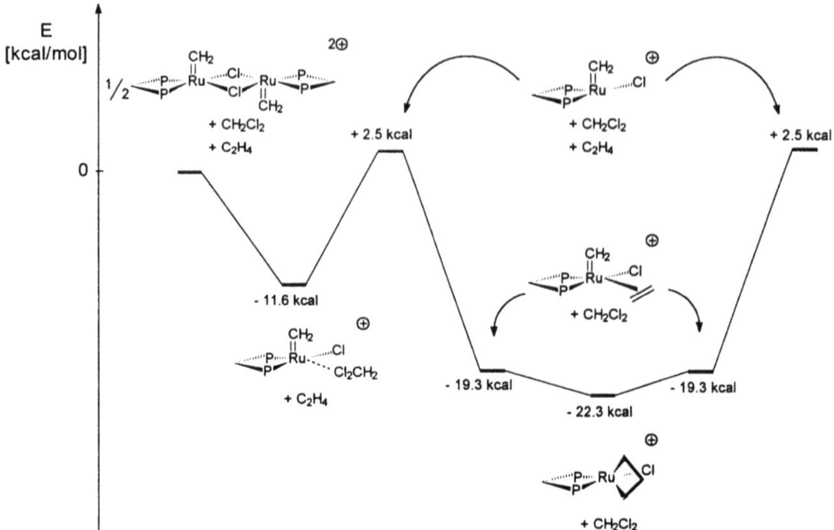

Fig. 14. Energy profile for the computed catalytic cycle for olefin metathesis (B3LYP/LANL2DZ)

The energy barrier towards the formation of the 14 electron metallacyclobutane species from its precursor turns out to be rather small (2.6 kcal mol^{-1}). In Fig. 15 the geometry optimized transition state (NIMAG=1) for this reaction path is shown.

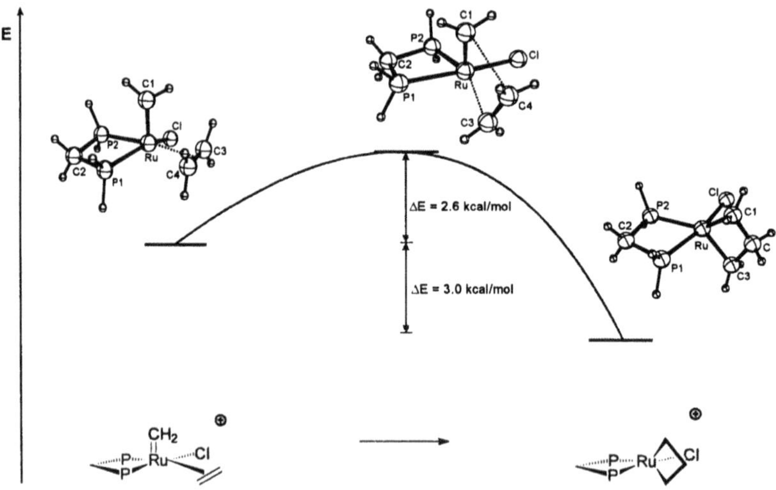

Fig. 15. Geometry optimized structures (B3LYP/LANL2DZ) for the ethylene complex, ring closure transition state and ruthenacyclobutane

In the olefin complex the methylene plane is oriented nearly orthogonal to the P1-Ru-Cl plane, so that optimal backbonding from the metal fragment to the carbene carbon acceptor orbital is possible. The C3-C4 axis of the olefin, which in this complex predominantly functions as a donor ligand with little backbonding from the metal fragment into π^*, is approximately orthogonal to the C1-Ru unit. In the transition state the methylene as well as the ethylene ligands rotate to form the metallacycle intermediate. As a result of the rotation of the methylene unit the P1-Ru-Cl angle becomes larger due to reduced backbonding.

The results of EH calculations are very similar. For the model system $[(\kappa^2\text{-}H_2PCH_2PH_2)ClRu{=}CH_2(\eta^2\text{-}C_2H_4)]^+$ the energy surface for the simultaneous rotation of the methylene unit and the ethylene ligand was calculated (Fig. 16).

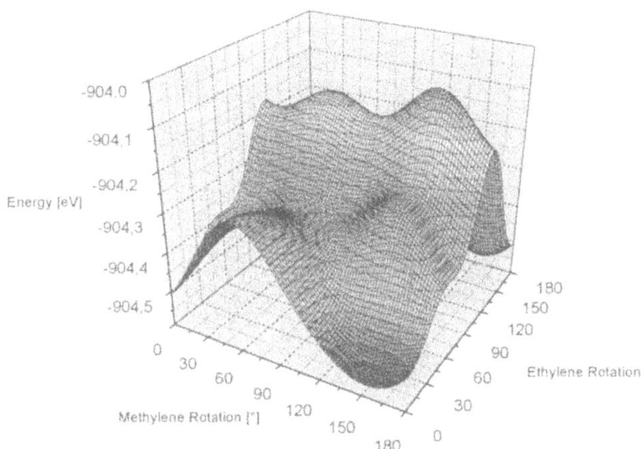

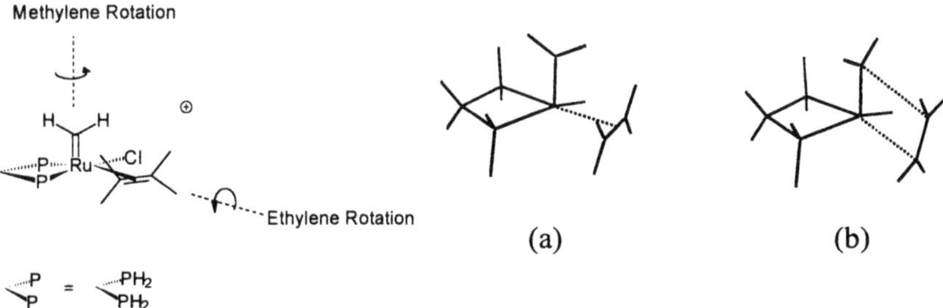

Fig. 16. EH calculation for $[(\kappa^2\text{-}H_2PCH_2PH_2)ClRu{=}CH_2(\eta^2\text{-}C_2H_4)]^+$ with simultaneous rotation of the ethylene and methylene unit: three dimensional energy surface with (a) minimum, (b) "transition state" on the way to the metallacyclobutane intermediate

In Fig. 16 the geometry of the complex with the lowest energy is displayed in (a), completely in accordance with the DFT optimized structure. In addition, the "transition state" (b) in EH calculation was found 5.1 kcal mol^{-1} less stable than (a) which is also in good agreement with the results from DFT investigations.

Although further aspects of polymer characterization will not be discussed here, it should be mentioned parenthetically, that the analysis of cyclooctene and norbornadiene ROMP polymers formed under the influence of the cationic complexes $[(\kappa^2\text{-dtbpm})(\mu\text{-Cl})Ru{=}CHR]_2[OTf]_2$ show a bimodal molecular weight distribution with a minor low molecular weight component, indicating the presence of two active species. The role of the dinuclear as opposed to mononuclear cations as catalitically active species therefore remains to be deciphered in detail (Scheme 7).

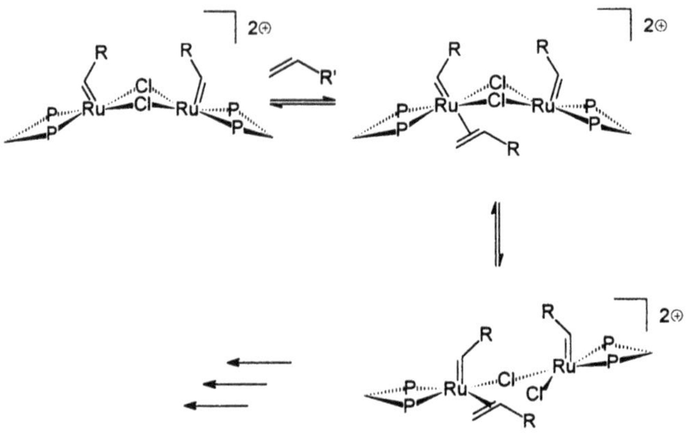

Scheme 7. Possible role of the dinuclear dicationic species in olefin metathesis

These results encourage further experimental work concerning the mechanism of olefin metathesis with the novel complexes $[(\kappa^2\text{-dtbpm})(\mu\text{-Cl})Ru{=}CHR]_2^{2+}$. With our present results at hand, mechanistic studies with gas phase analytical methods are a challenging possibility, because the cationic nature and the stability of our systems make them appear perfectly suited for e.g. electron spray tandem mass spectrometry studies.

5 Comments and Conclusion

The facile synthesis of stable neutral and cationic Grubbs type carbene complexes with a sterically crowded, strong electron donating diphosphinomethane ligand, which enforces *cis* coordination of the phosphines at the ruthenium center adds a completely new facet to the chemistry of this class of compounds. Halide abstraction from neutral precursor complexes, facilitated by the chelating ligand's strong *trans* influence upon the Ru-Cl bonds, rather than phosphine dissociation as claimed for *trans*-L_2Cl_2Ru=CHR Grubbs type complexes, leads to a remarkable increase in reactivity. Obviously, P_2-chelation also should carry a potential of employing chiral phosphines in metathesis.

Current efforts to tune the catalytic properties of these systems focus on a broad modification of the ligand set, counterion and carbene structure in order to outline the full potential of these systems in catalysis.

Acknowledgements: This work was supported by Studienstiftung des deutschen Volkes (Fellowship to M.A.O.V.), Deutsche Forschungsgemeinschaft (Graduate College Fellowship to S.M.H.), Fonds der Chemischen Industrie, BASF AG and Heraeus.

6 References

1. a) *Topics in Organometallic Chemistry Vol. 1: Alkene Metathesis in Organic Synthesis* (Ed.: A. Fürstner), Springer, Berlin, **1998**; b) M. Schuster, S. Blechert, *Angew. Chem. Int. Ed.* **1997**, *36*, 2037.
2. R. R. Schrock, J. S. Murdzek, G. C. Bazan, J. Robbins, M. DiMare, M. Oregan, *J. Am. Chem. Soc.* **1990**, *112*, 3875.
3. S. T. Nguyen, L. K. Johnson, R. H. Grubbs, *J. Am. Chem. Soc.* **1992**, *114*, 3974.
4. J. S. Kingsbury, J. P. A. Harrity, P. J. Bonitatebus Jr., A. H. Hoveyda, *J. Am. Chem. Soc.* **1999**, *121*, 791.
5. P. Schwab, R. H. Grubbs, J. W. Ziller, *J. Am. Chem. Soc.* **1996**, *118*, 100.
6. a) Z. Wu, S. T. Nguyen, R. H. Grubbs, J. W. Ziller, *J. Am. Chem. Soc.* **1995**, *117*, 5503; b) E. L. Dias, S. T. Nguyen, R. H. Grubbs, *J. Am. Chem. Soc.* **1997**, *119*, 3887.
7. S. Chang, L. Jones II, C. Wang, L. M. Henling, R. H. Grubbs, *Organometallics* **1998**, *17*, 3460.
8. a) E. L. Dias, R. H. Grubbs, *Organometallics* **1998**, *17*, 2758; b) A. Fürstner, K. Langemann, *J. Am. Chem. Soc.* **1997**, *119*, 9130.
9. a) J. Huang, E. D. Stevens, S. P. Nolan, J. L. Petersen, *J. Am. Chem. Soc.* **1999**, *121*, 2674; b) T. Weskamp, F. J. Kohl, W. Hieringer, D. Gleich, W. A. Herrmann, *Angew. Chem. Int. Ed.* **1999**, *38*, 2416; c) M. Scholl, T. M. Trnka, J. P. Morgan, R. H. Grubbs, *Tetrahedron Lett.* **1999**, *40*, 2247 d) M. Scholl, S. Ding, C. W. Lee, R. H. Grubbs, *Org. Lett.* **1999**, *1*, 953.
10. T. Weskamp, W. C. Schattenmann, M. Spiegler, W. A. Herrmann, *Angew. Chem. Int. Ed.* **1998**, *37*, 2490; see also Corrigendum: *Angew. Chem. Int. Ed.* **1999**, *38*, 262.
11. M. Ulmann, R. H. Grubbs, *Organometallics* **1998**, *17*, 2484.
12. S. M. Hansen, F. Rominger, M. Metz, P. Hofmann, *Chem. Eur. J.* **1999**, *5*, 557.
13. S. M. Hansen, M. A. O. Volland, F. Rominger, F. Eisenträger, P. Hofmann, *Angew. Chem. Int. Ed.* **1999**, *38*, 1273.
14. C. Hinderling, C. A. Adlhart, P. Chen, *Angew. Chem. Int. Ed.* **1998**, *37*, 2685.
15. O. M. Aagaard, R. J. Meier, F. Buda, *J. Am. Chem. Soc.* **1998**, *120*, 7174.
16. P. Chen, private communication.
17. J. A. Tallarico, P. J. Bonanitatebus, M. L. Snapper, *J. Am. Chem. Soc.* **1997**, *119*, 7175.

18. S. T. Nguyen, R. H. Grubbs, *J. Am. Chem. Soc.* **1993**, *115*, 9858.
19. W. Stüer, J. Wolf, H. Werner, P. Schwab, M. Schulz, *Angew. Chem. Int. Ed.* **1998**, *37*, 3421.
20. a) M. L. Christ, S. Sabo-Etienne, B. Chaudret, *Organometallics* **1994**, *13*, 3800; b) T. E. Wilhelm, T. R. Belderrain, S. N. Brown, R. H. Grubbs, *Organometallics* **1997**, *16*, 3867.
21. a) S. M. Hansen, *Zur Komplexchemie und Metallorganik des Rutheniums: Der Weg zu Neuen Katalysatoren für die Olefinmetathese*, Dissertation, Universität Heidelberg, **1999**; b) The fascinating chemistry of **6** and **7** will be reported in detail elsewhere: manuscript in preparation.
22. For DFT calculations the empirically parametrized B3LYP method within the Gaussian 94 (Revision E.2) package was used: a) M. J. Frisch, G. W. Trucks, H. B. Schlegel, P. M. W. Gill, B. G. Johnson, M. A. Robb, J. R. Cheeseman, T. A. Keith, G. A. Petersson, J. A. Montgomery, K. Raghavachari, M. A. Al-Laham, V. G. Zakrzewski, J. V. Ortiz, J. B. Foresman, J. Cioslowski, B. R. Stefanov, A. Nanayakkara, M. Challacombe, C. Y. Peng, P. Y. Ayala, D. J. Fox, J. S. Binkley, D. J. Defrees, J. Baker, J. P. Stewart, M. Head-Gordon, C. Gonzales, J. A. Pople, Gaussian Inc., Pittsburgh PA, **1995**; b) A. D. Becke, *Phys. Rev.* **1988**, *A38*, 3098; c) C. Lee, W. Yang, R. G. Parr, *Phys. Rev.* **1988**, *B37*, 785; d) S. H. Vosko, L. Wilk, M. Nusair, *Can. J. Phys.* **1980**, *58*, 1200; e) P. J. Hay, W. R. Wadt, *J. Chem. Phys.* **1985**, *82*, 299; f) P. J. Hay, W. R. Wadt, *J. Chem. Phys.* **1985**, *82*, 284; g) T. H. Dunning, P. J. Hay in *Modern Theoretical Chemistry*, H. F. Schaefer III (Hrsg.), Plenum: New York **1976**, 1.
23. In these calculations a large 6-311 + G(2d,2p) basis set was used for the main group atoms. The ruthenium basis for the Hay and Wadt ECP was expanded by uncontracting one s, one p and one d function and by adding a diffuse d function and two f functions contracted from three primitive f functions.
24. a) J. L. Hérisson, Y. Chauvin, *Makromol. Chem.* **1971**, *141*, 161; b) F. N. Tebbe, G. W. Parshall, D. W. Ovenall, *J. Am. Chem. Soc.* **1979**, *101*, 5074; c) T. J. Katz, S. J. Lee, N. Acton, *Tetrahedron Lett.* **1976**, 4247; d) R. H. Grubbs in *Comprehensive Organometallic Chemistr, Vol. 8.* (Ed.: G. Wilkinson), Pergamon Press, New York, **1982**.
25. For the following DFT calculations the empirically parametrized B3LYP method within the Gaussian 98 (Revision A.5) package was used: Gaussian 98, Revision A.5, M. J. Frisch, G. W. Trucks, H. B. Schlegel, G. E. Scuseria, M. A. Robb, J. R. Cheeseman, V. G. Zakrzewski, J. A. Montgomery, Jr., R. E. Stratmann, J. C. Burant, S. Dapprich, J. M. Millam, A. D. Daniels, K. N. Kudin, M. C. Strain, O. Farkas, J. Tomasi, V. Barone, M. Cossi, R. Cammi, B. Mennucci, C. Pomelli, C. Adamo, S. Clifford, J. Ochterski, G. A. Petersson, P. Y. Ayala, Q. Cui, K. Morokuma, D. K. Malick, A. D. Rabuck, K. Raghavachari, J. B. Foresman, J. Cioslowski, J. V. Ortiz, B. B. Stefanov, G. Liu, A. Liashenko, P. Piskorz, I. Komaromi, R. Gomperts, R. L. Martin, D. J. Fox, T. Keith, M. A. Al-Laham, C. Y. Peng, A. Nanayakkara, C. Gonzalez, M. Challacombe, P. M. W. Gill, B. Johnson, W. Chen, M. W. Wong, J. L. Andres, C. Gonzalez, M. Head-Gordon, E. S. Replogle, J. A. Pople, Gaussian, Inc., Pittsburgh (PA) **1998**.
26. (a) R. Hoffmann, *J. Chem. Phys.* **1963**, *39*, 1367; b) R. Hoffmann, W. N. Lipscomb, *J. Chem. Phys.* **1962**, *36*, 2197; c) Off-diagonal elements H_{ij} were calculated by using the weighted H_{ij} formula: J. H. Ammeter, H.-B. Bürgi, J. C. Thibeault, R. Hoffmann, *J. Am. Chem. Soc.* **1978**, *100*, 3686; d) The programs used were FORTICON8 and CACAO: J. Howell, A. R. Rossi, D. Wallace, K. Haraki, R. Hoffmann, Quantum

Chemical Program Exchange, QCPE No. 334; d) C. Mealli, D. M. Proserpio, *J. Chem. Ed.* **1990**, *67*, 399; e) Atomic parameters for C, H, P and Cl were standard values. The H_{ii} values for ruthenium were derived from a SCCC calculation for *trans*-[$(PH_3)_2Cl_2Ru{=}CH_2$] with Ru charge iteration parameters from Calzaferris's EH package: G. Calzaferri, M. Brändle, ICONC & INPUTC, Quantum Chemistry Program Exchange, QCMP 116. The Ru H_{ii} values (wave function parameters) used were: H_{ii} Ru(4d)=-12.2 eV (ξ_1=5.378, ξ_2=2.303, c_1=0.5573, C_2=0.6642); H_{ii} Ru(5s)=-9.26 eV (ξ=2.078); H_{ii} Ru(5p)=-5.47 eV (ξ=2.043).

Enforced Coordination Geometries – Preorganization, Catalysis and Beyond

Peter Comba

Anorganisch-Chemisches Institut, Universität Heidelberg, INF 270, D-69120 Heidelberg, Germany
e-mail: comba@akcomba.oci.uni-heidelberg.de

Abstract.

Stabilities, reactivities and electronic properties of transition metal coordination compounds are related to the number and type of donor groups and to their geometric arrangement. Carefully designed rigid ligands may be used to enforce specific coordination geometries, reactivities and stabilities. Applications range from metal ion discrimination and stereoselective coordination of substrates to compounds with unique spectroscopic and electrochemical properties, to low molecular weight structural and spectroscopic models of metalloproteins and to catalytically active coordination compounds.

Introduction

All molecular properties are the result of the corresponding structures; for transition metal coordination compounds this involves the metal centers and their oxidation states, the donor atoms and their substitution pattern and the geometric arrangement of the donors around the metal ions. Structures of coordination compounds are the result of a compromise between metal ion preference and ligand enforced geometries. In general, torsional modes are weaker than angle bending, and this is weaker than bond stretching; the geometry around metal ions is more flexible than the ligand backbones. This emerges from vibrational spectroscopy, structural analyses and force fields used successfully for the computation of structures, thermodynamic and spectroscopic properties of transition metal compounds [1-3]. The weakest modes in ligands are rotations around single bonds, and very rigid ligands generally have backbones which are reinforced by rings or multiple bonds. Therefore, with very rigid ligands, specific and unusual geometries may be enforced to metal ions, and this may lead to compounds with unusual properties. Rigid ligands with specific topologies have been used to influence reactivities, thermodynamic and electronic properties, and key examples include compounds in the entatic state (that is, energized to enhance the reactivity [4, 5]), metal ion selective ligands, for example for metal ion separation processes [6], and the stabilization of specific oxidation and electronic states [3].

Empirical, semi-empirical and quantum mechanical calculations have been used to compute ground and transition state structures and the corresponding properties of transition metal coordination compounds. Empirical force field calculations (molecular mechanics) have been found to produce structural data with relatively high accuracy [2, 7, 8], and the efficient structure optimization allows for a thorough conformational search [9]. Various methods have been used to compute molecular properties from the refined structures, and these include single point MO calculations, ligand field calculations, spectra simulations, as well as correlations based on strain energy differences from molecular mechanics calculations [2, 10]. All these methods may and have been used for the design of new compounds with specific properties and for the determination and interpretation of structural features (e.g. structures in solution).

In this short account I will summarize some of our own recent results and plans in this area.

Tuning the ligand field and reduction potentials

Cobalt(III) hexaamines generally are yellow (first dd transition at approximately 21000 cm^{-1}; 475 nm), they have relatively low reduction potentials (around 0.0 V, vs. SHE), and the cobalt-nitrogen bonds are relatively short (approximately 1.96 Å [11]), they vary only in small limits and are relatively strong [3]. Two hexaamine ligands have recently been reported, which enforce very short cobalt-nitrogen bonds (strong ligand field, negative reduction potential) in one, and very long cobalt-nitrogen bonds (weak ligand field, high reduction potential) in the other case.

The strongest ligand field excerted by purely aliphatic amines, for cobalt(III) and many other first row transition metal ions, is achieved with a bis-pendent amine derivative of cyclam, *trans*-diammac, the zinc/acid-reduction product of a readily available compound, obtained by a copper(II)-induced template reaction (Figure 1a) [12]. The cobalt(III) compound of *trans*-diammac has Co-N distances of 1.937 Å and 1.946 Å, the first dd band at 22170 cm^{-1} (450 nm) and a reduction potential of –0.63 V (vs. SHE) [3, 13-16]. Metal complexes of *trans*-diammac may occur in three different conformations (Figure 1b); their relative stability depends on the metal ion size, and this has been analyzed on the basis of force field calculations and verified by a series of X-ray structural analyses [2, 3, 15, 16]. Due to differences in the tilt angle θ (deviation from 90° of the angle between the axis through the two primary amine donors and the mean plane through the metal center and the four macrocyclic amine donors), there are significant spectroscopic variations. A combination of force field (MM) and ligand field (AOM) calculations (the MM-AOM method) has been used to compute the corresponding spectroscopic properties, and this was the basis for a thorough analysis of solution structures, e.g. for iron(II) and iron(III) compounds (both low spin) [17, 18]. The conformations of the in-plane, macrocyclic five-membered chelate rings of the coordinated *trans*-diammac ligand, that is, any of the three isomers in Figure 1b, may be enforced by substitution of the ethylene bridges. Fully preorganized ligands may be obtained, when pure specific isomers of bis-(cyclohexane-1,2-diamine)-copper(II) salts are used in the template reaction (see Figure 1c). Some of the corresponding copper(II) precursors have been isolated [19].

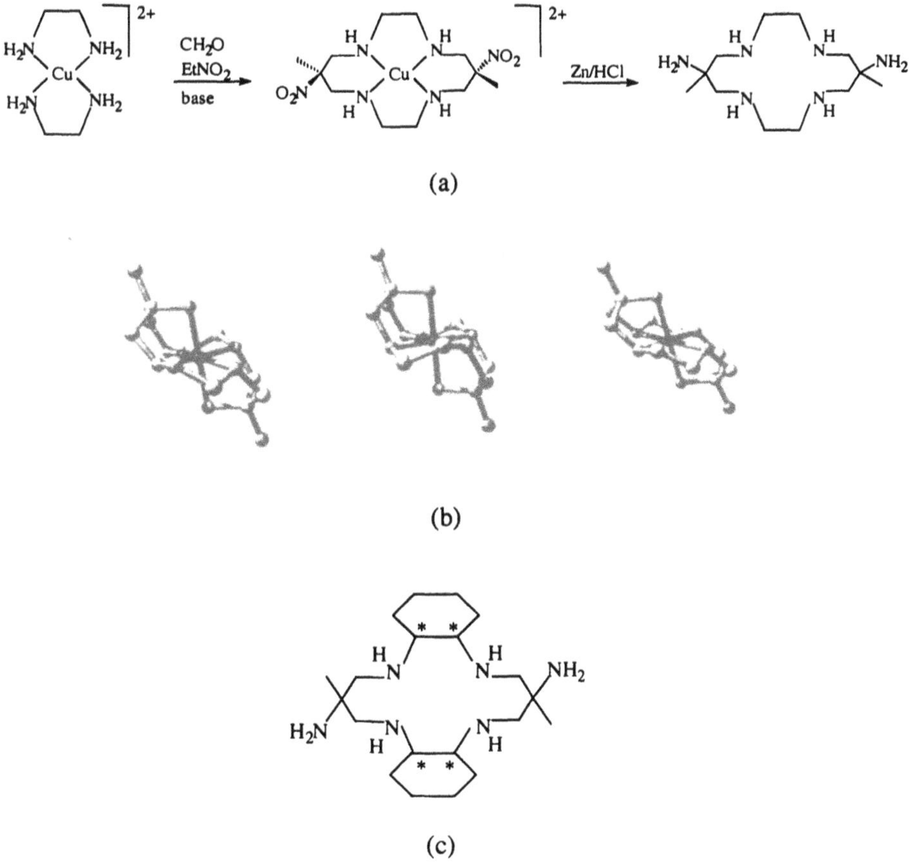

(a)

(b)

(c)

Fig. 1. (a) The template synthesis of the hexaamine ligand diammac [12]; (b) computed structures of the three isomers of $[M(\textit{trans}\text{-diammac})]^{n+}$; (c) the diammac derivative based on cyclohexane-1,2-diamine [19]

The weakest ligand field for a cobalt(III) hexaamine has been observed in one of the two stable conformers of a cobalt(III) compound with a hexaamine cage ligand (Figure 2) [20]. The other stable isomer has properties which are as expected for hexaamine cobalt(III) compounds, and its structure has been determined by X-ray crystallography (Figure 2a): the three lateral six-membered chelate rings have, due to the two methyl substituents, twist conformations, and their orientation is oblique (ob) with respect to the C_3-axis through the metal center and the two adamantane-derived caps; the observed Co-N distances are 1.990 Å; the compound is yellow (first dd transition at 20833 cm^{-1}; 487 nm); the reduction potential is 0.0 V (vs. SHE) [3, 20, 21]. The solid state structure of the second isomer has not been determined, and it has properties which are unprecedented for cobalt hexaamines: the compound is blue with a first dd transition at 16670 cm^{-1} (600 nm, that is, an energy difference of 4160 cm^{-1} (0.5 eV; 50 kJ/mol) with respect to the yellow conformer; the reduction potential is +840 mV (vs. SHE), that is, an energy difference of 80 kJ/mol (0.8 eV) with respect to the structurally characterized isomer [20, 21]. A reorientation of the lateral, six-membered chelate rings to lel conformations (parallel with respect to the C_3 axis, twist conformations conserved) was expected to yield a larger cavity (stabilization of the reduced form, weaker ligand field), and this was confirmed by force field calculations (structures and relative stabilities) and by the computation of the ligand field spectra (MM-AOM) and of the reduction potentials (MM-Redox) [21], with methods, which have been validated and described before [3, 10, 14, 15, 18, 22]. Details are given in Figure 2 and Table 1.

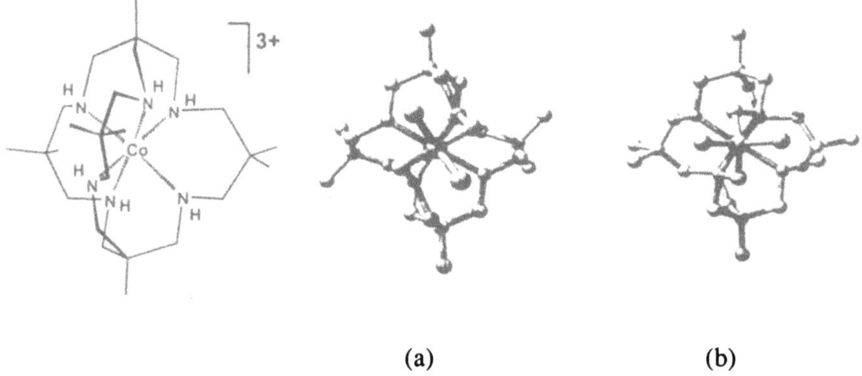

Fig. 2. Computed structures of the two stable conformers ((a) ob_3, (b) lel_3) of a cobalt(III) compound with a hexaamine cage ligand [21]

parameter		ob_3 (yellow)		lel_3 (blue)	
		obs	calc	obs	calc
Co^{III}- N	(Å)	1.990	1.985		2.046
Co^{II} - N	(Å)		2.124	2.235	2.224
strain energy (Co^{III})	(kJ/mol)		279	282	
strain energy (Co^{II})	(kJ/mol)		238	177	
E°	(V)	0.0	-0.1	0.84	0.89
dd absorption	(nm)	487	481	600	612
		350	344	370	416

Table 1. Experimentally determined and computed properties of the two isomers of the cobalt hexaamine compound shown in Figure 2 [20, 21]

Square planar geometry of copper(II) tetraamines (4- instead of the usual (4+1)- or (4+2)-coordination) has been achieved with a sterically demanding ligand, which efficiently shields the axial sites (see Figure 3) [23-25]; this leads to a significant shift of the dd transitions to higher energy, and the four-coordinate isomer of the copper(II) tetraamine compound in Figure 3 is orange (and not blue or purple as usually would be assumed) [25, 26]. Similar observations have been made with other copper(II) tetraamines [27]. Due to the expected inverse correlation between axial and in-plane bonding in tetragonal d^9 copper(II) compounds [28, 29], ligands which enforce a strong in-plane ligand field generally lead to long axial bonds and relatively high energy dd transition [25, 26].

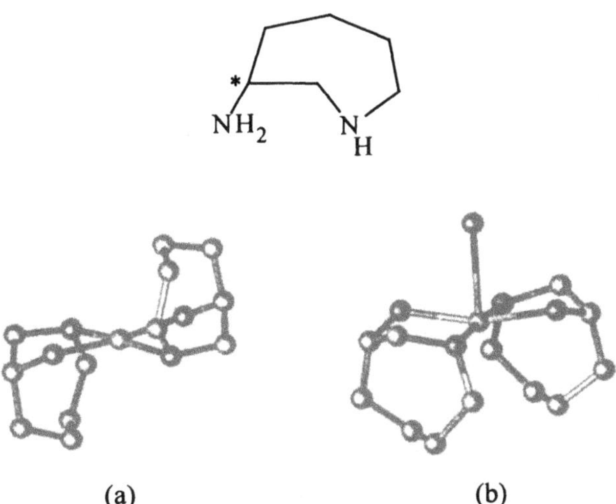

Fig. 3. Computed structures of (a) $[Cu(R\text{-}ahaz)(S\text{-}ahaz)]^{2+}$ and (b) $[Cu(S\text{-}ahaz)_2(OH_2)]^{2+}$ [25]

One possibility to achieve strong in-plane ligand fields is to enforce a planar geometry with a macrocyclic tetraamine with a relatively small hole size. A cyclen derivative with two reinforced chelates (see Figure 4a) was proposed to fulfill these requirements [30, 31]; limitations in preparative methods have so far prevented the exploration of this exciting chemistry. Another highly preorganized tetraaza-macrocyclic ligand has just become available (see Figure 4b), and preliminary modeling studies indicate that the structures of the metal-free and to copper(II) coordinated ligand are very similar [32]. Note, that a further restriction of the flexibility of the metal-free ligand might be achieved by substitution of the ethylene bridges, as described for *trans*-diammac above (see Figure 1c). In addition to the interesting ligand field properties, highly preorganized ligand systems such as these are predicted to be very selective complexation reagents (that is, they lead to very stable complexes with a small range of metal ions), and to have interesting electron transfer properties (reduction potentials, electron transfer rates) [5].

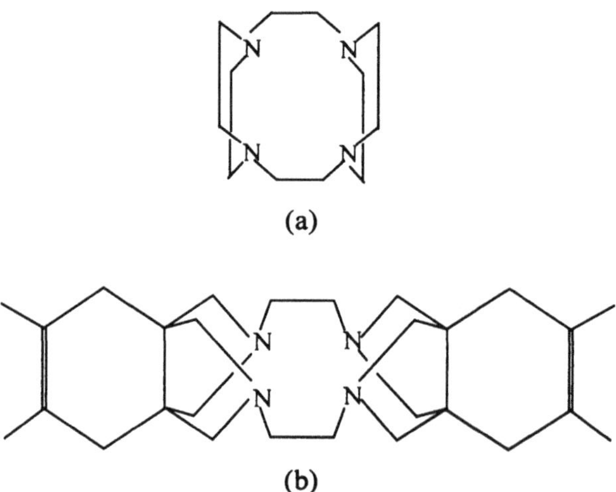

Fig. 4. Two highly preorganized tetraazamacrocyclic ligands which enforce a strong in-plane ligands field [30-32]

Reversible oxygen binding, oxygen activation and oxidation catalysis by copper(I/II) systems

Oxygen transport and oxidation catalysis are important processes in nature and industrial technology [33-37]. Hemocyanin and tyrosinase are two particularly well characterized examples [38-41], and low molecular weight model compounds of these enzymes have been studied extensively, in order to understand their structural and spectroscopic properties, to mimic their reactivity and eventually design, prepare and study active model compounds, which might be used for processes of industrial interest [33-38, 42-49]. Three binding modes for dioxygen to dicopper(I) units have been described, and model compounds of all of these have been studied extensively (Figure 5) [33-35, 38, 42-49].

$$2[Cu^{I}L] + O_2 \rightleftharpoons [(Cu^{II}L)_2(O_2)]$$

(a)

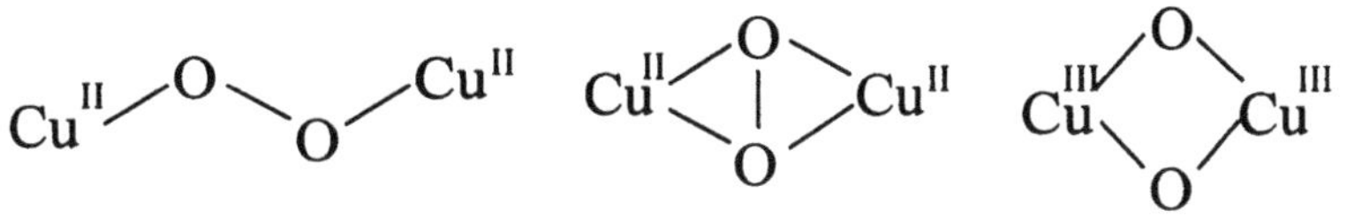

(b)

Fig. 5. (a) The oxygenation reaction of copper(I) (charges omitted); (b) binding modes for dioxygen to dicopper compounds

A general problem of low molecular weight model compounds in this (and other) areas is a relatively low stability of the active site models, which is primarily due to the missing protection by the protein backbone and difficulties to efficiently preorganize the ligand systems. For example, a variety of end-on (μ-peroxo)-dicopper(II) compounds have been reported [42, 43], but generally these are only stable at very low temperature (typically –50°C), and only one solid state X-ray structure has been reported so far [43]. Based on extensive studies on the kinetics and thermodynamics of the oxygenation reaction, some considerably more stable end-on (μ-peroxo)dicopper(II) compounds have recently been reported [45-48], and the main features for the stabilization involve (see also Figure 5a): preorganization of the dinuclear product by carefully designed dinucleating ligands [44-46, 48]; destabilization of the copper(I) precursor by the choice of an appropriate solvent [45, 47] or by a ligand that prefers the oxidized form of the copper site [5, 49]; stabilization of the copper-oxygen bond by enforcing a coordination geometry which leads to a strong interaction between the relevant copper and oxygen orbitals [48].

Tetradentate bispidine derivatives are easily accessible, very rigid, tetradentate ligands (Figure 6) [50, 51]. Apart from a bispidine-derived tetradentate ligand with a different topology, and which enforces planar geometry [52], their coordination chemistry was largely unexplored until recently [53]. Molecular modeling indicates that ligands such as those in Figure 6 are indeed very rigid and generally lead to square pyramidal or octahedral structures with two non-equivalent bonds to the two tertiary amine donors [53, 54]. This was confirmed by a series of structurally characterized compounds, including mono- and dinuclear manganese(II) [55, 56], iron(II) [57], cobalt(II) [53], copper(I) [49] and copper(II) [48, 58] compounds; the rigidity of the ligand was confirmed by its experimentally determined structure

[49]; note, that recently we also observed seven-coordinate species with manganese(II)-bispidine chromophores [56].

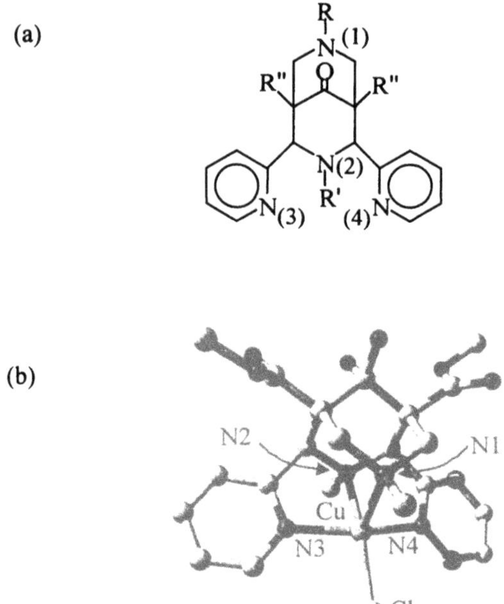

Fig. 6. (a) A bis-pyridine-substituted bispidine-type tetradentate ligand (R = R' = CH_3, R" = $COOCH_3$); (b) representation of the experimentally determined structure of a five-coordinate copper(II)-bispidine compound (acetalization of the carbonyl group upon coordination).

Bispidine-type ligands, similar to those shown in Figure 6a, but with six-membered chelate rings involving the pyridine donors and the metal ion, have rather small cavity sizes and enforce a distorted tetrahedral coordination geometry [53, 54, 59]. Hexadentate derivatives with four pyridine donors have also been prepared, and modeling studies, as well as experimental structures of a series of transition metal coordination compounds indicate that the cavity size of this ligand is rather large [60].

For the dicopper-peroxo chemistry with donors of the type described in Figure 6a it emerges that (i) the bispidine-type diaminodipyridinyl cavity is ideally preorganized for copper(II) compounds, with an enforced square pyramidal coordination geometry with the two pyridine donors trans to each other, one of the tertiary amines (N(2) in Figure 6a) and a monodentate ligand trans to N(2) (e.g. Cl^-, see Figure 6b, or a bridging ligand in a dinuclear compound, e.g. peroxide, see below) completing the plane, and with a significantly elongated bond to the second tertiary amine (N(1) in Figure 6a); this tetragonal geometry, which stabilizes the Jahn-Teller labile copper(II) ground state, has also been observed for other metal complexes with this ligand [53, 55, 57]; (ii) small substrates (e.g. peroxide) generally are coordinated in the basal plane, trans to the tertiary amine donor N(2), that is, there may be strong σ-interactions (Cu d_{x2-y2}) and possible π-interactions (Cu d_{xz}, Cu d_{xy}) between the donor and some copper orbitals; (iii) dinucleating ligands, where N(1)-R (or N(2)-R') is part of diamine, which links two bispidine-type cavities, are readily available, that is, two cavities may be oriented in such a way that the structure of the precursor is well preorganized for the dicopper(II) peroxo product; (iv) the rigidity of the ligand, together with the high degree of preorganization for copper(II), imply that the copper(I) precursor is destabilized, that is, in an entatic state [5]. All these features have been tested experimentally [48, 49, 58]. The basic tetradentate bispidine-type ligand (see Figure 6a) leads to an end-on (μ-peroxo)dicopper(II) compound which is stable up to about –20°C. A qualitative spectroscopic analysis indicates that the chromophore has a square pyramidal structure with a significant distortion towards trigonal bipyramidal (N(2)-Cu-O in the computed structure (MM) is 154°, compared to N(2)-Cu-Cl of 165° in the chlorocopper(II) compound (X-ray), compare Figures 6 and 7) [48]. Therefore, a considerable contribution to the stabilization of this peroxo compound might be due to a destabilization of the copper(I) precursor with respect to the (peroxo)dicopper(II) product [49], see below.

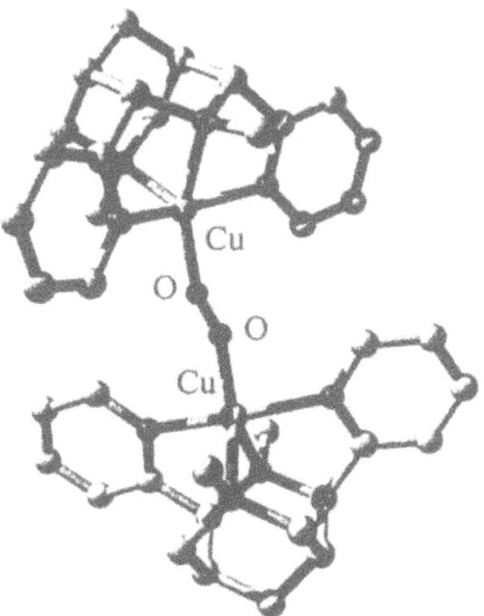

Fig. 7. Representation of the computed structure of the end-on (μ-peroxo)dicopper(II) compound of the mononucleating bispidine-type ligand (substituents R" and carbonyl group (see Figure 6) omitted)

The electronic stabilization, due to a square pyramidal geometry of the (μ-peroxo)dicopper(II) product with the copper—oxygen bond in the basal plane is further supported by studies involving the bis-α-methylpyridine-substituted derivative of the bispidine-type ligand: structural analyses of two copper(II) compounds indicate that, due to steric strain, coordination to the equatorial site is destabilized (see Figure 8), and oxygenation of the corresponding copper(I) precursor does not lead to a stable μ-peroxo intermediate [49].

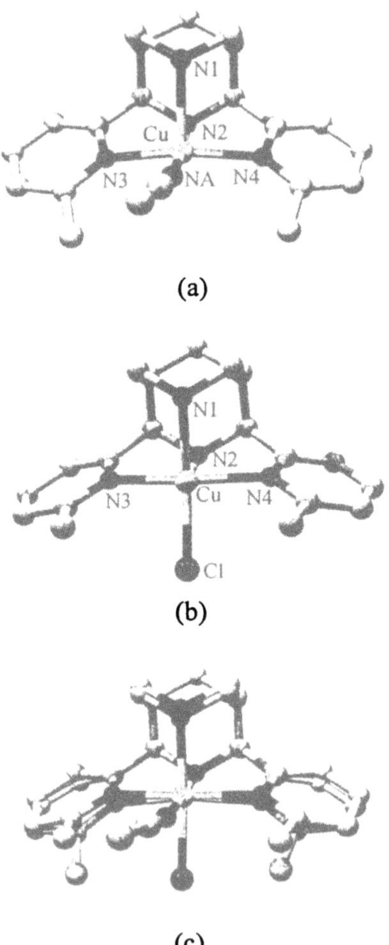

Fig. 8. Plots of the experimentally determined structures of copper(II) compounds with a bis-α-methylpyridine-substituted bispidine derivative (substituents R, R', R" and carbonyl group (see Figure 6) omitted); (a) acetonitrile complex; (b) chlorocomplex; (c) overlay plot of (a) and (b)

Modeling studies (MM) [44] indicated that a dinucleating bis-bispidine-type ligand with an ethyl bridge between the two tertiary amines N(1) (see Figure 6a) would lead to a minimum of strain, induced to the (μ-peroxo)dicopper(II) product, and a highly preorganized dicopper(I) precursor [48, 49]. Oxygenation of the dicopper(I) compound with an ethyl spacer group indeed leads to a very stable peroxo product, and this was characterized spectroscopically [48, 58]; with a half-life of 50 min at 298 K this is one of the most stable (μ-peroxo)dicopper(II) compounds known to date. The computed structure indicates that this compound is less distorted towards trigonal bipyramidal geometry than the unbridged compound (see Figure 9a; N(2)–Cu–O: 169°). The high degree of preorganization is visualized in overlay plots of this computed structure with experimental structural data of bridged and mononuclear copper(II) bispidine compounds (Figure 9b,d). The increased stability of the copper–oxygen bond, due to the square pyramidal chromophore, also emerges from the spectroscopic analysis: the copper–oxygen vibration (Raman) is moved to significantly higher energy, while the oxygen–oxygen bond is weakened, and, on a qualitative basis, this is also supported by a shift of the charge transfer transition in the UV-vis spectrum [48]. A possibility to further increase the stability of the peroxo dicopper(II) product is to restrict the rotation around the single bonds of the diamino-1,2-ethane fragment. A cyclohexane derivative of the ligand might lead to significant further stabilization; this is supported by preliminary modeling studies, and a corresponding ligand has now been prepared and characterized [61].

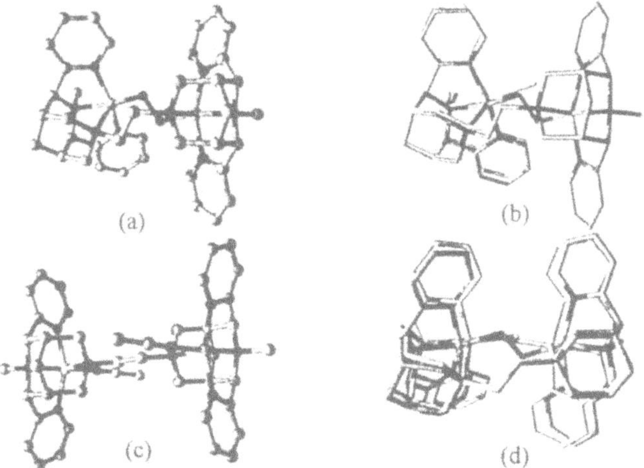

Fig. 9. Stabilization of the dicopper(II) peroxo compound with a highly preorganized dinucleating bis-bispidine-type ligand (substituents R', R" and carbonyl group (see Figure 6) omitted); (a) computed structure of the peroxo-dicopper(II) complex; (b) overlay of (a) with the experimentally determined structure in Figure 6b; (c) plot of the experimentally determined dicopper(II) structure with the dinucleating ligand (bis-acetonitrile complex); (d) overlay of (a) and (c) (torsional angles involving the ethyl spacer group have been adjusted in (c))

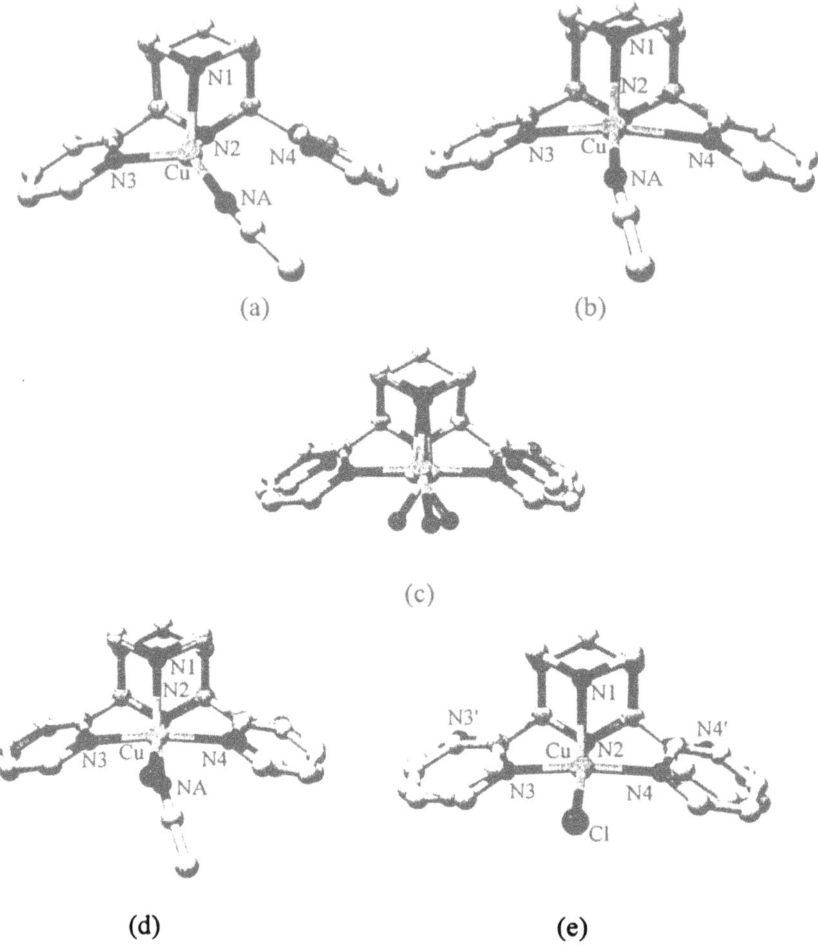

Fig. 10. Plots of experimentally determined structures of the mononucleating bispidine-type ligand and its copper(I) complexes (substituents R, R', R" and carbonyl group (see Figure 6) omitted); (a) four-coordinate copper(I) complex; (b) five-coordinate copper(I) complex; (c) overlay of the two enantiomers of (a) and (b); (d) overlay of (b) and the corresponding copper(II) complex (Figure 6b); (e) overlay of the metal-free ligand and the copper(II) complex (Figure 6b)

From the structural analyses of copper(I) compounds with the mononucleating bispidine-type ligand (Figure 6a) it follows that the copper(I) precursor compounds are indeed destabilized, that is, in an energized (entatic) state [3]. Two isomers have been isolated: a yellow, four-coordinate species with only one coordinated pyridine donor, and a red, five-coordinate compound, both with a coordinated acetonitrile molecule (see Figure 10) [49]. An overlay plot of the two structures (Figure 10c) indicates that rotations of approximately 20° around the C—C single bonds, involving the pyridine donors, and a translation of the copper center along the pyridine nitrogen—pyridine nitrogen axis leads to an interconversion of the two possible four-coordinate isomers, with the five-coordinate species as an intermediate structure. Accordingly, this compound shows a fast dynamic process in solution (^{1}H-NMR spectroscopy) [49]. The rigidity of the bispidine ligand, the high degree of preorganization for coordination to a copper(II) center and the energization of the copper(I) precursor are further supported by overlay plots of the copper(II) and copper(I) structures (chloro complex and five-coordinate, red isomer, respectively, Figure 10d) and of the copper(II) compound with the metal-free ligand (Figure 10e) [49].

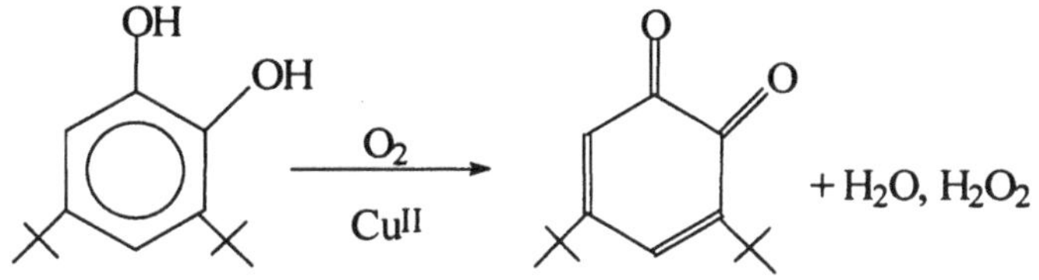

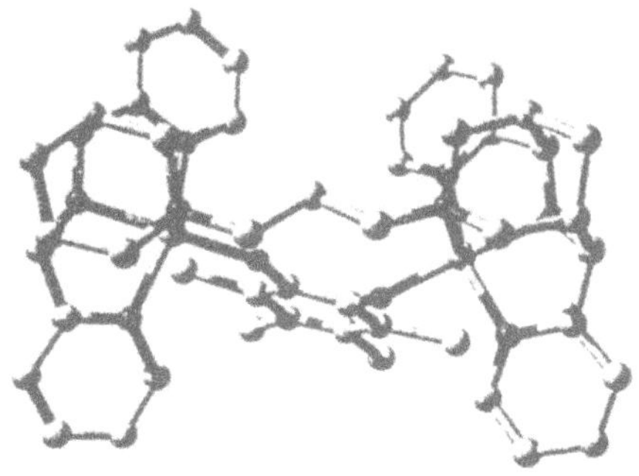

Fig. 11. (a) Catechol oxidation; (b) plot of the experimentally determined structure of a catechol-brigded dicopper(II) compound (propylene-bridged bis-bispidine (substituents R', R" and carbonyl group (see Figure 6) omitted); tetrachlorocatechol)

Preliminary experiments indicate that copper(II)-bispidine compounds catalyze the oxidation of catechol derivatives (see Figure 11; note, that various mechanisms have been observed in dependence of the copper(II) catalyst, and this has not yet been analyzed for our systems) [62, 63]. A significant variation of the catalytic activity with the nature of the bridge of the dinucleating ligand indicates that the catalytically active species has a bridging catecholate, and this is also supported by spectrophotometric studies that involve the copper(II) catalyst and variable amounts of a deactivated (chloro-substituted) catechol derivative, and by the experimentally determined solid state structure of a catechol-derivative bridged dicopper(II) compound with a dinucleating bispidine ligand (Figure 11b) [64]. Structures of copper(II) bispidine compounds with monodentate and bidentate catecholate have now also been obtained [64], and these are all structural types which have been proposed in the tyrosinase/catechol oxidase cylce [65].

Our recent studies with bispidine-type ligands have been successful and fascinating primarily because these ligands are extraordinarily rigid and highly preorganized for copper(II). At present, we are further investigating the coordination chemistry of this type of ligand with other metal ions. A key but not the only feature for possible catalytic activities in other systems is a structural match or mismatch between the ligand and the corresponding metal center or metal-substrate fragment.

An interesting question is, what the properties of transition metal compounds of the *anti*-isomer of the basic bispidine-type ligand in Figure 6a might be, compared to those of the readily available *syn*-isomer. The predicted structures (Figure 12) indicate that electronically, these compounds might be quite different and that the enforced tetragonal geometry (Jahn-Teller distortion) is lost. However, the *anti*-isomer (Figure 12b) is chiral, and this might lead to other interesting applications. Unfortunately, we have not yet been able to isolate this ligand in the desired configuration [66].

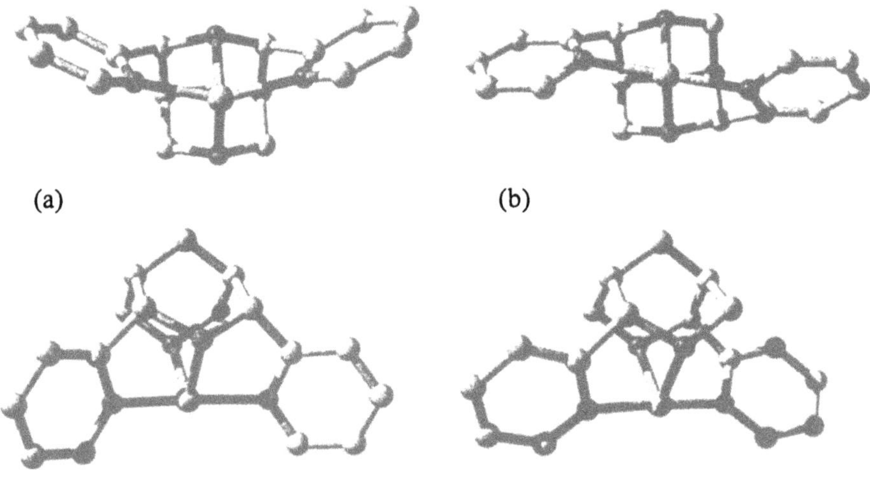

Fig. 12. Computed structures of tetracoordinate complexes with (a) a *syn*-bispidine derivative and (b) an *anti*-bispidine derivative

Acknowledgement

Our own studies in this area are supported by the German Science Foundation (DFG), the Fond of the Chemical Industry (FCI) and the VW-Stiftung. I am grateful for that, for the excellent work done by my coworkers, whose names appear in the references to our work, and to Brigitte Saul, Marlies von Schoenebeck-Schilli and Karin Stelzer for their help in preparing the manuscript.

References

1. Comba P (1993) Coord. Chem. Rev. 123: 1
2. Comba P, Hambley TW (1995) Molecular Modeling of Inorganic Compounds. VCH, Weinheim
3. Comba P (1999) Coord. Chem. Rev. 182: 343
4. Williams RJP (1995) Eur. J. Biochem. 234: 363
5. Comba P (2000) Coord. Chem. Rev. accepted
6. Comba P (1999) Coord. Chem. Rev. 185: 81
7. Boeyens JCA, Comba P (2000) Coord. Chem. Rev. accepted
8. Cundari TR (1998) J. Chem. Soc., Dalton Trans.: 2771
9. Bartol J, Comba P, Melter M, Zimmer M (1999) J. Comput. Chem. 20: 1549
10. Comba P (1999) In: Howard JKA, Allen FH (eds) Implications of Molecular and Materials Structure for New Technologies . Kluwer, Dordrecht p 87
11. Orpen AG, Bramner L, Allen FH, Kennard O, Watson DG, Taylor R (1989) J. Chem. Soc., Dalton Trans.: S1
12. Comba P, Curtis NF, Lawrance GA, Sargeson AM, Skelton BW, White AH (1986) Inorg. Chem. 25: 4260
13. Bernhardt PV, Lawrance GA, Hambley TW (1989) J. Chem. Soc. Dalton Trans. : 1059
14. Comba P, Sickmüller AF (1997) Inorg. Chem. 36: 4500
15. Bernhardt PV, Comba P (1993) Inorg. Chem. 32: 2798
16. Bernhardt PV, Comba P (1991) Helv. Chim. Acta 74: 1834 (1992) Helv. Chim. Acta 75: 1645
17. Comba P (1994) Inorg. Chem. 33: 4577
18. Comba P, Börzel H, Pritzkow H, Sickmüller A (1998) Inorg. Chem. 37: 3853
19. Bernhardt PV, Comba P, Elliot BL, Lawrance GA, Maeder M, O'Leary MA, Wei G, Wilkes EN (1994) Aust. J. Chem. 47: 1171
20. Geue RJ, Hanna J, Höhn A, Qin CJ, Ralph SF, Sargeson AM, Willis AC (1997) In: Issied SS (ed) Electron Transfer Reactions . American Chemical Society, New York p 137
21. Comba P, Sickmüller AF (1997) Angew. Chem. 109: 2089; Angew. Chem. Int. Ed. Engl. 36: 2006
22. Comba P, Jakob H (1997) Helv. Chim. Acta 80: 1983
23. Saburi M, Miyamura K, Morita M, Mizoguchi Y, Yohikawa S, Tsuboyama S, Sakurai T, Tsuboyama K (1987) Bull. Chem. Soc. Jpn 60: 141

24. Saburi M, Miyamura K, Morita M, Yoshikawa S, Tsuboyama S, Sakurai T, Yamazaki H, Tsuboyama K (1987) Bull. Chem. Soc. Jpn. 60: 2581
25. Comba P, Hambley TW, Hitchman MA, Stratemeier H (1995) Inorg. Chem. 34: 3903
26. Comba P (1998) In: Gans W, Boeyens JCA (eds) Intermolecular Interactions. Plenum Press, New York p 97
27. Comba P, Hilfenhaus P, Nuber B (1997) Helv. Chim. Acta. 80: 1831
28. Gazo J, Bersuker IB, Garaj J, Kabesokavá M, Kohout J, Langfelderová H, Melník M, Serátor M, Valach F (1976) Coord. Chem. Rev. 19: 253
29. Hathaway BJ, Hodgson PG (1973) J. Inorg. Nucl. Chem. 35: 4071
30. Wainwright KP (1980) Inorg. Chem. 19: 1396
31. Wainwright KP, Ramasublou A (1982) J. Chem. Soc., Chem. Commun. : 277
32. Comba P, Schiek W (1999) work in progress
33. Kitajima N (1994) Adv. Inorg. Chem. 39: 1
34. Karlin KD, Tyeklar Z, Zuberbühler AD (1993) In: Reedijk J (ed) Bioinorganic Catalysis . Marcel Dekker Inc., New York p 261
35. Tolman WB (1997) Acc. Chem. Res. 30: 227
36. Sheldon RA, Kochi JK (1981) Metal-Catalyzed Oxidations of Organic Compounds. Academic Press, New York
37. Que LJ (1993) In: Reedijk J (ed) Bioinorganic Catalysis . Marcel Dekker Inc., New York p 347
38. Karlin DK (1993) Bioinorganic chemistry of copper. Chapman & Hall, London New York
39. Magnus KA, Ton-That H, Carpenter JE (1994) Chem. Rev. 94: 727
40. Hazes B, Magnus KA, Bonaventura C, Bonaventura J, Danter Z, Kalk KH, Hol WGJ (1993) Protein Sci. 2: 597
41. Volbeda A, Hol WGJ (1989) J. Mol. Biol. 209: 249
42. Karlin KD, Kaderli S, Zuberbühler AD (1997) Acc. Chem. Res. 30: 139
43. Tyeklar Z, Jacobson RR, Wei N, Murthy NN, Zubieta J, Karlin KD (1993) J. Am. Chem. Soc. 115: 2677
44. Comba P, Hilfenhaus P, Karlin KD (1997) Inorg. Chem. 36: 2309
45. Karlin KD, Lee D-H, Kaderli S, Zuberbühler AD (1997) J. Chem. Soc., Chem. Commun. : 475
46. Bol JE, Driessen WL, Ho RYN, Maase B, Que jr L, Reedijk J (1997) Angew. Chem. 109: 1022-1025 Angew. Chem. Int. Ed. Engl. 36: 998
47. Becker M, Heinemann FW, Schindler S (1999) Chem. Eur. J. 5:3124
48. Börzel H, Comba P, Katsichtis C, Kiefer W, Lienke A, Nagel V, Pritzkow H (1999) Chem. Eur. J. 5: 1716
49. Börzel H, Comba P, Hagen KS, Katsichtis C, Pritzkow H (1999) Chem. Eur. J., in press
50. Samhammer A, Holzgrabe U, Haller R (1984) Arch. Pharm. 322: 557
51. Holzgrabe U, Ericyas E (1992) Arch. Pharm. (Weinheim, Germany) 325: 657
52. Hosken GD, Hancock RD (1994) J. Chem. Soc. Chem. Commun. : 1363
53. Comba P, Nuber B, Ramlow A (1997) J. Chem. Soc. , Dalton Trans. : 347
54. Comba P, Okon N, Remenyi R (1999) J. Comput. Chem. 20:781
55. Comba P, Kanellakopulos B, Katsichtis C, Lienke A, Pritzkow H, Rominger F (1998) J. Chem. Soc., Dalton Trans. : 3997
56. Comba P, Pritzkow H, Xiong Y publ. in preparation
57. Börzel H, Comba P, Hagen K publ. in preparation
58. Börzel H, Comba P, Pritzkow H, Schindler S publ. in preparation

59. Comba P, Lienke A, Pritzkow H work in progress
60. Börzel H, Comba P, Lienke A, Maas O, Pritzkow H publ. in preparation
61. Comba P, Kerscher M, Pritzkow H work in progress
62. Kitajima N, Koda T, Iwata Y, Moro-oka Y (1990) J. Am. Chem. Soc. 112: 8833
63. Berreau LM, Mahapatra S, Halfen JA, Houser RP, Young Jr. VG, Tolman WB (1999) Angew. Chem., Angew. Chem. Int. Ed. Engl. 111: 180; Angew. Chem. Int. Ed. Engl. 38: 207
64. Börzel H, Comba P, Pritzkow H work in progress
65. Karlin KD, Zuberbühler, AD (1999) In: Reedijk J, Bouwman E (eds), Bioinorganic Catalysis, Marcel Dekker Inc., New York p 469.
66. Comba P, Lienke A, Schiek W work in progress

Supramolecular Functions of Designed Transition Metal Ion Complexes

Paolo Scrimin[a], Paolo Tecilla[b] and Umberto Tonellato[a]

[a]Università di Padova, Dipartimento di Chimica Organica and Centro CNR Meccanismi di Reazione Organiche, Via Marzolo,1 I - 35131 Padova, Italy; [b]Università di Trieste, Dipartimento di Scienze Chimiche, Via L. Giorgieri 1, I- 34127 Trieste, Italy.
E-mail: P.S.: scrimin@mail.chor.unipd.it; P.T.: tecilla@dsch.univ.trieste.it; U.T.: tonum@mail.chor.unipd.it

1. Introduction

In the last two decades, the area of supramolecular chemistry [1] has developed at an impressive rate, first focussing on the creation of new molecules and new assemblies through noncovalent bonds to create remarkable molecules and supermolecules. However, chemists are not only the architects of wonderful static molecular structures but they are naturally interested to make them work or properly assemble to realize new molecular engines, whatever their functions may be. So, recently much attention is being directed toward the functions and applications of the supramolecular objects [2]. Much impetus struck novel fast developing areas, such as that of material science, as well old ones but with a new light and from a different perspective. The latter aspect is quite relevant in the area of life sciences. Supramolecular interactions form the basis of processes that occur in biology and one of the problems from the start was related to the possible reductive convergence to one novel branch of the "old" biomimetic chemistry [3]. Instead, chemists used the supramolecular concepts to understand biological events, such as the aggregation of lipids and membrane formation which had been largely neglected, and, by joining their efforts with biochemists, they helped to fill the cultural gap between chemistry and biochemistry with exciting perspectives. However, the affinity between supramolecular sciences and life sciences is unavoidably strong and it is in particular suggestive of the many functions that supramolecular systems can perform. Nature binds, assembles, and organizes with extreme efficiency molecules of biological relevance through noncovalent intermolecular forces and proper use of energetic and other features, in order to perform a variety of basic functions essential to life.

During the years in our laboratory we investigated many supramolecular systems with the aim of realizing reactive or catalytic species. More recently, much of our research work has been focussed on complexes of transition metal ions designed to perform different functions other than catalysis such as translocation and recognition with signal (sensing) and some of the relevant results indicated below will be presented and discussed.

2. Receptors and aggregate systems.

Following the canonical definitions of J.-M. Lehn [4] a supramolecular system is formed by the interaction with binding and selection of substrate(s) by a given receptor system, as outlined in Figure 1. Therefore, molecular receptors play a key role and their identification among the natural products (such as cyclodextrins) or, most often, their design if they are synthetic structures is a major challenge for the realization of supermolecules.

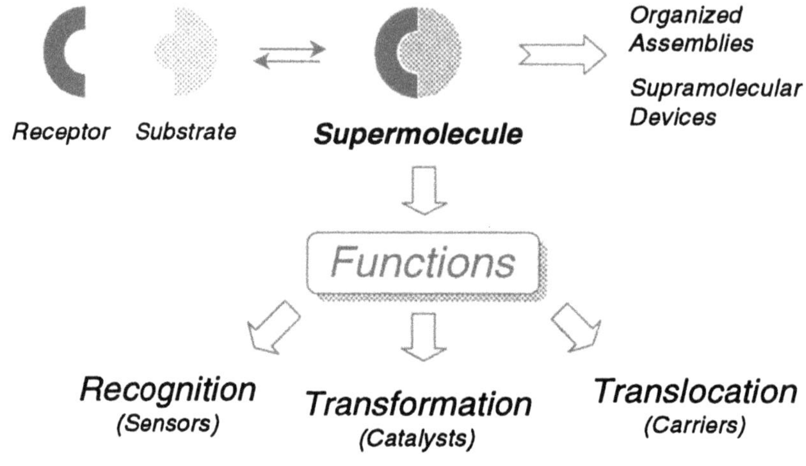

Fig. 1.

Molecular receptors should feature binding moieties whose nature, positioning and strength of binding must be complementary to those in the substrate. In the early days when attention was mainly focussed on crown ethers, cryptands and cage-type receptors, which do not have self-complementary functionalities, the definition of a molecular receptor was seemingly straightforward. However, as the research on supramolecular chemistry evolved toward more sophisticated systems designed to perform complex functions that require the introduction of spacers to allow separation of binding moieties which are complementary to each other, the design of artificial receptors had to face a number of different problems. As a matter of fact it appears [5] that much synthetic work has been wasted due to the possibility, partly underestimated, of self-quenching of binding moieties, self-association that may prevent interaction or binding with the substrate and hence solubility in the solvent of choice.

An apparently unrelated problem we had to face from a conceptual point of view, was the following. May aggregates [6], such as micelles, vesicles or membranes, be considered supramolecular receptors? Strictly speaking aggregates are not receptors. Although, at least in aqueous solutions, they efficiently associate quite a variety of substrates, they bind them with little or no recognition thus violating one of the main requisite of a proper receptor. Yet the aggregates are formed by exploiting weak non covalent interactions between the lipophilic parts of the so-called surfactants components and, as such, they may be envisaged as supramolecular systems. They bring together binding sites but, normally, they do not arrange them in suitable patterns.

Looking for a bridge between surfactant and supramolecular chemistry was our subliminal drive due to the fact that we had been working in the field of surfactant chemistry with the aim of realizing biomimetic-type catalysts [7]. Over the years, we devoted much attention to the realization of functionalized micellar or vesicular systems as catalysts of the hydrolytic cleavage of esters or amides with remarkable results. There was much claim of analogies between functionalized aggregates and enzymic reactions. The analogy, however is mostly restricted to formal kinetic aspects [8] and is, in general, unjustified since very little, if any, substrate selectivity or stereospecificity was observed and except for a few cases, the turnover rate was too poor to justify the claim that functionalized aggregates are catalysts in their own right [9]. Mechanistic investigation of these systems made it clear that the kinetic benefits observed were almost entirely due to concentration effects and, in the case of ionic aggregates, to electrostatic effects [10].

3. Transition Metal Ligands Complexes as a Source of Supramolecular Systems and Functions.

At one point, we [11] as well as other research groups [12] thought that the functionalized aggregate system could turn out to much more effective catalysts for the hydrolysis of carboxylate or phosphate esters if they were made, or could incorporate, complexed transition metal ions to give assemblies termed metalloaggregates. Needless to say, the idea of investigating metalloaggregates was mainly inspired by the mode of action of metalloenzymes such as the Zn(II) containing carboxypeptidase A. Of course, well known general arguments stimulated the use of these metal ions. Their effects involve: a) Lewis acids catalysis; b) charge neutralization; c) activation of nucleophilic functions or coordinated water molecules by substantially decreasing their pK_a, and, hence, nucleophilic catalysis; d) assistance in leaving group departure. However, although it has been largely overlooked, one of the most relevant roles of the metal ion may be the template effect *i.e.* the capability of the ion bound to a properly designed ligand to bind also a given substrate with a chelating subunit so that reactant and substrate are not only brought into proximity but somehow organized or oriented. Since accumulation and organization are main properties of molecular receptors to realise supramolecular systems, metalloaggregates may be viewed as receptors of their own kind.

In recent years our research activity developed following the idea that properly designed transition metal ion complexes may perform the canonical supramolecular functions such as selective catalysis, recognition (with signalling, i.e. sensing) and translocation of ionic or neutral organic species across natural or so-called bulk membranes. From this point of view *i.e.* that of the supramolecular functions, the results of our activity will be overviewed and commented in the next sections.

4. Reactivity and Catalysis.

Reactivity and catalysis rank among the major functional properties of supramolecular systems. There was much hope that receptors bearing reactive groups may bind suitable substrates, react with them, release products and rapidly restore the reagent for a new cycle. Again, biological events are challenging insofar as enzymes represent the highest expression of effective and selective chemical catalysis. Yet in spite of the many efforts and the very large number of new supramolecular blocks and arrays very few effective catalysts have been reported in the literature [13] and this is certainly a source of disappointment.

Rather than review the many systems reported in the literature or investigated in our laboratory, let us focus the attention on reactants and catalysts that may be comparable for the type of process that they are involved in and for the idea at the basis of their realization. We here restrict the comparison to the reactivity of metalloaggregates [11, 14] and designed metalloreceptors intended to catalyze the hydrolysis of esters of amino acids.

Our work started with the synthesis and investigation of hydroxy-functionalized lipophilic pyridine based ligands such as **1**, **2** to cite the prototype of a large variety of compounds that may form micelles in aqueous solution. They are rather strong ligands for a number of transition metal ions, in particular Cu(II), and their complexes do also form aggregates or are easily included in aggregates of inert surfactants like cetyltrimethylammonium bromide (CTABr). Following a scrutiny of quite a number and types of substrates, activated esters of carboxylic or phosphoric acids, the most impressive results [11] were obtained in the case of α-amino acid esters and the substrate of choice for the assessment of the efficacy of the systems was the *p*-nitrophenyl picolinate (PNPP). The observed rate constants actually measured under accessible conditions are, in some cases, over a million-fold larger than those observed in pure buffer and the accelerations are quite remarkable even if compared to those observed in the presence of Cu(II) ion alone, which is a good catalyst. The efficacy of the metalloaggregates toward other types of substrates is generally very modest except in the case of aggregates containing the Cu(II) complex of **1** [15]. The main modes of reaction in the case [11, 14] of α-amino acid esters are indicated in Scheme 1 as path **a** and path **b**. Path **a** involves: (i) formation of a ternary complex (ligand/metal ion/substrate); (ii) pseudo-intramolecular attack of the activated hydroxyl on the carbonyl of the ester resulting in its acylation; (iii) metal ion mediated hydrolysis of the acylated intermediate. Path **b**, involving nucleophilic attack by a coordinated water molecule, may compete with **a** which, however, largely prevails in aggregates.

The deacylation of the intermediate following path **a** is fast enough to ensure a sizable turnover rate to the metalloaggregate which thus may be defined as a true catalyst. Moreover, employing micellar Cu(II) complexes of pure enantiomers of chiral ligands such as **3-4** and the *p*-nitrophenyl esters of phenylalanine, phenylglycine or leucine as substrates we observed a remarkable enantioselectivity [16] the rate ratios being in the range 10-30 and over 50 under very special conditions [17]. Thus, metalloaggregates of the type here indicated are substrate selective and enantioselective catalysts.

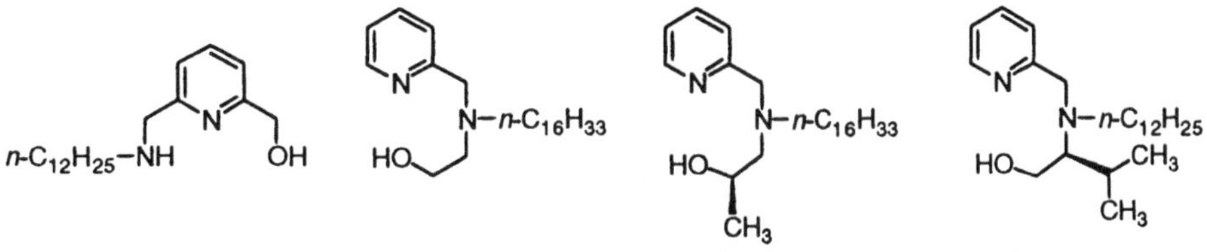

1 2 3 4

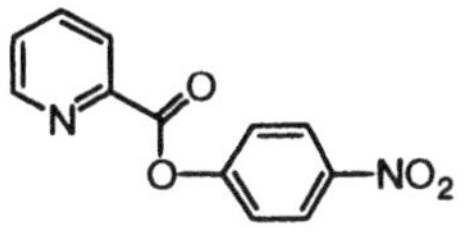

PNPP

$C_{16}H_{33}\overset{+}{N}(CH_3)_3\ \overset{-}{Br}$

CTABr

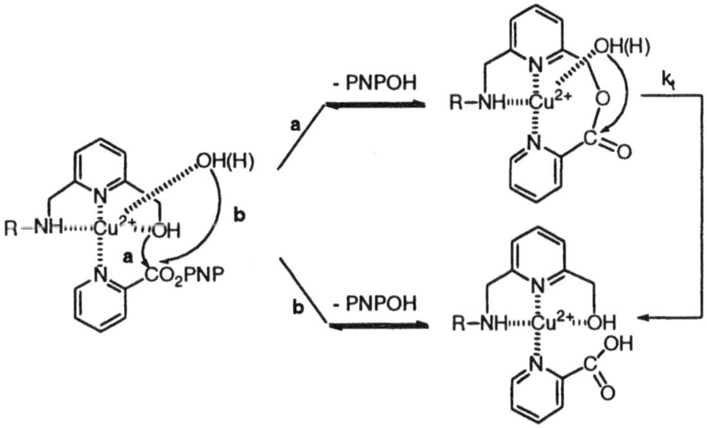

Scheme 1.

Stimulated by these results, we turned our attention to supramolecular systems that may act as metallocatalysts through a similar mode of action implying the formation of ternary complexes but in a much more rigid structure than that of a micellar aggregate. First we synthesized and investigated [18] functionalized β-cyclodextrin derivatives such as **5**. These are canonical molecular receptors providing a cavity where a proper substrate may be included and also a 2-hydroxymethyl pyridine linked to the 3-position of one of the glycopyranose rings via a thioether bond. Such receptor binds a Cu(II) ion and, in principle, it may form a reactive ternary complex with an α-amino acid ester somehow included in the cavity and protruding its relatively hydrophilic cleavable portion to interact with Cu(II). It turned out to be a failure. As a matter of fact, the pending metal ion complexed 2-hydroxymethyl pyridine moiety may react with unbound substrates much more easily than with any included one. Only in the case of the very special substrate, PNPQPh, the proper positioning and productive geometry was partly attained but the acceleration observed was quite modest.

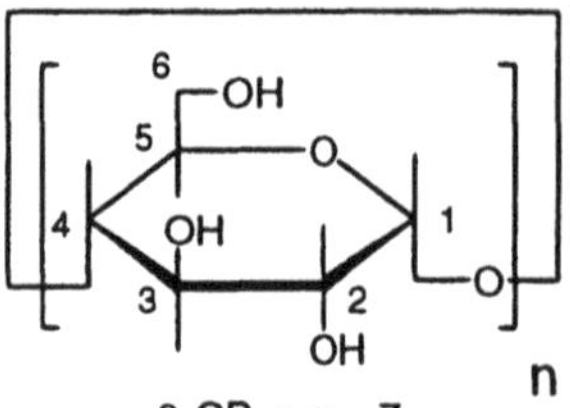

β-CD : n = 7 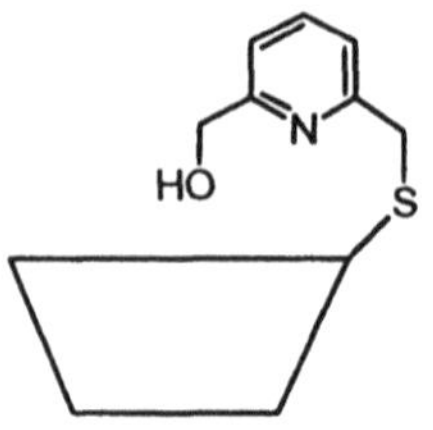 5 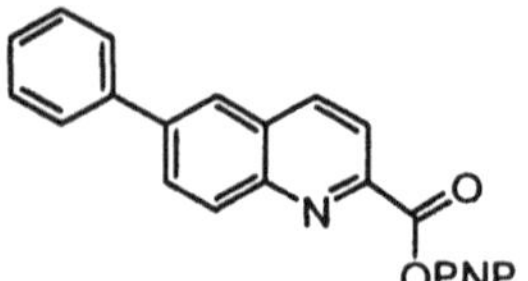 PNPQPh

We hoped that organization, recognition and transformation could be more easily achieved with a different type of receptors such as the macrocycle **6** and the open chain counterpart **7** [19]. The structures of these designed receptors feature two 2-6 dimethylaminopyridine moieties (separated by diphenylmethane spacers) in the macrocyclic **6** and one in the open structure of **7** as chelating subsites for Cu(II); in the case of **7** two amino groups are placed at the extremities in such a way that upon chelating a second metal ion the whole ditopic complex system can close up to define a (pseudo)-macrocycle. Both systems contain two Cu(II) ions and an hydrophobic cavity; moreover, in the case of **7**, two trimethylammonium residues have been attached to ensure solubility in aqueous solutions. In the cases of **6** and **7**, there are no nucleophilic functions in the proximity of the bound metal ions; the design was based on the assumption that an amino acid ester of proper size could be accomodated in the cavity in such a way that while its amino group is coordinated to one Cu(II) ion, the cleavable part of the ester could be brought in the proximity of the second resident ion where an activated water molecule could act as the nucleophile for the hydrolysis as indicated in Figures 2. Both systems did work as expected in the case of β-amino acid esters, such as that of β-alanine, while they were inert in the case of α-amino acids esters such as that of leucine, too short to properly fit into the cavity. Although clear evidence of mode of action was obtained also by comparison with half-model structure binding only one metal ion, the acceleration observed for the hydrolysis of β-amino acid esters as measured in a 1:1 water DMSO mixture in the case of **6** and in water in that of **7** were quite modest (by less than one order of magnitude). One of the reasons for such a relative inefficacy is probably due to the fact that the two metal ions bound into the macrocyclic or pseudomacrocyclic supermolecules, do not point toward the interior of the cavity in their most stable conformations but rather impose a twisted structure to the cyclic structures to minimize electrostatic repulsion [19].

The above examples, openly restricted to the case of supramolecular potential metallocatalysts of the cleavage of amino acid esters, on one hand show how difficult and frustrating may be the design of proper receptors and, on the other hand, how remarkable results can be achieved on relying on a non rigid, apparently unorganized system like that of simple aggregates. As recently pointed out by Sanders [20], by thinking of the extreme efficacy of enzymes developed by nature during millions of years, it may be illusory to think that a chemist may realize effective catalysts in a few decades in spite of the indisputable progresses thanks also to the onset of supramolecular chemistry. Quite likely, the substantial failures reported (to say nothing of the untold ones) are probably due to a mistaken strategy which has been mainly based on our "fear" of entropy so that the design of receptors for a catalytic action was based on the search of rigidity rather than on the overall flexibility of the whole reacting system.

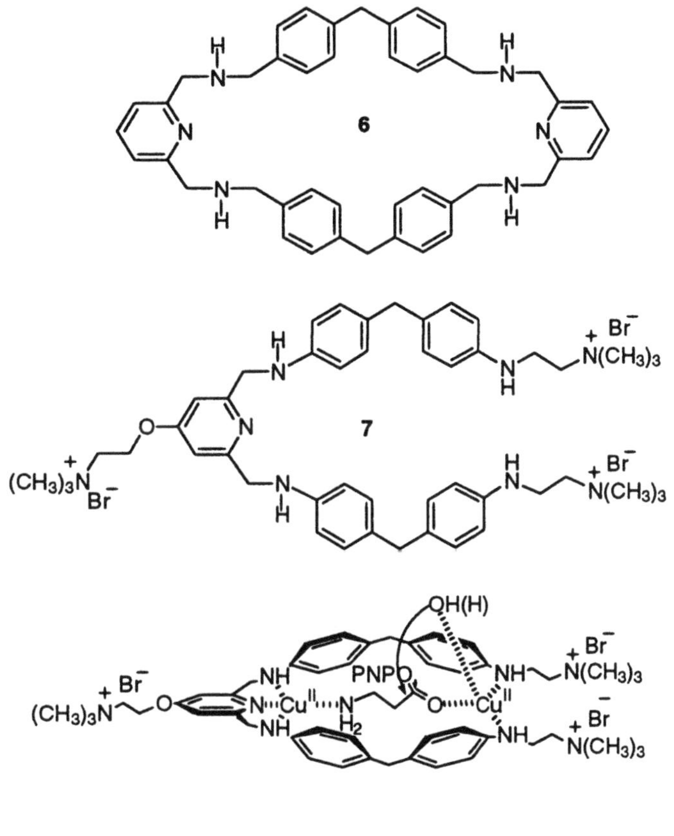

Fig. 2.

5. Translocation

The realization of receptors that bind selectively organic and inorganic substrates could turn them into molecular carriers and induce selective transport across membranes. Indeed, transport has been rightly claimed to represent one of the basic functions of supramolecular species [4]. Much work has been devoted to such functional feature with at least one very important result: eventually chemist devoted their due attention to the realm of natural membranes not only investigating their peculiar structure but also exploring and realising the promotion and control of the permeation and transport across the amphiphilic or lipophilic molecular constituents of the membrane layers. Due to the complexity of many natural membranes, many studies have been addressed to simpler models ranging from vesicles of synthetic (usually double-tailed) or natural surfactants to the so-called "bulk" membranes, *i.e.*, organic solvents separating two aqueous phases [21]. Model membranes are useful insofar as they may allow the definition of the factors playing a role in the translocation [22] and bulk membranes are clearly related to the achievement of the goal of the selective extraction and translocation of organic or inorganic species. Transport, as a supramolecular function, is mediated by a carrier molecule which at some points of the process act as a molecular receptor by binding and releasing a given substrate at the interface.

Taking advantage of our experience with ligands employed in the metalloaggregates investigated as catalysts, we recently explored the ability of simple N-mono- or dialkylated dipeptide ligands such as **8** to act as selective and effective carriers for the transport of transition metals across a liquid chloroform bulk membrane [23]. The system is quite effective in the transport of Cu(II) from a buffer source to a 0.1 M HCl receiving phase across a bulk chloroform membrane using the setup illustrated in Figure 3. The efficiency depends on the easy formation of the complex with the metal ion, **9**, which is neutral (and hence more soluble in the organic solvent) due to the deprotonation of the peptide (with an apparent pK_a close to 5) and carboxylic acid functions, at the source-chloroform interface and of the facile decomplexation at the acidic receiving phase. Under the pH conditions used the carriers are much less effective in the transport of Zn(II) and Ni(II) (by factors larger than 10^3 and 10^4, respectively, using as a carrier **8a**) due to the fact that only Cu(II) among other transition metal ions can form neutral complexes at the pH value of the source phase.

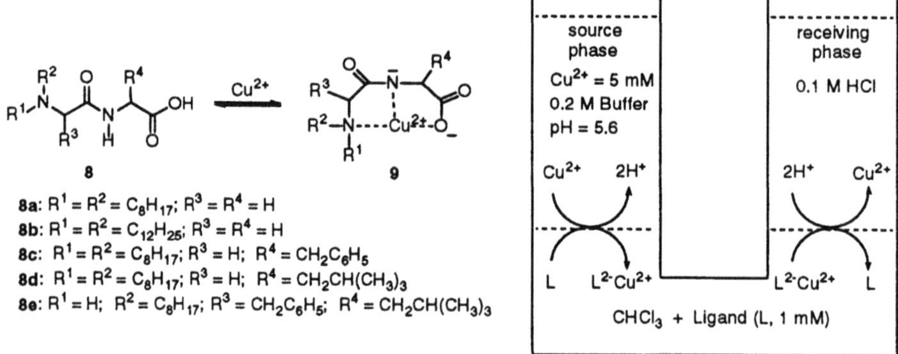

Fig. 3.

Our early idea [24] was to employ as carriers 1,2-diamino ethane derivatives bearing on one or both nitrogen atoms long paraffinic chain as carriers of both Cu(II) and α-amino acids taking advantage of the fact that they may form ternary complexes at the source and release them at the receiving phase containing EDTA as the complex is disrupted. The co-transport is effectively achieved and its effectiveness depends on the lipophilicity of the amino acid and on the affinity constant of the carrier for the metal ion which should be as close as possible to that of the amino acid transported. We also employed chiral ligands to look for enantioselectivity in the co-transport of enantiomeric amino acids but such effects were small if any under the conditions used.

The task with lipid bilayers, *i.e.*, closer models of biological membranes, is far more complex. Vesicles are sphere-shaped aggregates made of a bilayer of lipids trapping in its interior a water pool. The membrane made of the bilayers is generally tightly packed providing an effective barrier for the permeation of polar species such as water and ionic species [25]. We synthesized and investigated designed amphiphilic species [26, 27], both cationic and anionic, such as **10** and **11**, the latter one in collaboration with the group of Professor R.A. Moss at Rutgers University, N.J., U.S.A., which are functionalized with reactive moieties (*p*-nitrophenyl esters). Because of the presence of the labile functional group they may be catalytically hydrolyzed in the presence of Cu(II) (**10**) or Eu(III) (**11**) ions. When the proper metal ion was added to a solution of preformed vesicles below the phase transition temperature, only a fraction (around 60%) of the cleavable membrane components are hydrolyzed indicating that the ions interact only with the ester residing on the outer leaflet of the vesicle. On the other hand, above the phase transition temperature when "melting" of the paraffinic chains occurs and the vesicles loose their rigidity, the esters are fully hydrolyzed and this was shown to occur not because of the permeation of the metal ion across the bilayer but rather because the amphiphilic components commute from one layer to the other one in a flip-flop type of motion so exposing themselves to the external ions that act as catalysts of their hydrolysis. Thus the membranes of the vesicles are not permeable to metal ions such Cu(II) or Eu(III) and this is independent from electrostatic factors.

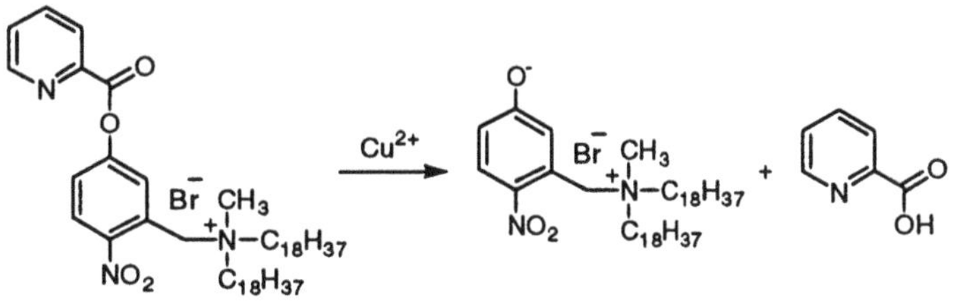

Cu^{2+}

10

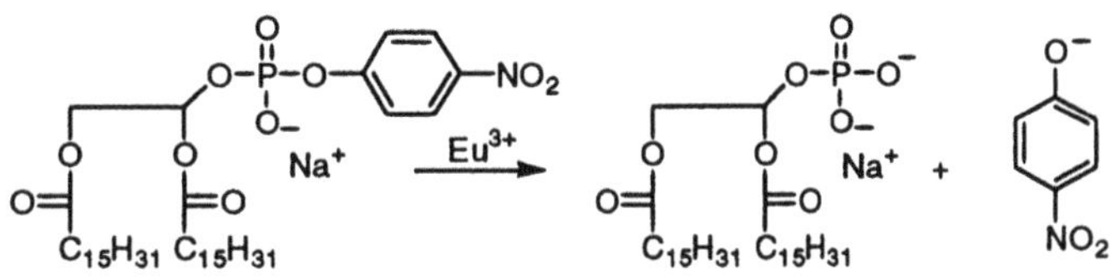

Eu^{3+}

11

More recently, exploiting the very same principle, we have prepared vesicle-forming lipids which upon Ln(III)-catalyzed hydrolysis are converted into single chain surfactants [28]. In this way the Ln(III) induces, hydrolyzing the lipids, vesicle decapsulation with consequent release of their content. This process may find application in the controlled delivery of drugs.

Then we turned our attention to the permeability of biological membranes and to the possibility to control it. Nature often employs amphipatic polypeptides which form channels or pores across the bilayer. Some of them, such as the peptaibols [29], are peptides which are easily trapped in the bilayer of the membrane in an helical conformation and if they are long enough to span the entire length of the membrane can form channels by clustering and exposing the hydrophobic part of the helix of the lipids. Helical conformation, proper length and an hydrocarbon chain connected to the N-terminus of the peptide are apparently prerequisites for the formation of channels and hence for permeability. In collaboration with the group of Professor C. Toniolo at the University of Padova, we realized [30] synthetic templates of general structure **12**, made of peptides containing helicogenic α-amino isobutyric acid (Aib) connected to tris(2-aminoethyl)amine (Tren). The Tren moiety strongly binds transition metal ions such as Zn(II) and the formation of the complex induces a reversible change from an extended to a calix-like conformation as shown in Figure 4. Our idea that the polypeptide templates **12** could form channels in the extended form and impair their formation when in the calix-like form, too short to span the length of the membrane, apparently proved correct. In fact, some of the templates **12**, in the absence of metal ions, do promote the release of carboxyfluorescein trapped inside unilamellar liposomes made of a mixture of phosphatidycholine and cholesterol but such release is strongly inhibited by addition of a stoichiometric amount of Zn(II). The efficacy of the system can be restored when the metal ion is removed from the Tren platform by addition of EDTA, a stronger ligand than Tren. Thus the properly designed polypeptide templates are effective in inducing the leakage of liposomes, at least in the case of the dye employed and control of their permeability can be achieved upon addition of Zn(II) quite likely via the modulation of the conformation of the templates.

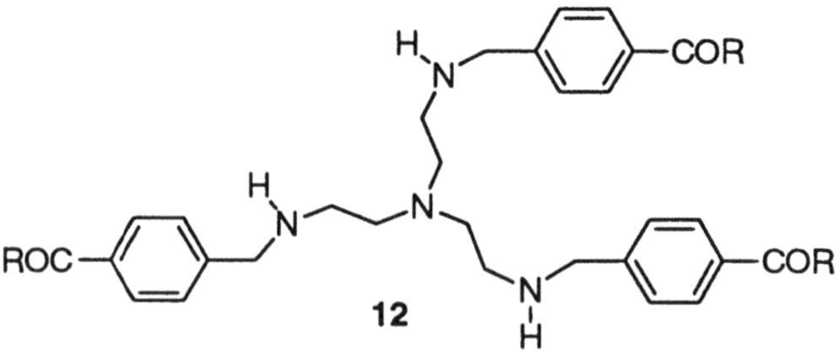

12

12a : R = GlyLeuAibGlyGlyLeuAibGlyIleLeuOMe
12b: R = GlyLeuAibGlyIleLeuOMe
12c : R = LeuAibGlyIleLeuOMe
12d : R = AibGlyIleLeuOMe
12e : R = GlyLeuAibNHiPr

$C_7H_{11}-C(=O)-$AibGlyLeuAibGlyGlyLeuAibGlyIleLol Trichogin A IV

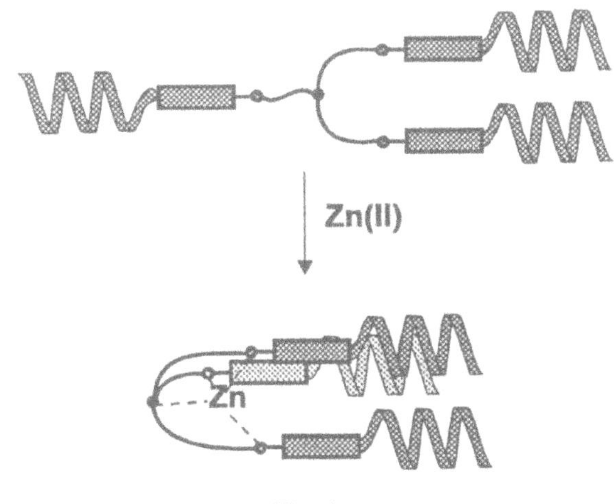

Fig. 4.

6. Chemical Sensing.

The translation of supramolecular recognition properties of a receptor has brought about the realization of sensors that allow for the continuous measurement of physical or chemical parameters [31]. Chemical sensing is part of an acquisition process in which some information on the composition of a system is obtained in real time and translated in magnified electric signals. Recognition and amplification are the functions of a chemical sensor.

Recently, we turned our attention to transition metal ion sensors and, following a commonly used strategy, we realized a system made of a recognition site (a selective metal-chelating molecule) and a readout moiety (a fluorophore) covalently linked trough a spacer [32]. We used a rather versatile ligand platform, the *all-cis* 1,3,5-triamino 2,4,6-trimethoxy cyclohexane to which a 9-anthracenyl moiety was covalently linked to one of the amino nitrogens, **13**. The main features of this chemosensor may be summarized as shematically shown in Figure 5: a) at pH = 5 the sensor is fluorescent and upon binding Cu(II), a quenching metal ion, the fluoresence decreases sharply also in the presence of other metal ions; b) at pH = 7, the sensor is not fluorescent but upon selectively binding Zn(II), a non quenching metal ion, the fluorescence is restored; c) the Zn(II) complex at pH = 7, on the other hand, may bind organic substrates such as some dicarboxylic acids or nucleotides which may switch off the fluorescence of the ternary complex.

Studies of the system are still under way to better understand the mode of action but the results, so far obtained and consolidated, point to a very versatile chemosensor which upon changing the pH of the solution reveals the presence of either Cu(II) or Zn(II) and the complex of the latter ion may detect organic molecules of specific structure.

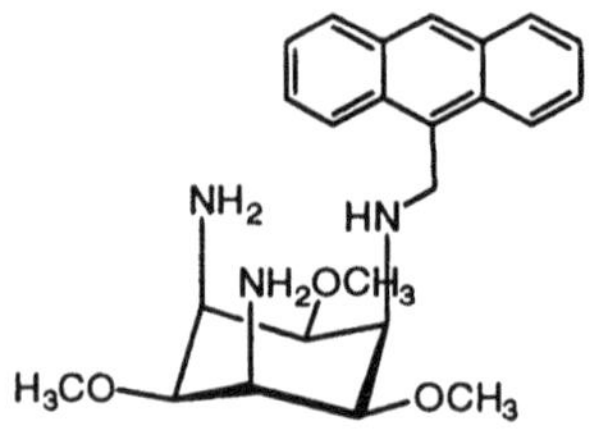
NH2
HN
NH2OCH3
H3CO
OCH3

13

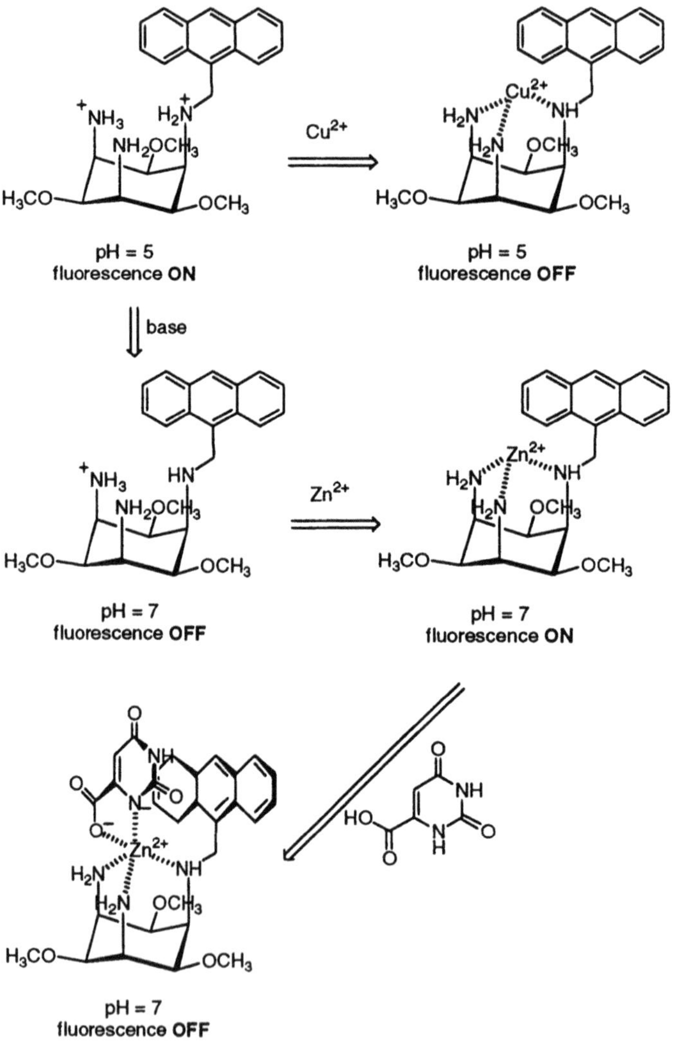
+NH3
H2N+
NH2OCH3
H3CO
OCH3
pH = 5
fluorescence ON
Cu2+
Cu2+
H2N
NH
H2N
OCH3
H3CO
OCH3
pH = 5
fluorescence OFF
base
+NH3
HN
NH2OCH3
H3CO
OCH3
pH = 7
fluorescence OFF
Zn2+
Zn2+
H2N
NH
H2N
OCH3
H3CO
OCH3
pH = 7
fluorescence ON
O
NH
O
N
O
HO
O
N
H
O
NH
O
N
O
O-
Zn2+
H2N
NH
H2N
OCH3
H3CO
OCH3
pH = 7
fluorescence OFF

Fig. 5.

Linking the recognition site to the fluorophore on one hand ensures proximity of the sensor components but, on the other hand, requires a proper design and synthetic efforts. A different approach is based on the self-assembly of the ligand and the signalling device and this can be achieved through specific interaction such as hydrogen bonding or by, less specific but more easily obtained, hydrophobic interactions inside a surfactant aggregate. We have recently reported the first example [33] of such a system in which a ligand and a fluorophore spontaneusly assemble in a CTABr aggregate and act as a very sensitive and selective sensor for Cu(II) ions. The ligand, a liphophilic derivative of a glycylglycine dipepdite, forms micelles with CTABr in a 1:2 ratio. The hydrophobically driven binding of the fluorescent probe ANS to such micelles results in the formation of the supramolecular sensor schematically illustrated in in Figure 6. Upon binding of Cu(II) ions, the fluorescence emission on ANS is strongly quenched allowing its detection in the micromolar concentration range. Due to its mode of binding, similar to the one described for compounds **8**, the ligand is selective toward Cu(II) ions and other metal ions such as Zn(II), Ni(II), Mn(II), Co(II), Pb(II), do not interfere with the observed quenching. A great advantage in using the self-assembling strategy is the possibility of changing the components to set up a combinatorial approach to the realization of the most effective sensor. Once again the use of flexible, less organized systems ensures several advantages and allows the formation of devices able to perform the designed task by simply mixing the different components.

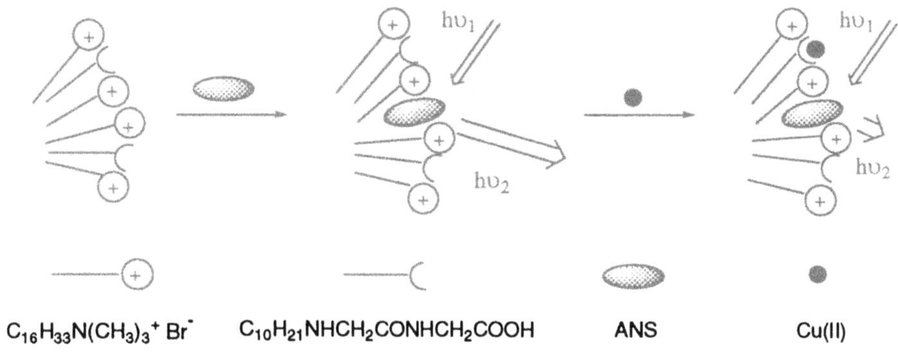

Fig. 6.

7. Concluding remarks

The issue of supramolecular functions has been here used to follow the thread of our research activity. However this is not only intended as a sole artifact: supramolecular chemistry, mostly mastered by the academics, is not only addressed to the creation of remarkable molecules or supermolecules of great beauty but is increasingly directed (and should be more so) to the realisation of species that do useful work and lead to a variety of applications and to a decisive impact in new interdisciplinary fields. At the beginning of the third millenium, it is predictable [34] the emergence of supramolecular *science and technology* as a multidisciplinary domain providing a fertile ground for the creativity of scientists. Applications are in sight and this hopefully will be a big reward for the many scientists looking back to an experience that has always been stimulating, providing them with continuous intellectual challenges and the capacity to survive failures [35].

Acknowledgements.

The authors thank the many coworkers, whose names appear in the references which contributed to the work, for their intellectual contributions (and Mr. Enzo Castiglione for technical assistance). The work has been supported, over the years, by MURST, the Italian Ministry of the University and Scientific Research under the frameworks of various projects and, recently, of the "Supramolecular Devices Project", by the Italian C.N.R., recently within the "Sensori Fluorescenti Supramolecolari" project, and by the European Economic Community for fellowships.

References.

1. J.-M. Lehn, *Supramolecular Chemistry. Concepts and Perspectives*; **1995,** VCG, Weineheim,; D. J. Cram, *Science*, **1988**, *240*, 760; C. J. Pedersen, *J. Am. Chem. Soc.*, **1967**, *89*, 7017.
2. For recently published general accounts, see: *Supramolecular Science, Where it is and Where it is going,* **1999**, R. Ungaro and E. Dalcanale eds, Kluwer Acad Publ, Dohrdrecht.
3. R. Breslow, *Chem. Soc. Rev.* **1972**, *1*, 553. For comprehensive books, see: H. Dugas *Bioorganic Chemistry,* **1996**, Springer, New York, 3rd ed,; W. Kaim and B. Shwederski, *Bioinorganic Chemistry: Inorganic Elements of Life,* **1994**, Wiley, Chichester.
4. J.-M. Lehn, *Angew. Chem. Int. Eng. Ed.*, **1988,** *27*, 89.
5. A. P. Davis, **1999**, in ref.2, pp. 125-146.
6. J .H. Fendler , *Membrane Mimetic Chemistry,* **1982**, Wiley, New York
7. U. Tonellato in *Solution Chemistry of Surfactants*, **1979**, K. Mittal ed, vol 2, Plenum, New York, pp 541-553; L. Anoardi, R. Fornasier, U. Tonellato *J. Chem. Soc., Perkin II*, **1981**, *260*, 1; U. Tonellato, *Colloids and Surfaces*, **1989**, *35*, 121; R. A. Moss, R. C. Nahas S. Ramaswani, and W. J. Sanders, *Tetrahedron Lett.*, **1975**, 4435; R. A. Moss, R. C. Nahas, S. Ramaswani, *J. Am. Chem. Soc.*, **1977**, *99*, 627; R. A. Moss, T. J. Lukas, and R. C. Nahas, *J. Am.. Chem. Soc.*, **1978**, *100*, 5920.
8. M. Menger and C. E. Portnoy, *J. Am. Chem. Soc*, **1967**, *89*, 4698; K. Y. Yatsimirski, , K. Martinek , and I. Berezin, *Tetrahedron*, **1971**, *27*, 2855; L. S. Romsted in *Micellization, Solubilization, and Microemulsions*, **1977** K. Mittal ed, vol. 2, p. 509, Plenum, New York; C. A. Bunton and G. Savelli, *Adv. Phys. Org. Chem.*, **1986**, *22*, 213.
9. M. Menger and L. G. Whitesell, *J. Am. Chem. Soc.* **1985**, *107*, 707; R. A. Moss, K. W. Alwis, J.-S. Shin, *J. Am.. Chem. Soc.* **1984,** *106*, 2651.

10. R. Fornasier and U. Tonellato, *J. Chem. Soc., Faraday I*, **1980**, *76*, 1301.
11. (a) R. Fornasier, D. Milani, P. Scrimin, and U. Tonellato, *Gazz. Chim. Ital.*, **1986**, *116*, 55; (b) R. Fornasier, P. Scrimin, P. Tecilla, and U. Tonellato, *J. Am. Chem. Soc.*, **1989**, *111*, 224; (c) P. Scrimin, P. Tecilla, and U. Tonellato in *Organic Reactivity and Biological Aspects*, **1995**, B. T. Golding, R. G. Griffin, and H. Maskill eds., The Royal Society of Chemistry, pp 223-231.
12. L. L. Melhado and C. D. Gutsche *J. Am. Chem. Soc*, **1978**, *100*, 1850; W. Tagaki, K. Ogino, *Top. Curr. Chem.* **1985**, *128*, 144 and *J. Am. Chem. Soc.*, **1987**, *111*, 5086.
13. For recent overviews, see: J. K. M. Sanders, *Chem. Eur. J.*, **1998**, *4*, 1378 and, in ref. 2, pp 273-286.
14. U. Tonellato, *Pure & Appl. Chem.*, **1998**, *70*, 1961 and references therein.
15. P. Scrimin, P. Tecilla , and U. Tonellato, *J. Org. Chem.*, **1991**, *56*, 199.
16. M. C. Cleij, P. Scrimin, P. Tecilla, and U. Tonellato, *Langmuir*, **1996**, *12*, 2956; P. Scrimin, P. Tecilla, and U. Tonellato *J. Org. Chem.*, **1993**, *58*, 3025 and *Tetrahedron*,**1995**, *51*, 527; M. C. Cleij, P. Scrimin, P. Tecilla, and U. Tonellato, *Tetrahedron*, **1997**, *53*, 357.
17. F. Bertoncin, F. Mancin, P. Scrimin, P. Tecilla, and U. Tonellato, *Langmuir*, **1998**, *14*, 975.
18. R. Fornasier, E. Scarpa, P. Scrimin, P. Tecilla, and U. Tonellato, *J. Incl. Phenom*, **1992**, *14*, 230.
19. P. Scrimin, P. Tecilla, U. Tonellato and U, Vignaga, *J. Chem. Soc., Chem. Commun.*, **1991** 449; P. Scrimin, P. Tecilla, and U. Tonellato, G. Valle and A. Veronese, *Tetrahedron*, **1995**, *51*, 527; P. Scrimin, P. Tecilla, and U. Tonellato, *Eur. J. Org. Chem.*, **1998**, *1*, 1143.
20. J. K. M. Sanders, , in ref. 2, **1999**, pp 273-286.
21. T. M. Fyles and W. F. Van Straaten-Nijenhuis, in *Comprehensive Supramolecular Chemistry*, **1996**, D. N. Reinhoudt ed., vol. 10, Elsevier, Oxford, UK, pp 53-77; F. De Fong and H. C. Visser, *ibidem*, pp 13-51.
22. F. M. Menger and J. J. Lee, *J. Org. Chem*, **1993**, *58*, 1909.
23. M. C. Cleij, P. Tecilla, U. Tonellato, and P. Scrimin, *J. Org. Chem.*, **1997**, *62*, 5592.
24. P. Scrimin, U. Tonellato, and N. Zanta, *Tetrahedron Lett.*, **1988**, *29*, 4967; P. Scrimin, P. Tecilla, and U. Tonellato, *Tetrahedron*, **1995**, *51*, 217.
25. D. D. Lasic, *Liposomes: from Physics to Applications*, **1993**, Elsevier, Amsterdam.
26. G. Ghirlanda, P. Scrimin, P. Tecilla, and U. Tonellato, *J. Org. Chem.*, **1994**, *59*, 18.
27. P. Scrimin, P. Tecilla, R. A. Moss, and K. Bracken K., *J. Am. Chem. Soc.*, **1998**, *120*, 1179.
28. P. Scrimin, S. Caruso, N. Paggiarin, and P. Tecilla, *Langmuir*, in press.
29. E. Benedetti, A. Bavoso, B. Di Blasio, V. Pavone, C. Pedone, C. Toniolo, and G. M. Bonora, *Proc. Natl. Acad. Sci. U.S.A.*, **1982**, *79*, 7951.
30. P. Scrimin, A. Veronese, P. Tecilla P, U. Tonellato, V. Monaco, F. Formaggio, M. Crisma, and C. Toniolo, *J. Am. Chem. Soc.*, **1996**, *118*, 2505.
31. For overviews, see: *Fluorescent Chemosensors for Ion and Molecule Recognition*, **1993**, A. W. Czarnik ed, ACS Symp. Ser., 538; A. P. de Silva, H. Q. N. Gunaratne, T. Gunnlaugsson, A. J. M. Huxley, C. P. McCoy, J. T. Rademacher, and T. E. Rice, *Chem. Rev.*, **1997**, *97*, 1515.
32. L. Fabbrizzi, M. Licchelli, F. Mancin, M. Pizzeghello, A. Taglietti, P. Tecilla, and U. Tonellato, unpublished results.

33. P. Grandini, F. Mancin, P. Tecilla, P. Scrimin, and U.Tonellato, *Angew. Chem. Int. Ed. Engl.*, **1999**, *38*, 3061.
34. J. M. Lehn in ref.2, **1999**, pp 287-304.
35. D. N. Reinhoudt, J. F. Stoddard, and R. Ungaro, *Chem. Eur. J.*, **1998**, *4*, 1349.

Photoinduced Electron Transfer: Perspectives in Organic Synthesis

Angelo Albini, Maurizio Fagnoni and Mariella Mella

Department of Organic Chemistry, University of Pavia, via Taramelli 10, I-27100 Pavia, Italy
albini@chifis.unipv.it

Abstract. Electron transfer between organic molecules, almost non existent in the ground state, is common occurrence when an excited state is involved. The innovative characteristics of (photoinduced) single electron transfer between organic molecules are discussed in the frame of the general mechanism of organic chemistry. The redox step generates a pair of oppositely charged radical ions under exceptionally mild conditions in neat organic solvent by irradiation. These hitherto little investigated species show a rich and varied chemistry. Representative examples are illustrated, in particular photoinduced radical conjugate alkylation using unconventional radical precursors. The versatility of the photochemical method for radical ions generation suggests that it may give a significant contribution to the development of organic synthesis in the next years.

1. Photochemical Reactions

1.1. Chemical reactivity of excited states.

It has often been asserted that photochemistry adds a new dimension to chemistry. Indeed, new chemical reactions involving the excited states are added to 'common' ground-state chemistry. It is important to notice, however, that photochemical reactions not only are a new addition to the known repertoire, but are also qualitatively different from the those occurring in the ground state. Some key aspects are pointed out in the following. The first one concerns energy. Electronically excited states of organic molecules are 2 to 4 eV higher in energy than the corresponding ground states and thus photochemical reactions start from an energetic much higher than that of thermal reactions (see Fig. 1). This is not *per se* an indication that an efficient reaction occurs, since the very fact that the energy is so high makes physical decay to the corresponding ground state - with or without emission of luminescence – a very fast process (rate constant $\geq 10^9 s^{-1}$ for singlet states, $\geq 10^6 s^{-1}$ for triplet states) However, and this is the second key aspect, excitation in most cases also involves a large electron redistribution. Thus excited states differ from ground state not only for their energy, but also for their electronic structure (bottom of Fig. 1). This change opens new reaction paths, some of which have a low activation energy enabling them to compete with physical decay. As a result, photochemical reactions are necessarily characterized by a much larger rate than any ground state reaction.

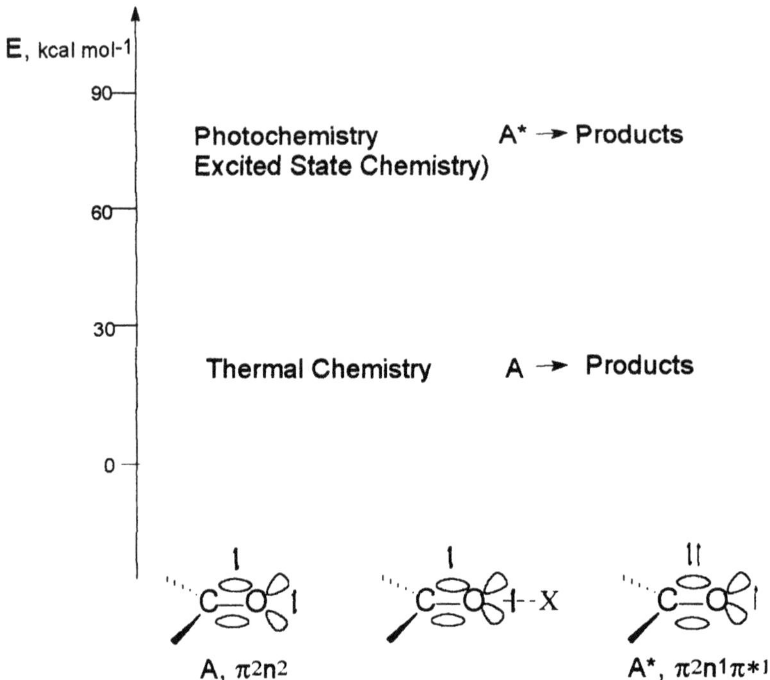

Fig. 1. Ground state chemistry and excited state chemistry take place in two different energy domains. Even more importantly, the electron distribution drastically changes upon excitation. This is shown here in the case of a ketone: the ground state is a weak electrophile, which can be activated by means of a Lewis acid (X). The nπ* excited state has the characteristics of an oxygen-centered radical.

Thus, the fact that a photochemical reaction is sufficiently fast to occur during the exceedingly short lifetime of the excited state is due to the large change in the electron distribution caused by electronic excitation, rather than to the high energy. As an important corollary, the reactivity in the excited states is generally unrelated to that of the ground state. On the other hand, also in ground state chemistry one induces an organic molecule to react more often through a suitable perturbation of the electronic structure than by supplying energy. As an example, a ketone is a the electronic structure than by supplying energy. As an example, a ketone is a weak electrophile, due to the polarization of the π C-O bond. Thus, in order to make such a molecule react at a useful rate one has either to use a very strong (usually charged) nucleophile or to increase the polarization of the C-O bond, e. g. by adding a Lewis acid (X in Fig. 1) and thus increasing the polarization.

Electronic excitation causes a stronger perturbation in the electron distribution than an acid, not in the sense of reinforcing the ground state polarization but by introducing a novel feature favorable to reactivity. The nπ* excited state of ketones is certainly not a C-electrophile (since there are now three electrons in the π space rather than two); rather, the determining change is due to the single electron remaining in the non bonding orbital of the oxygen atom, making the excited state comparable to an oxygen centered (electrophilic) radical. A radical chemistry is indeed observed with such species, which both abstract hydrogen from donors such as alcohols and add to alkenes at a rate in the order of $10^6\ M^{-1}\ s^{-1}$ or higher, a rate unparalleled in ground state chemistry. Photochemistry thus contributes a variety of peculiar chemical reactions, which are interesting for mechanistic studies and in many instances have also a significant preparative value. At least some of these reactions are becoming familiar with the practitioners of organic synthesis and are introduced in complex synthetic strategies [1-3].

1.2 Redox Processes in the Excited State.

As it appears from the Fig. 1, photochemical reactions are rationalized on the basis of the electronic structure of the excited state involved, exactly as it is usually done with ground state processes, and are obviously typical of each chromophore. It is perhaps not as generally appreciated, however, that there is a general property that all excited states share independently from their structure. This may be considered as a further new dimension that photochemistry offers and amounts to the fact that excited states are *both* strong oxidants and strong reducing agents.

Photochemists have been long familiar with the Rehm-Weller equation [4], stating that the standard free energy change for the redox equilibrium in eq. 1

$$D + A \rightleftarrows D^{+\cdot} + A^{-\cdot} \quad (1)$$

is equal to that in the ground state subtracted of the excitation energy when an excited state, either D* or A*, is involved (the last term of the equation accounts for coulombic attraction and is small in polar solvents).

$$\Delta G° = -nF\ [E°(A/A^{-\cdot}) - E°\ (D^{+\cdot}/D)] - \Delta E_{exc} - e_0^2/\varepsilon_a \quad (2)$$

Since, as mentioned above, ΔE_{exc} corresponds to a large amount of energy (2 to 4 eV) the change with respect to the ground state causes a major variation in the position of the equilibrium (see Fig. 2) [5-7]. As shown in the figure, organic molecules are generally reduced at a negative potential vs NHE, but excited states (see the formulae at the bottom) are reduced at a largely positive potential.

In terms of localized orbitals, this new behavior of excited state can rationalized by the fact that such species are both easily oxidized, since they transfer the electron that has been promoted to an antibonding orbital by excitation, and are likewise easily reduced by adding an electron to the bonding or non bonding orbital which is half filled in the excited state (see Fig. 3).

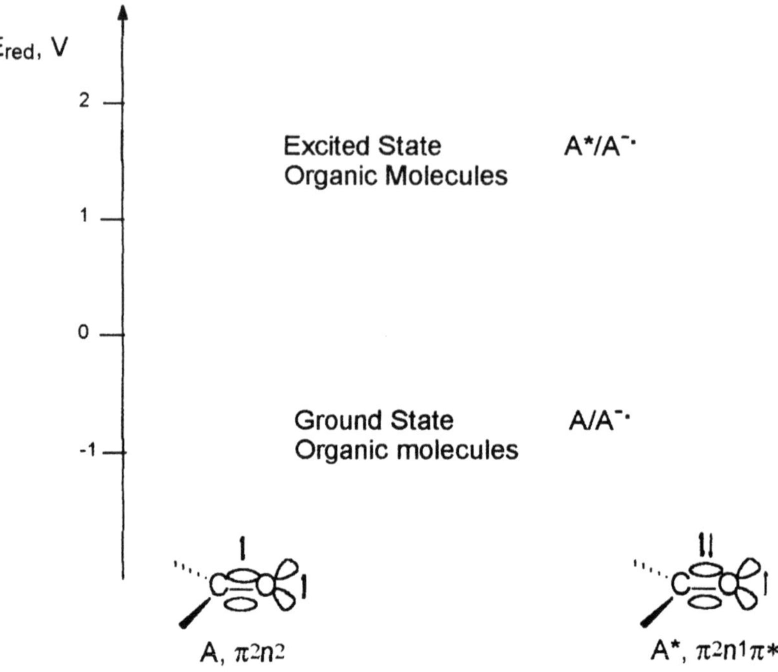

Fig. 2. Excited state are strong reducing and oxidizing agents. As an example, a ground state ketone is reduced at a largely negative potential vs NHE, the excited state at a largely positive potential

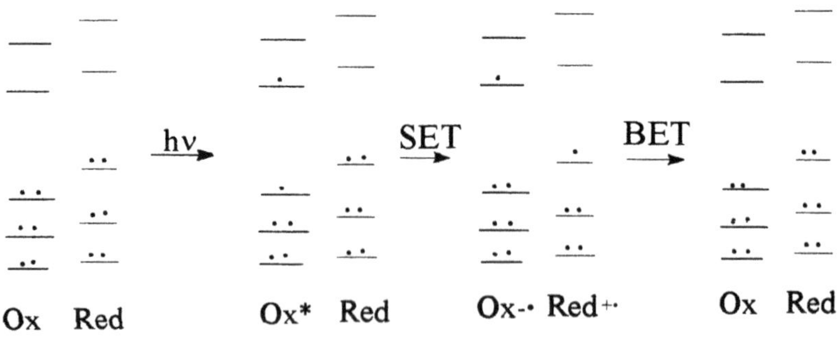

hν SET BET

Ox Red Ox* Red Ox-· Red+· Ox Red

Fig. 3. Envisaging single electron transfer with localized orbitals: SET after excitation and back electron transfer (BET) regenerating the starting materials.

2. Redox Processes in Organic and in Inorganic Chemistry

Single electron transfer between organic molecules is a unique possibility that photochemistry affords, since redox processes are extremely common between inorganic species (eq. 3) but not at all between organic molecules (eq. 1) [8, 9].

$$M^{m+} + N^{n+} \rightleftarrows M^{(m+1)+} + N^{(n-1)+} \qquad (3)$$

Indeed, redox equilibria take a very small place in organic chemistry, both in teaching and in research, contrary to what happens in inorganic chemistry. Inorganic reactions can be roughly classified in two groups, Lewis acid/base reactions (where a pair of electrons is shared to form a covalent bond or a to build up a complex) and redox reactions (where electrons are exchanged in any number, e. g. one as in eq. 3).

Most organic reactions involve the interaction between a Lewis acid and a Lewis base, or with more usual terms between an electrophile and a nucleophile. The different terms indicate that in reactions between organic molecules the sharing of a pair of electrons is the rate determining step. Accordingly, the mechanism of organic reactions is usually envisaged by using the typical curved arrows that evidence how pairs of electrons are displaced in the process of bond making and cleaving.

$$E \quad :Nu \longrightarrow E{-}Nu$$

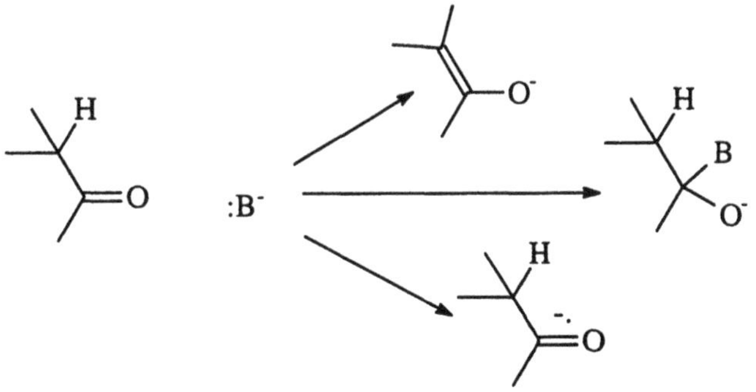

Fig. 4. A base :B⁻ may react with a ketone as a Brønsted base, as a nucleophile and as a reducing agent.

This way of envisaging reactions by displacing a pair of electrons is certainly a simplification (in fact, electrons do move in any case one by one, compare e.g. ref. [10]) but is convenient for both teaching and research. On the contrary, reactions involving single electron transfer not connected to the formation of a bond are uncommon. This is due to the fact that organic molecules are neither easily reduced nor easily oxidized. The potentials relative to such equilibria are largely negative or, respectively, largely positive, and thus redox processes occur only under drastic conditions.

There is an analogy between a base and a reducing agent, since both contribute electrons, but also an overwhelming difference. This can be shown by turning again to the example of ketones (see Fig. 4). When a strong base is added, this may act either as a Brønsted base by abstracting a proton or as a Lewis base (= nucleophile) by adding to the carbon atom. Both such processes occur only with quite strong bases because of the inherent low reactivity of organic molecules due to the strength of the covalent bonds and the scarce polarization. Furthermore there are reactions which lead to reduction of the ketone moiety as the overall result, but from the point of view of the mechanism are again nucleophilic additions, e.g. the reduction of ketones by a hydride.

It is difficult to find a reagent that one would call a base and has an available electron pair so high-lying as to reduce the ketone (lower path in Fig. 4). On the other hand, sodium metal has a highly-lying electron and behaves as an effective one-electron reducing agent leading to the formation of the ketone radical anion (see Fig. 5). In such a process electron exchange is not accompanied by bond formation or cleavage and is similar to the typical redox process of inorganic chemistry (see eq. 3). The blue color of benzophenone radical anion is well known to synthetic organic chemists since it is used as an indicator to check that a solvent is anhydrous when boiled over sodium or another reducing agent [11].

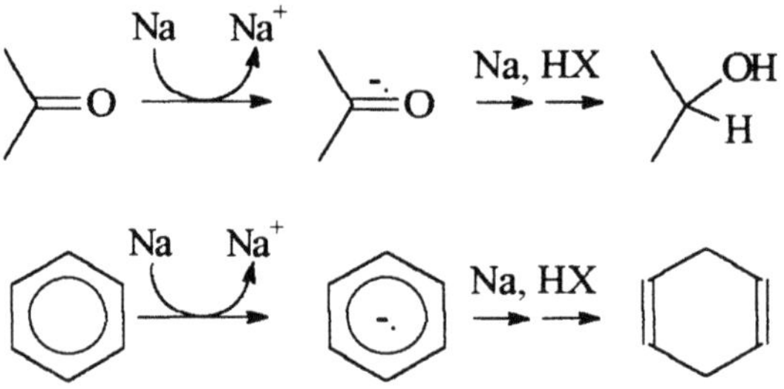

Na Na$^+$ Na, HX OH H

Na Na$^+$ Na, HX

Fig. 5. Single electron transfer to an organic molecules occurs when a strong reducing agent such as sodium metal is used.

Related one-electron reductions are well known, e. g. the Birch reduction of aromatic hydrocarbons (Fig. 5) [12]. In both cases, an *odd* electrons species results in the first step, the radical anion of the aromatic or of the ketone, and only in a later step this is protonated and further reduced leading again to an even electron product. In every case these reactions require very strong reagents, such as alkali metals, certainly not organic donors.

Likewise, in electrophilic substitution or addition an electrophile having a vacant orbital accepts an electron pair from a nucleophile (e. g. an alkene, see Fig. 6, compare with the reaction as a Brønsted acid in the first line) and forms a new bond. Here again, other paths are possible, e.g. some aromatic electrophilic substitutions appear to involve single electron oxidation of the substrate to the corresponding radical cation [13] (see again Fig. 6, where the process is also generalized for an organic molecule M and an oxidant Ox), and only in a following step a bond is formed and one comes back to an even electron species. Also in this case however, such a mechanism is considered to be the exception rather than the rule.

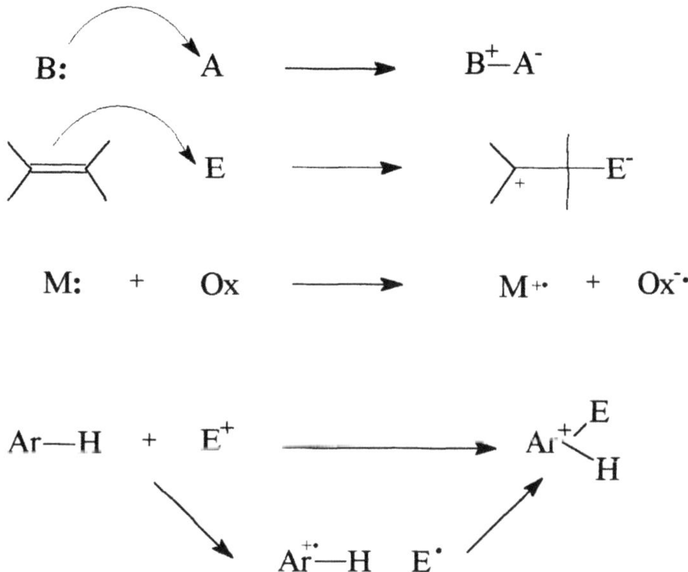

Fig. 6. A species A having an empty orbital acting as a Brønsted acid, as an electrophile and as an oxidizing agent.

The discussion above shows that reducing or oxidizing an organic molecule requires the use of a powerful inorganic reagent see e.g. the reaction of sodium with naphthalene or a ketone in Fig. 5, or to that of $NO_2^{+\cdot}$ (=E) with very electron-rich aromatics, see Fig. 6. On the contrary, redox processes involving only organic molecules are rare (actually these are known in a very few cases [14]) because thermodynamics precludes them. The situation in inorganic chemistry is different, since many elements form stable compounds with various oxidation number and there are examples of redox equilibria at any potential. A monodimensional diagram of redox potentials is densely filled (see a few examples in Fig. 7a), and it is always easy to find a pair of reagents for which the ΔG for a redox process is negative. As an example, the figure shows that permanganate is a convenient choice for oxidizing chloride to chlorine. Thus, reactions of the type shown in eq. 3 are common among inorganic species.

On the contrary, organic molecules are stable in their closed-shell configuration, and the corresponding reduced (radical anions) or oxidized species (radical cations) are strongly destabilized. Fig. 7b is the analogue of Fig. 7a for organic chemistry. With very few exceptions, the redox processes are grouped in two regions, oxidations and reductions, separated by an empty gap (the shaded area in Fig. 7b).

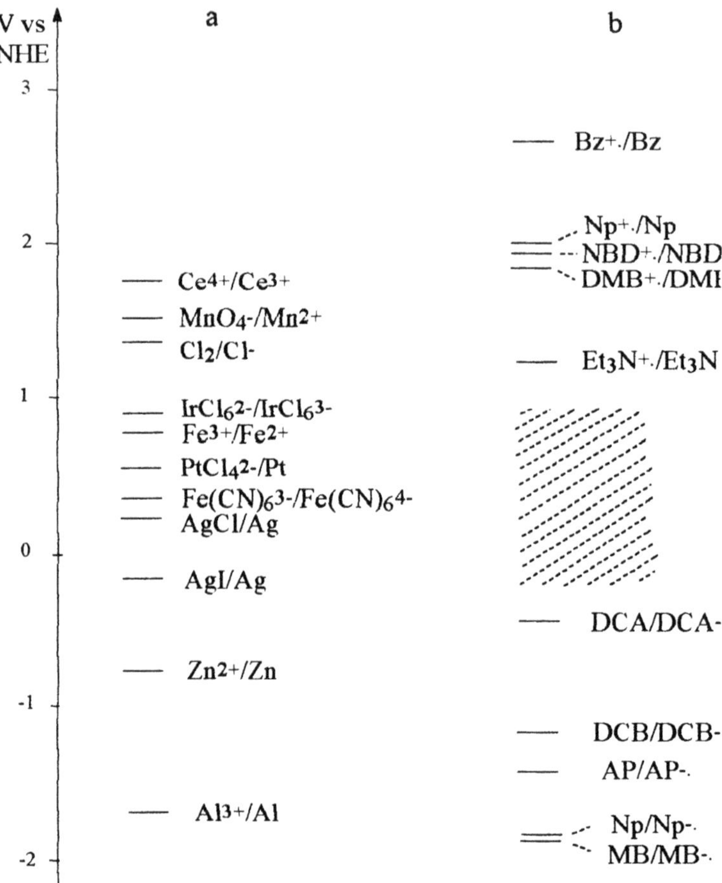

Fig. 7. Redox potentials: a) of representative inorganic species. b) of representative organic molecules: benzene (Bz), naphthalene (Np), norbornadiene (NBD), triethylamine (NEt_3), 9,10-dicyanoanthracene (DCA), 1,4-dicyanobenzene (DCB), acetophenone (AP), methyl benzoate (MB).

Oxidations occur only at a very positive potential, and reductions at a quite negative potential (as shown in the Figure, naphthalene, Np, is both a weak oxidizing and reducing agent and this holds in general organic molecules). This restricts the choice of possible redox processes, in practice requiring an inorganic reagent.

3. Redox Processes involving Excited Organic Molecules

There is a single possibility to change this situation, and this is to bridge the gap of Fig.7b by moving an electron upwards, viz to promote the molecules to their excited states, as discussed in Sec. 1.2. Fig. 3 shows how a SET process which is thermodynamically prohibited in the ground state becomes possible for the exited oxidant. As an example, Fig. 8 shows that molecules such as esters, ketones and nitriles, that are classed as 'acceptors' but certainly not as oxidants since their E_{red} vs NHE is negative, become strong oxidants (E_{red} largely positive) when excited. The same occurs with reducing agents. The gap between reductions and oxidations apparent in Fig 7b has now been overcome and taking a ground state molecule it is generally possible to find a convenient electronically excited partner for oxidizing it (the same holds for reducing ground state molecules, not shown in the Figure).

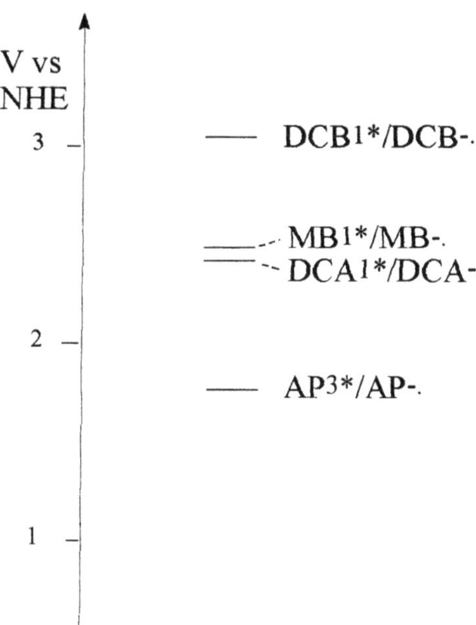

Fig. 8. The reduction potential of the acceptors shown in Fig. 7 is moved upwards upon excitation.

As it appears in Fig. 8, the leap upward upon excitation is substantial. Once again, organic molecules are quite stable and displacing an electron from a bonding to an antibonding orbital, viz electronically exciting the molecule from the ground to the excited state, involves a large amount of energy. One has to use a energetic photon (E_{exc}, organic molecules absorb in the UV, and thus this term is > 2 eV) and this is translated into a strong change in the redox potential [$E_{red}(M^*) = E_{red}(M)+E_{exc}$] (see eq. 2). This makes excited organic molecules more powerful oxidizing or reducing agents than most inorganic reagent.

As Fig. 8 shows, excited ketones are oxidants as strong as the most powerful inorganic oxidants, e.g. Ce(IV), but excited aromatic esters or nitriles are superior. In the ground state, ketones are no oxidants, since the lowest unoccupied orbital ($\pi^*_{C=O}$) lies relatively low in energy (lower e.g. than a $\pi^*_{C=C}$ orbital but certainly not enough to receive an electron from any *organic* molecule. Ketones can be reduced e.g. electrochemically or by a strong inorganic reagent, e.g. Na or SmI_2. In the excited state, however, ketones oxidize moderate organic donors such as amines [15] (see Fig. 9).

Summing up, it is sufficient to shine light upon the solution to make redox reactions, exceptional in organic chemistry as long as only ground states are considered, as common as in inorganic chemistry, and this happens under really mild conditions, in an organic solvent with no other reagent added. The limitation is that, as it appears from Fig. 3, a *pair* of oppositely charged radical ions is formed, and the most obvious, and always very fast, ensuing reaction is SET in the opposite direction (back electron transfer, BET). In this way, the original substrates are recovered and light is degraded to heat.

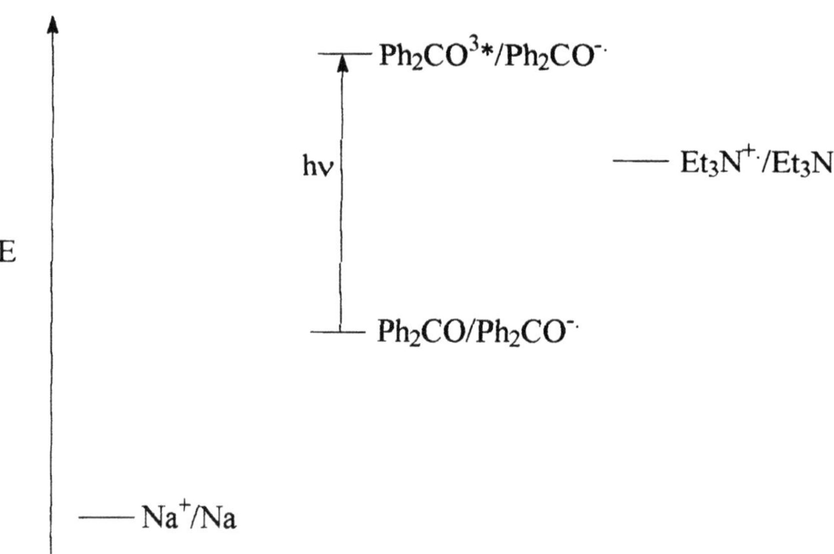

Fig. 9. Ground state benzophenone is reduced by sodium metal, while triethylamine is a sufficiently good reducing agent for the triplet state.

4. Photochemical Generation of Radical Ions.

Thus, charge separation may be only transient and in that case one has to make recourse to an appropriate technique for revealing it. As an example, the formation of the brightly colored radical anion by sodium reduction of benzophenone can be obtained in the excited state by means of moderate donors, e.g. amines, as shown in Fig. 9. The same visible absorption is indeed formed, though only as a transient. Thus it can be revealed only by using a fast kinetic technique such as flash photolysis, since then BET regenerates the starting molecules as shown in Fig. 3.

However, this shortcoming can be overcome either by incorporating the donor/acceptor pair into a supramolecular structure allowing for subsequent vectorial electron transfer, as it happens in photosynthesis, or through a chemical reaction. If one of the radical ions formed in the SET step is rapidly destroyed, the lifetime of the remaining one becomes much longer and in the limiting case persisting species are obtained. As an example, the radical anions of 9,10-dicyanoanthracene (DCA) or of 2, 6, 9, 10-tetracyanoanthracene show an extended absorption in the visible. This is formed by reduction of the excited singlet state either as a transient disappearing within microseconds when the concomitantly formed radical cation has no reaction available, and thus BET is the only process [16], or as a persisting species (days) when the oxidized product is rapidly consumed through an independent path [17].

5. Organic Synthesis via Photoinduced Electron Transfer.

5.1 Characteristics of the Method.

Apart from producing nice colors, the possibility that one (or both) or the radical ions chemically react fast enough to overcome BET opens up new strategies for organic synthesis [18-20]. Indeed, a photoinduced redox process leads to highly reactive species such as radical ions (eq. 4) in a way that is essentially independent on conditions, and thus allows for a large choice in the experimental parameters in order to direct the ensuing reaction of such species.

$$RX + A^* \rightarrow RX^{+\cdot} + A^{-\cdot} \qquad (4)$$

Other methods for carrying out single electron transfer induced organic syntheses are being developed in the meantime, notably via electrochemistry [21] and through catalytic reactions [22]. Two general advantages that a photoinduced redox process has over alternative methods are: 1) that the photochemical process can be carried out in neat organic solvents, not requiring the use of a conducting salt as in electrochemistry (eq. 5) and not encountering the problems of limited solubility typical of the use of inorganic compounds (eq. 6)

$$RX + M^{n+} \rightarrow RX^{+\cdot} + M^{(n-1)+} \quad (5)$$

$$RX - e^- \rightarrow RX^{+\cdot} \quad (6)$$

and 2) that the active redox reagent is the excited state formed by light absorption and present at a very low steady state concentration; this avoids competition with over-reduction or with over-oxidation (e. g., see eq. 7) which is typical of the other methods, where the first formed radical ions are formed either in the vicinity of an electrode or in the presence of a significant concentration of the inorganic reagent.

$$RX^{+\cdot} - e^- \rightarrow RX^{++} \quad (7)$$

5.2 Applications.

The scope and the versatility of photoinduced electron transfer for organic synthesis have been only marginally explored up to now. Some potentialities that have emerged are indicated below (see Fig. 10).

• Carrying out a reaction under mild conditions. The Birch reduction of aromatics requires sodium, but in the excited state this can be obtained by using a mild reagent such as sodium borohydride [23].

• Introducing a new reactivity. Alkenes and dienes are nucleophiles due to the high-lying HOMO; however, taking away an electron from this orbital makes such a species an electrophile and indeed a diene radical cation behaves as the dienophile in a Diels-Alder reaction [19f, 24]. This is an instance of Umpolung obtained not through complexation or protonation as usually done in thermal chemistry, but through a photoinduced redox process.

• Generating reactive intermediates. As mentioned above, organic molecules are poor oxidizing or reducing agents, and the energy of a radical ion is often comparable to that of covalent bonds. As a result, such species may undergo fragmentation of a σ bond, thus allowing the generation of neutral radicals and ions in a selective way [18, 19a-d, 20]. Cleavage of strong bonds (e.g. C-Si, C-C, C-H) can thus be selectively obtained, so that e.g. alkyl radicals can be generated under mild conditions from a variety of precursors, whereas the usual method involving atom transfer is necessarily limited to weak bonds.

Our work has concentrated mainly on the last group of reactions and we have exploited photoinduced electron transfer sensitization for a carrying out radical conjugate alkylation under unusual conditions.

Radical conjugate alkylation has been largely developed in the last two decades and has been shown to be a valuable tool in organic synthesis [25, 26]. Two methods have been developed.

The first one is a chain method based on atom abstraction. In this case the radical precursor must have a easily abstractable atom, which in practice restricts the range of substrates to alkyl iodides or bromides, and it is required that a chain carrier that has a sufficient affinity for the halogens is used.

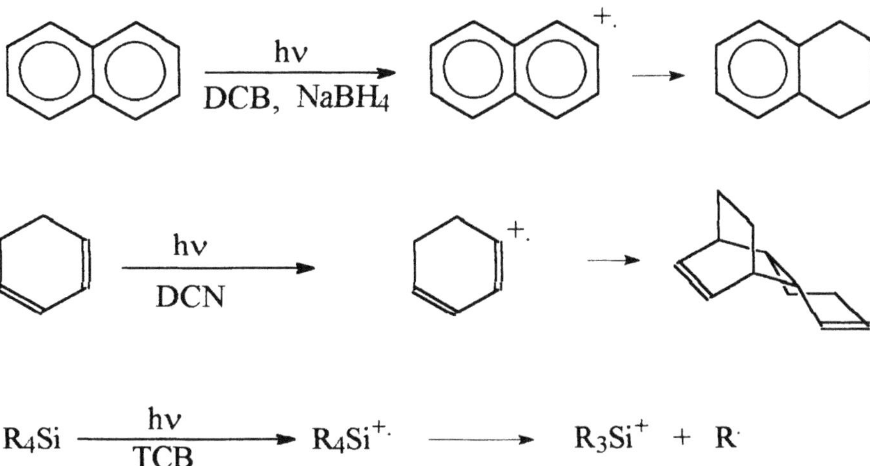

Fig. 10. Examples of new reactivity introduced by using photogenerated radical cations. 1. Enhancing the reactivity: mild reduction of naphthalene after electron transfer to DCB. 2. Umpolung, as in the reaction of the cyclohexadiene radical cation as an electrophile. 3. Fragmentation to give useful intermediates, as with the radical cation of a tetralkylsilane after electron transfer to excited 1,2,4,5 tetracyanobenzene (TCB).

Such a role is most often fulfilled by a tin radical, thereby introducing a further limitation in view of the toxicity of tin derivatives (see Fig. 11).

The latter one is a non chain redox method (see Fig. 12). In the oxidative variation this makes use of a strong oxidant, such as a Mn(III) or a Ce (IV) derivative for the formation of the radical cation of the radical precursor [27], which then cleaves and gives the key radical. In practice, such a method is limited to relatively good electron donors, such as (tautomeric) enols and is liable to the possibility that the radical is over-oxidized to the cation. At any rate, the adduct radical is generally oxidized, and the final product arises from the cation by addition of a nucleophile, thus consuming two equivalents of the oxidizing agent and leading to a product different from that obtained from the chain method.

In the photosensitized method, however (considered here in the oxidative version, the same concept applies to the symmetric reductive variation), a first oxidative step leads to the radical cation and, from it, to the key alkyl radical (see Fig. 13). This adds to the electrophilic alkene and the adduct radical is reduced by the persisting radical anion of the sensitizer. Thus the process is stoichiometric in light, but catalytic as far as the sensitizer is concerned (thus allowing the use of a limited amount of the latter). The process need to be accurately engineered, since the first electron transfer step depends on the reducing potential of the excited state of the sensitizer, while the latter (and reverse) ET step depends on the reducing potential of the corresponding ground state [28]. Aromatic ketones and esters have been shown to be effective sensitizers, and can be chosen in every case in such a way that they are tailored to the specific system used. By using this method a variety of conjugate alkylations have been carried out, by using unusual radical precursors such as tetralkylstannanes, but also tin-free reagents, such as silanes, ethers or acetals [29, 30]. A few examples are shown in Fig. 14.

$$In^{\cdot} + R'_3SnH \longrightarrow InH + R'_3Sn^{\cdot}$$

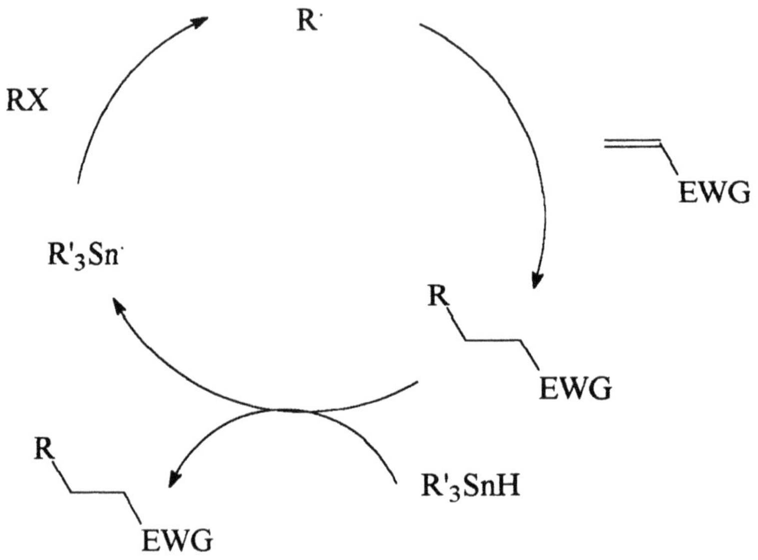

R·
RX
EWG
R'3Sn·
R
EWG
R
EWG
R'3SnH

Fig. 11. Radical conjugate addition to electron-withdrawing (EWG) substituted alkenes. The radicals are generated via the atom extraction method.

Furthermore, SET processes have a role in enzymatic reactions. Photoinduced electron transfer offers a versatile method for studying the chemistry or organic radical ions and thus to evaluate their participation. Here, as in general with this method, the advantage is that radical ions are generated in a 'clean' way, in neat solution, no addition of aggressive reagents or conducting salts being required, contrary to what occurs in thermal redox processes or in electrochemistry. Thus, another possible application concerns model studies of enzymatic redox processes.

In another field again, radical ions generation and chemistry is an area of obvious significance for material science and where again photochemistry has a role.

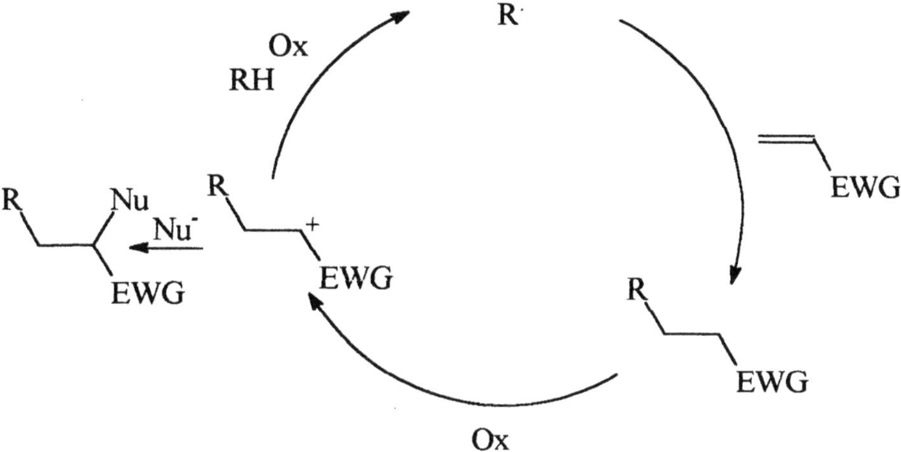

Fig. 12. Conjugate alkylation via radical generated by oxidation. Mn(III) or Ce(IV) compounds are often used as the oxidizing agent. Two equivalents of these reagents are used, since also the adduct radical is oxidized.

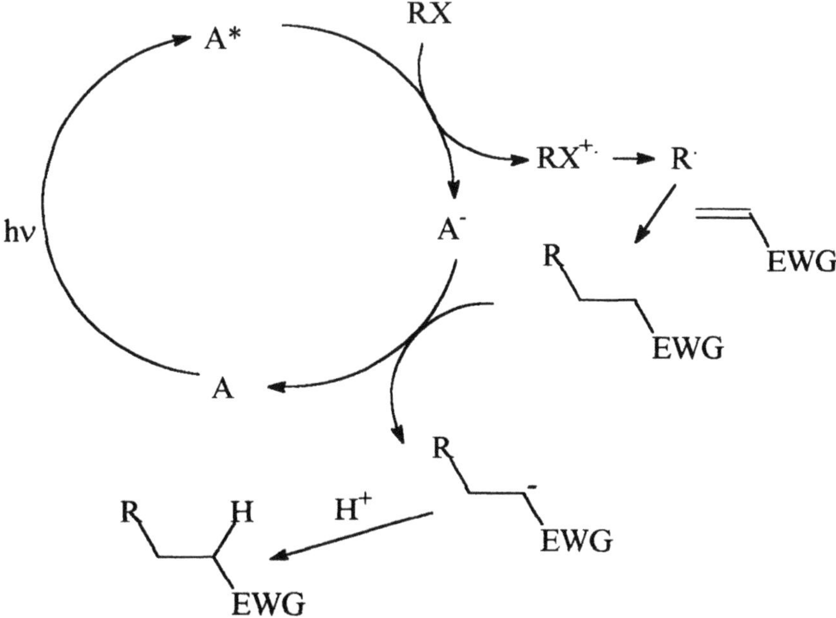

Fig. 13. Conjugate alkylation via radical generated by photoinduced electron transfer and radical cation fragmentation. The radical adduct is reduced by the radical anion of the sensitizer.

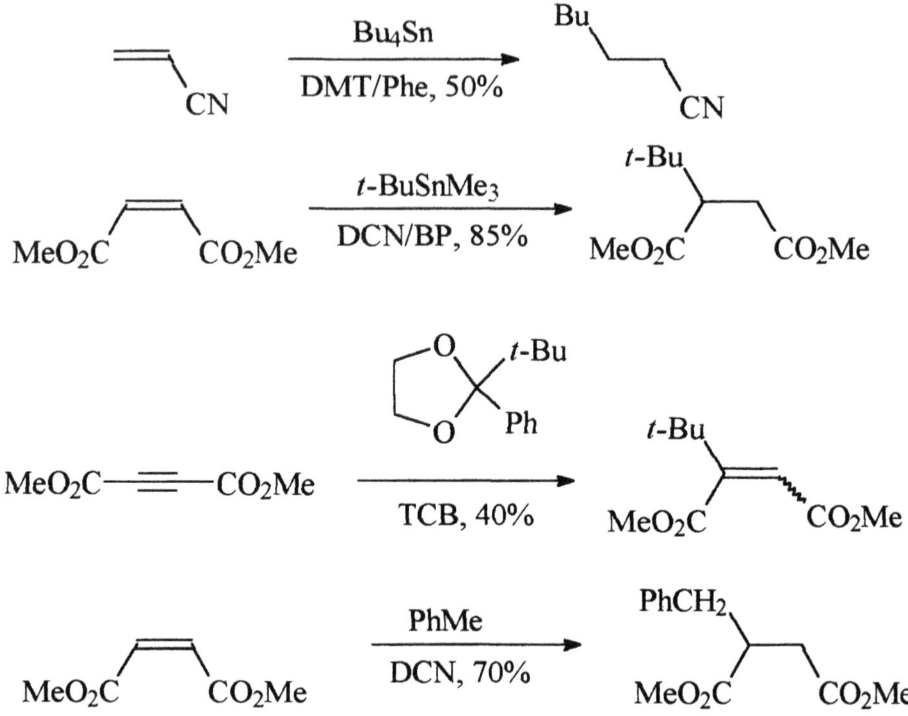

Fig. 14. Examples of conjugated alkylation via radicals generated via photoinduced electron transfer (to aromatic nitriles or esters, which function as non consumed sensitizers, see Fig. 12) and radical cation fragmentation.

A variation that further extends the versatility of the method is based on the use of a heterogeneous rather than a homogeneous sensitizer. This may be done by using a polymer incorporating the chromophore to be excited [31], or better still by using a semiconductor powder rather than an organic molecule. Titanium dioxide has a well established role for the photoinduced recovery of polluted water, since light absorption promotes an electron to the conduction band leaving a hole in the valence band (see Fig. 15). Under this condition oxygen is reduced to the superoxide anion and water oxidized to hydroxyl radical which initiate the decomposition of many organic compounds, and in particular of pollutants.

The same principle can be applied to a solution in an organic solvent. Although the electron-hole recombination on the semiconductor surface is very fast, adsorbed substrates may undergo electron transfer. When the radical cation of the donor is easily fragmented, another method of generating radical via photoinduced electron transfer is realized. This can be trapped by an electron withdrawing substituted alkene (which can function also as the initial acceptor A) and conjugate alkylation results.

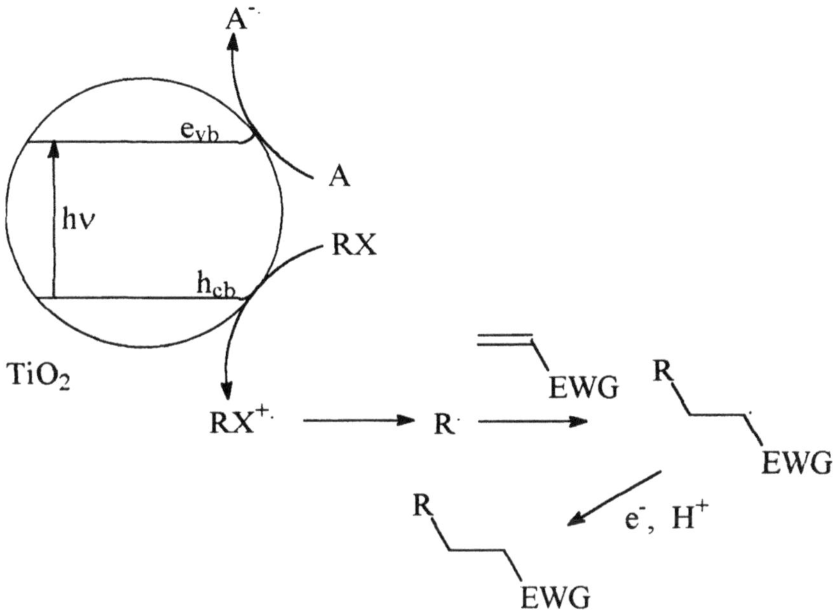

Fig. 15. Light absorption promotes an electron from the valence to the conduction band in a semiconductor such as titanium dioxide. Electron and hole can be trapped by substrates adsorbed on the surface. This is another way for carrying out a photsensitized radical alkylation.

It has turned out that this principle can be applied with reasonable preparative results and is complementary to the use of a homogeneous sensitizer since some radical precursors are more effective with this method [32]. Furthermore, it can be conveniently applied to the use of solar light rather than of a lamp [33].

6. Conclusions

Electronic excitation adds single electron transfer reactions to the mainframe of organic chemistry, where they had been previously considered a curiosity of little consequence.

Other methods for producing radical ions are likewise being developed in these years, but the peculiar mild conditions and versatility of the photoinduced electron transfer guarantee to this method an important role. It can be anticipated that, in the general frame of contemporary research pointing to a cleaner ('green') chemistry, activation of organic molecules via (photo) induced electron transfer will give a significant contribution, in parallel to activation by metal complexes, to organic synthesis. Their role may be crucial for the activation of strong bonds, e. g. C-H bonds, under mild conditions.

Furthermore, recognizing that redox processes easily occur in the excited state, may be useful from the didactic point of view, since this suggests that a more unitary view can be attained of some general concepts on chemical reactions that have been independently developed in different fields of chemistry. It is hoped that the above discussion, though certainly oversimplified in many respects, has been convincing in this respect.

Acknowledgement. Partial support of the our work in this field by MURST, Rome, and CNR, Rome, is gratefully acknowledged.

7. References.

[1] J. Mattay, A. Griesbeck, Eds. *Photochemical Key Steps in Organic Synthesis*, VCH, Weinheim, 1998.
[2] W. H. Horspool, P. S. Song, Eds. *CRC Handbook of Organic Photochemistry and Photobiology*, CRC Press, Boca Raton, 1995.
[3] W. Horspool, D. Armesto, *Organic Photochemistry: a Comprehensive Treatment*, Ellis Horwood, New York, 1992.
[4] D. Rehm, A. Weller, *Isr. J. Chem.*, **1970**, *8*, 259.
[5] M. A. Fox, M. Chanon, Eds. *Photoinduced Electron Transfer*, Elsevier, Amsterdam, 1988.
[6] (a) G. J. Kavarnos, N. J. Turro, *Chem. Rev.*, **1986**, *86*, 401. (b) G. J. Kavarnos, *Top. Curr. Chem.*, **1991**, *156*, 21
[7] J. Mattay, Ed., *Top. Curr. Chem.*, **1990**, *156* ; **1990**, *158*; **1991**, *159*; **1992**, *163*; **1993**, *168*.
[8] L. Eberson, *Electron Transfer Reactions in Organic Chemistry*, Springer, Berlin, 1987.
[9] P. S. Mariano, Ed. *Advances in Electron Transfer Chemsitry*, JAI Press, Greenwich, CT, 1991-1998, Vol. 1-6.
[10] A. Pross, S. S. Shaik, *Acc. Chem. Res.*, **1983**, *16*, 363.
[11] W. Schlenck, E. Bergmann, *Liebigs Ann. Chem.*, **1928**, *464*, 1.
[12] D. Caine, *Org. React.*, **1976**, *23*, 1.
[13] (a) L. Eberson, R. Gonzalez-Luque, M. Merchan, F. Radner, B. O. Roos, S. Shaiks, *J. Chem. Soc., Perkin Trans. 2*, **1997**, 463. (b) L. Eberson, M. P. Hartshorn, F. Radner, M. Merchan, B. O. Roos, *Acta Chem. Scand.*, **1993**, *47*, 176.
[14] J. Ko, G. B. Schuster, *J. Am. Chem. Soc.*, **1977**, *99*, 6107.
[15] (a) S. G. Cohen, A. Parola, G. H. Parsons, *Chem. Rev.*, **1973**, *73*, 141. (b) J. Cossy, J. P. Pete, in ref. 9, Vol 5, p. 141.
[16] I. R. Gould, D. Ege, J. E. Moser, S. Farid, *J. Am. Chem. Soc.*, **1990**, *112*, 4290.
[17] (a) M. Freccero, M. Mella, A. Albini *Tetrahedron* **1994**, *50*, 2115. (b) M. A. Kellet, D. G. Whitten, I. R. Gould, W. R. Bergmark, *J. Am. Chem. Soc.*, **1991**, *113*, 358.
[18] (a) M. Mella, M. Fagnoni, M. Freccero, E. Fasani, A. Albini, *Chem. Rev.*, **1998**, *27*, 81. (b) A. Albini, M. Mella, M. Freccero, *Tetrahedron*, **1994**, *50*, 575. (c) A. Albini, E. Fasani, M. Mella, *Top. Curr. Chem.*, **1993**, *168*, 143.

[19] (a) U. C. Yoon, P. S. Mariano, *Acc. Chem. Res.*, **1992**, *25*, 233. (b) E. R. Gaillard, D. G. Whitten, *Acc. Chem. Res.*, **1996**, *29*, 292. (c) R. Popielartz, D. R. Arnold, *J. Am. Chem. Soc.*, **1990**, *112*, 3068. (d) F. D. Lewis, *Acc. Chem. Res.*, **1986**, *29*, 520. (e) I. R. Gould, S. Farid, *Acc. Chem. Res.*, **1996**, *29*, 520. (f) K. Mizuno, Y. Otsuji, *Top. Curr. Chem.*, **1994**, *169*, 301. (g) G. Pandey, *Top. Curr. Chem.*, **1993**, *168*, 175.
[20] (a) P. Maslak, *Top. Curr. Chem.*, **1993**, *168*, 1. (b) F. D. Saeva, *Top. Curr. Chem.*, **1991**, *156*, 59.
[21] C. F. Guertler, S. Blechert, E. Steckhan, *Synlett.*, **1994**, 141.
[22] T. Hirao, *Synlett.*, **1999**, 175.
[23] M. Yasuda, C. Pac, H. Sakurai, *J. Org. Chem.*, **1981**, *46*, 788.
[24] M. Mella, E. Fasani, A. Albini, *Tetrahedron*, **1991**, *47*, 3137.
[25] B. Giese, *Radicals in Organic Synthesis. Formation of Carbon-Carbon Bonds*, Pergamon, Oxford, 1986.
[26] D. P. Curran, in B. M. Trost, I. Fleming, *Comprehensive Organic Synthesis*, Pergamon, Oxford, 1991, Vol. 6, p. 714.
[27] (a) B. B. Snider, *Chem. Rev.*, **1996**, *96*, 339. (b) V. Nair, J. Mathew, J. Prabhakaran, *Chem. Soc. Rev.*, **1997**, *97*, 127.
[28] M. Fagnoni, M. Mella, A. Albini, *J. Am. Chem. Soc.*, **1995**, *117*, 7877.
[29] (a) M. Fagnoni, M. Mella, A. Albini, *J. Org. Chem.*, **1994**, *59*, 5614. (b) M. Fagnoni, M. Mella, A. Albini, *Tetrahedron*, **1995**, *51*, 859.
[30] M. Fagnoni, M. Mella, A. Albini, *J. Org. Chem.*, **1998**, *63*, 4026
[31] A. Albini, S. Spreti, *J. Chem. Soc., Chem. Commun.*, **1986**, 1426.
[32] L. Cermenati, M. Mella, A. Albini, *Tetrahedron*, **1998**, *54*, 2575.
[33] L. Cermenati, C. Richter, A. Albini, *J. Chem. Soc., Chem. Commun.*, **1998**, 805.

“Bioinspired” Metal Complexes of Macrocyclic $[N_4^{2-}]$ and Open Chain $[N_2O_2^{2-}]$ Schiff Base Ligands - a Link between Porphyrins and Salicylaldimines

Ernst-G. Jäger

Institut für Anorganische und Analytische Chemie, Friedrich-Schiller-Universität Jena, August-Bebel-Strasse 2, D-07743 Jena, Germany
E-mail: Cej@uni-jena.de

Abstract. Transition metals play an essential role as active sites of many enzymes. The complex catalytic performance of these biocatalysts should present a continuing challenge to chemists far into the third Millennium. Many attempts have been made in the past decades of the last century to develop new catalysts based on coordination compounds which mimic natural models. Initial success and the first technical applications involved particularly porphyrins and complexes of salicylaldimines. In this paper, we give an overview of chelate complexes of tetradentate Schiff base ligands with either a macrocyclic $[N_4]^{2-}$ or an open chain $[N_2O_2]^{2-}$ donor set derived from aliphatic 3-oxoaldehydes and diamines. These complexes represent a link between those of porphyrin type and those of salicylaldimine type and prove to have many properties and reactions in common with them. The high variability of the complexes' ligands with regard to the ring size, the extent of the π-electron system, and electronic as well as steric effects of peripheral substituents allow a broad variation of those properties decisive in catalytic performance; such as redox potentials, reactivity of axial coordination sites, and the spin state of the central atom. The redox couples $Ni^{III/I}$ as well as the equilibrium constants for the addition of axial ligands to the planar nickel(II) or the penta-coordinated organo-cobalt(III) complexes reflect the high sensitivity of the central atom to electronic effects from equatorial ligands. Some of the macrocyclic nickel complexes are good electrocatalysts for the reduction of carbon dioxide. Most of the discussion focuses on iron complexes, especially their reactivity with different axial ligands, the "push-pull" effects in adducts with mixed axial ligands and some special structural features. First observations of the catalysis of hydroquinone oxidation by an oxidase-like four-electron reduction of dioxygen show that - besides redox potentials, axial reactivity and spin state - the formation of oligonuclear units, stabilized by H-bridges between peripheral oxo-groups and/or axial ligands, seem to play an essential role in catalytic performance. H-bridges are obviously also responsible for the formation of "molecule based magnets" with specific solid-state structures and cooperative magnetic properties.

1. Macrocyclic metal complexes as active sites of biocatalysts - a challenge to chemistry of the third Millennium

In view of the non-renewability of fossil fuels as well as the increasing threat to life on earth through the greenhouse effect and warming of the atmosphere, the search for new sources of energy and new sources of basic materials for organic synthesis is one of the most important tasks and challenges at the beginning of the third Millennium. Processes of life give evidence for the fundamental possibility of using carbon dioxide for syntheses of a great variety of energy-rich organic fuels and valuable products (including carbohydrates, fats and proteins) through the use of sun light as energy source (scheme 1).

The life building performance of green plants, involving the stepwise strongly endergonic "up hill" reactions in photosynthesis or assimilation, is the fundament of virtually all life on earth and was the original source of all our fossil fuels. The highly complex processes of photosynthesis and respiration are dependent on certain typical features of living organisms. These include a high degree of differentiation and organization in space and time, special inter- and intracellular mechanisms for transporting substances and information and - last but not least - the involvement of enzymes as highly effective catalysts. Biochemical research during the past decades has demonstrated the central role of metals as active sites of such biocatalysts [1].

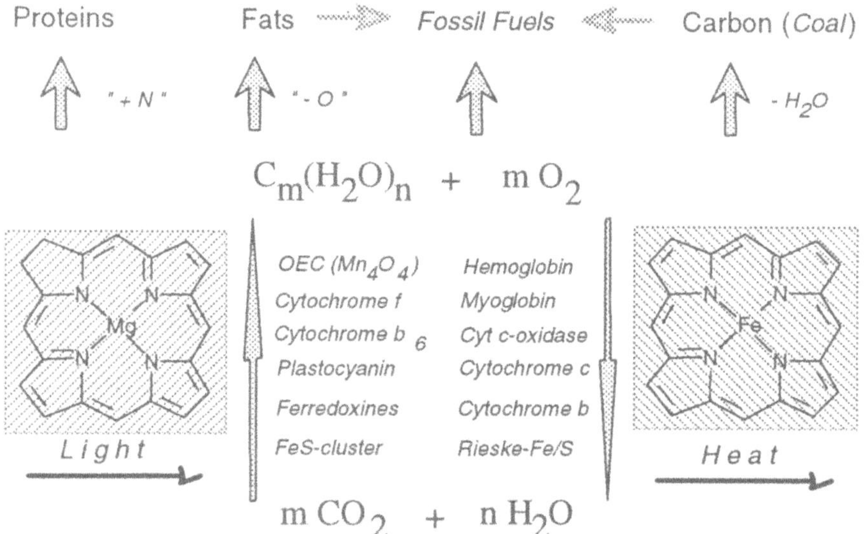

Scheme 1. Biological conversion of solar energy

Many individual steps of the cyclic transformation of energy and material in photosynthesis and respiration are controlled by coordinatively bound metal ions that make possible the redox steps over a range of potentials of more than 1.23 V (the equilibrium potential for the cleavage of water into hydrogen and oxygen). The transfer of oxidation equivalents from the photochemically formed radical cation of chlorophyll to water leading to the release of dioxygen require the high potential of a $[Mn_4O_4]$ cluster of the "Oxygen Evolving Complex" (*OEC*) [2]. The vectorial electron transfer and the redox reactions with decreasing potentials are mainly catalyzed by complexes of either copper [3] (*e.g.* in the charge transfer catalysts plastocyanin and azurin, as active site of many enzymes involved in the O_2 metabolism, including oxidases, oxygenases and superoxide dismutase) or iron [4] (*e.g.* as central atom of heme in hemoglobin, myoglobin, and cytochromes, as active site in non-heme iron proteins including the redox and charge-transfer catalysts based on iron-sulfur clusters). The redox active central atoms are often assisted by other metal ions, *e.g.* zinc or calcium, supporting the correct geometric arrangement of the catalytic center..

The most conspicuous and therefore perhaps the best known metal complexes in biology are the green leaf pigment chlorophyll and the red blood pigment of hemoglobin. Their essentiality to life was recognized long before their chemical identity was known. The typical macrocyclic structure of both complexes was elucidated by Willstätter and by Hans Fischer [5] who showed that the basic organic macrocycle chlorin of chlorophyll only differs from the porphin ring of heme by hydrogenation of one peripheral double bond. During the following decades further active centers of metalloenzymes have been characterized as macrocyclic complexes with typical "trace elements" as central atoms (scheme 2): In the early sixties, the active site of the anti-perniciosa factor, the coenzyme B_{12}, was realized to be an organo-cobalt complex [6] with a mono-anionic ligand derived from corrin, having a 15-membered inner great ring with a delocalized acyclic 14π electron system instead of the dianionic 16-membered 18π (Hückel-aromatic) inner ring of porphin. Only about fifteen years ago, the factor F 430, the active center of an enzyme involved in the last step of methane formation from carbon dioxide, was discovered as a nickel complex of so-called dihydrocorphin, a monoanionic macrocyclic ligand with a 16-membered inner ring (like porphin), but with a π-conjugation system extended over only two of the four six-membered chelate rings [7].

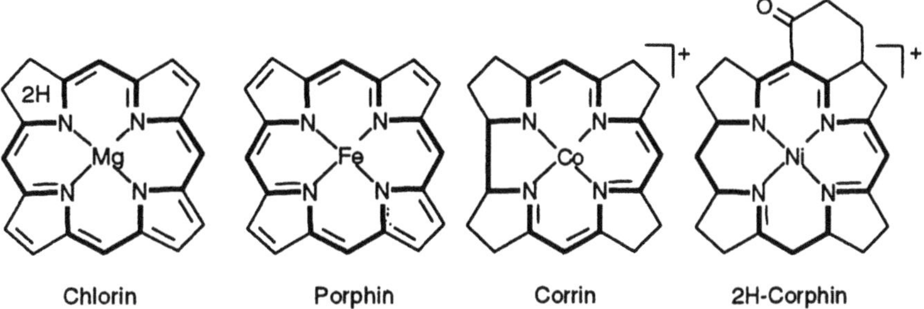

Scheme 2. Natural metal complexes with typical macrocyclic $[N_4]$ ligands

With these results it became increasingly evident that living organisms can use the ubiquitous elements iron, cobalt, and nickel in a specially suited environment of a macrocyclic ligand to achieve highly effective catalytic reactions. Similar technical processes need in most cases noble metals for this purpose, *e.g.* the higher homologues in the periodic table.

The best known example of these macrocyclic bio-complexes and that of the most general biocatalytic importance is doubtless the iron porphyrin of heme enzymes. Heme (with only slightly varied peripheral substitution, *cf.* scheme 3) is involved in the transport and storage of dioxygen in vertebrates (hemoglobin and myoglobin), in the four-electron reduction of dioxygen to water in the respiratory chain (together with copper in cytochrome c-oxidase), as a mono-oxygenase in the insertion of a *single* oxygen atom from dioxygen into organic substrates (cytochrome P450), in many side reactions of O_2 metabolism (catalase, peroxidase), in redox processes within the metabolism of inorganic S- or N-compounds (sulfite reductase, nitrite reductase, hydroxylamine oxidase) and additionally in many vectorial charge transfer processes (cytochromes b, c, d, f). Only a few years ago, the enzyme NO synthase (producing the highly important biomolecule NO by a five-electron oxidation of arginine with dioxygen) has been recognized as also containing heme as one of the active centers [8].

To learn the mysteries of the activity and selectivity of these catalytic centers in nature - through designing synthetic catalysts of similar performance under abiotic conditions - is one of the most interesting and challenging problems for future chemical research.

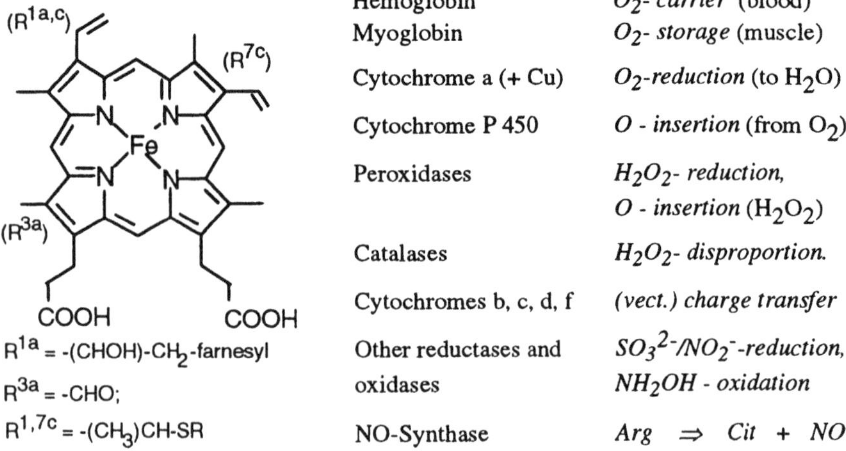

Scheme 3. Heme b (a, c in cytochromes a and c) as active site of biomolecules derived from iron porphyrins

2. Synthetic complexes designed after biological models

2.1. Porphyrins and salicylaldimines

The first attempts [9] to use natural or synthetic porphyrins as catalysts under abiotic conditions were not very successful. Only after a more detailed knowledge of structures and reaction mechanisms of heme enzymes obtained during the past three or four decades, has there been any substantial progress in this field. Considerable success was then achieved using porphyrin ligands [10] bearing specially designed substituents which can mimic in part the function of the protein.

More useful results for practical application have been found with artificial structures, differing considerably from that of the biological models. When the early investigations with natural porphyrins were under way, Pfeiffer and his coworkers developed the complex chemistry of salicylaldimines, a very powerful class of chelate ligands. The close relation of Pfeiffer's "salen" complexes to heme containing enzymes became clear when Tsumaki observed in 1938 [11] that salicylaldiminate complexes of cobalt (Cosalen) can add dioxygen reversibly and mimic the function of hemoglobin without a protein. Currently, Jacobsen and Katsuki have developed catalysts for enantioselective epoxidations [12a,b] based on a similar type of ligands. These manganese complexes (scheme 4) catalyze the O-transfer from hypochlorite (or other oxygen donors) in a similar way as the cytochrome P450-type enzymes do from molecular oxygen. While experiments involving the modeling of the P450 function using synthetic porphyrins are still at the stage of basic research, the Jacobsen/Katsuki catalysts have already gained technical ripeness. Even more recently, the development of single-component polyolefin catalysts based on salicyl aldimine complexes of nickel by Grubbs *et al.*[12c] have demonstrated again the potential of this class of ligands.

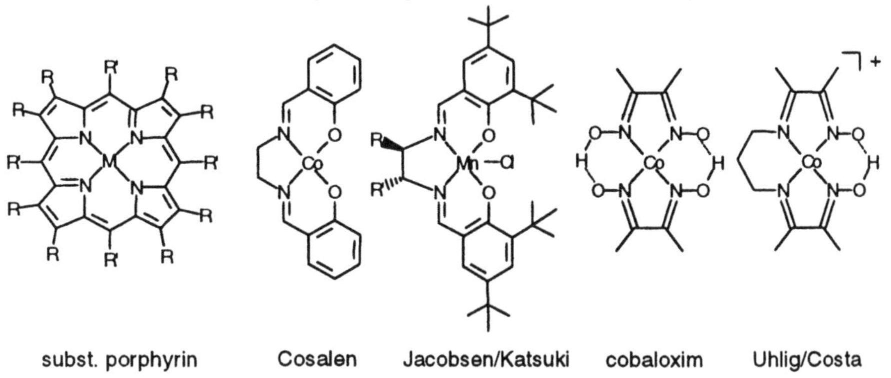

subst. porphyrin   Cosalen   Jacobsen/Katsuki   cobaloxim   Uhlig/Costa

Scheme 4. Selected "bioinspired" metal chelate complexes

Porphyrins (including their tetrabenzo-tetraaza derivatives, the phthalocyanines) and salen complexes are up to now the most intensively used metal complexes for mimicking the macrocyclic active sites of biocatalysts. Other examples are the "cobaloximes" by Schrauzer [13] and a cobalt complex derived from a ligand monoanion (designed first by Uhlig [14] and subsequently intensively studied by Costa [15]) for modeling the macrocyclic part of coenzyme B_{12}.

2.2. Complexes of macrocyclic imino-enamines and open-chain oxo-enamines

In the last decades, we developed a pool of transition metal complexes with tetradentate macrocyclic $[N_4]^{2-}$ [16-20] as well as open chain $[N_2O_2]^{2-}$ base ligands [21-25]. They are derived from 3-oxo-aldehydes instead of salicylaldehyde

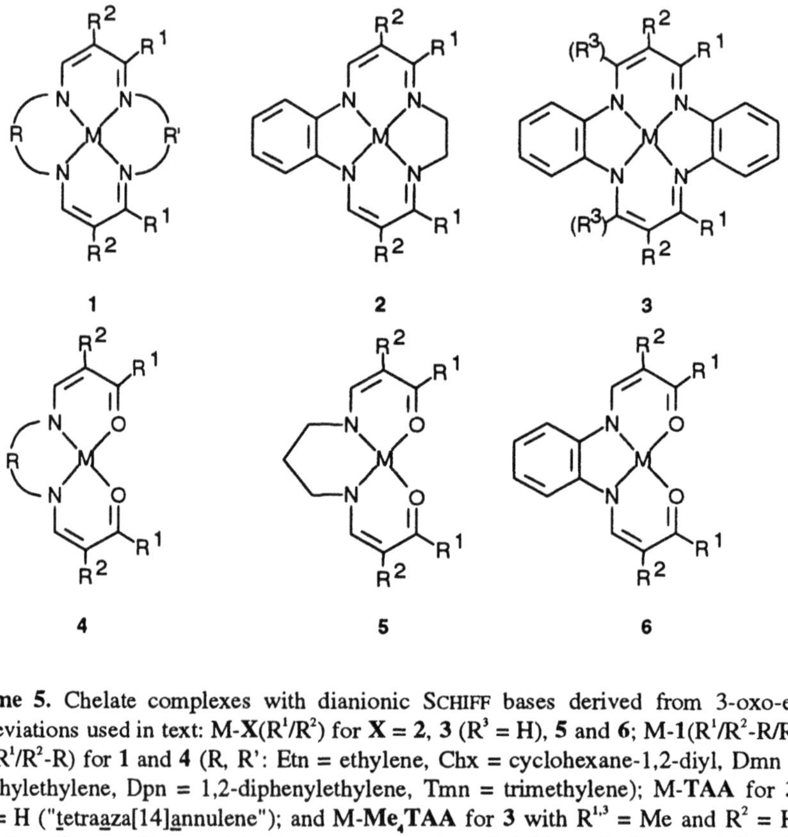

Scheme 5. Chelate complexes with dianionic SCHIFF bases derived from 3-oxo-enoles. Abbreviations used in text: M-**X**(R^1/R^2) for **X** = **2**, **3** (R^3 = H), **5** and **6**; M-**1**(R^1/R^2-R/R') and M-**4**(R^1/R^2-R) for **1** and **4** (R, R': Etn = ethylene, Chx = cyclohexane-1,2-diyl, Dmn = 1,2-dimethylethylene, Dpn = 1,2-diphenylethylene, Tmn = trimethylene); M-**TAA** for **3** with $R^{1,2,3}$ = H ("tetraaza[14]annulene"); and M-**Me$_4$TAA** for **3** with $R^{1,3}$ = Me and R^2 = H. The symbol **m (fe, co, ni)** is used to denote any complex. Axial ligands: Py = pyridine, HIm = imidazole.

(scheme 5). These complexes represent a link between those of porphyrins and salicylaldimines and have some characteristic structural features in common with the macrocyclic active sites of many biocatalysts:

1. The redox active central atom;

2. two empty or labile axial coordination sites;
3. an anionic, equatorial arranged, tetradentate chelate ligand;
4. at least two conjugate-unsaturated six-membered chelate rings;
5. in case **1** to **3** the $[N_4]$ donor set and the macrocyclic structure.

An advantage of these complexes is their broad variability. They can easily be synthesized with the biological relevant central atoms iron, cobalt, nickel, copper and - in part -manganese. The N,N-bridges determine the ring size and the extent of the π-electron system. The substituents R^1 (Me, ^{t}Bu, Ph, OEt) and R^2 (H, Me, Ph, COMe, COOEt, COPh, CN) can be used to control very efficiently the electronic features of the central atom and also the more hydrophilic or more lipophilic character of the ligands.

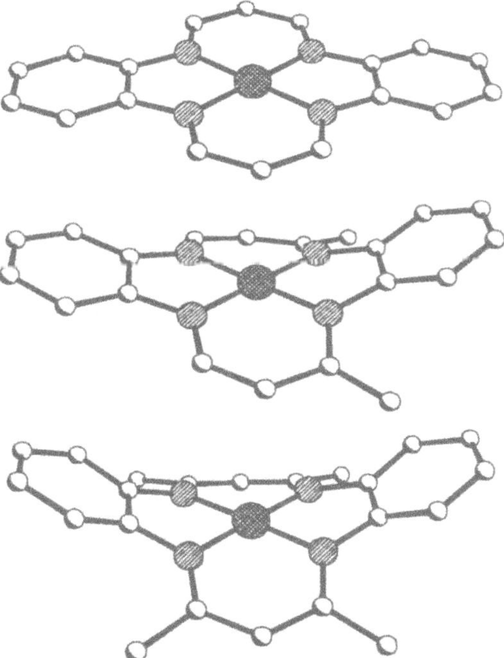

Fig. 1. Effect of ligand methylation on the saddle-shaped distortion of nickel complexes Ni-**TAA** (top) [28], Ni-**3**(Me/H) (middle) [29], and Ni-**Me$_4$TAA** (bottom) [30a].

The geometric features of the complexes are effected by the *N,N*-bridges and their repulsion with the substituents R^1 (and R^3). For example, the complexes of so called tetraaza[14]annulenes **3** show an increasing saddle-shaped distortion of the macrocycle with increasing substitution of H by Me in R^1 and R^3. This is displayed for the nickel complexes in Figure 1. This distortion (caused by repulsion between methyl-H and phenylene bridges) affects dramatically the properties of complexes: the resistance against strong acids decreases. The axial positions of the central atom are no longer equivalent and give rise to existence of additional isomers. The rotation barriers for axial ligands, especially heterocyclic bases, such as imidazole, Py *etc*, increase and these ligands can be bound only with a defined geometric orientation of their planes with respect to the $[MN_4]$ core. The rich structural flexibility of the ligand $\mathbf{Me_4TAA}^{2-}$ and the chemistry of its complexes have recently been reviewed [26, 27].

Steric effects can also be responsible for differences in the reactivity of the axial sites of complexes derived from stereoisomers of chiral ligands. The different orientation of the 1,2-cyclohexandiyl bridge in the nickel complexes *RR*- and *meso*-Ni-**4**(OEt/CN-Chx) (Figure 2) results for the *meso* isomer in a lower reactivity of the axial coordination sites, particularly with respect to the addition of bulky ligands at the *first* axial vacancy [25b].

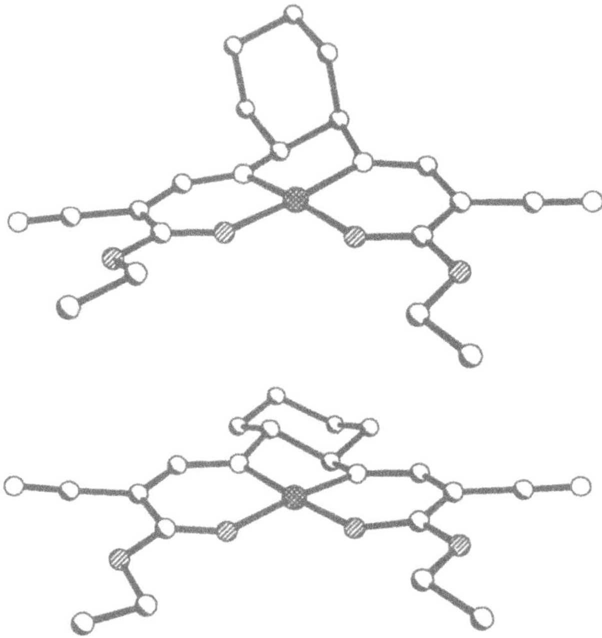

Fig. 2. Structures of *meso*-Ni-**4**(OEt/CN-Chx) (top) and *RR*-Ni-**4**(OEt/CN-Chx) (bottom)

3. Selected results of catalytic experiments

Our main interest with respect to the catalytic performance of these complexes is presently focused on the binding and activation of dioxygen and - for a few years - on the electrochemical reduction of carbon dioxide. For both reactions, the results point out a striking influence of the equatorial "controlling ligand" as well as the solvent which can act as axial ligand in many cases. Interesting catalytic properties have been observed in both cases. Only selected results are summarized here. A more detailed discussion is given in ref. [34, 36] (O_2 activation) and [40] (electrocatalytic CO_2 reduction).

3.1. Activation of dioxygen

Hydroquinone (H_2Q) in organic solvents was used as a model substrate to evaluate the activation of dioxygen in the presence of our metal complexes [31-34]. Quinone-/hydroquinone systems act in nature as a link between the mainly heme controlled part and the more "organic" catalyzed part of the respiratory chain.

$$\text{hydroquinone} + 1/2\ O_2 \xrightarrow[MeCN]{cat} \text{quinone} + H_2O \qquad (1)$$

Without any catalyst, no significant formation of quinone can be observed over hours or days. No significant catalytic activity was observed with most of the investigated complexes (including acetylacetonates, salen complexes, porphyrins and others [32a]). In special cases under special conditions, however, an efficient activation of dioxygen takes place.

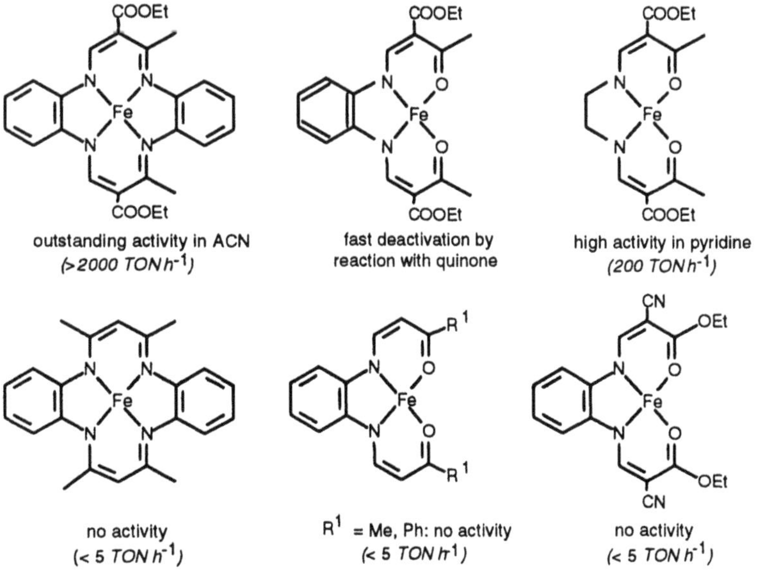

outstanding activity in ACN ($>2000\ TON\,h^{-1}$)

fast deactivation by reaction with quinone

high activity in pyridine ($200\ TON\,h^{-1}$)

no activity ($< 5\ TON\,h^{-1}$)

R^1 = Me, Ph: no activity ($< 5\ TON\,h^{-1}$)

no activity ($< 5\ TON\,h^{-1}$)

Scheme 6. Oxidase like activity of iron complexes with respect to the oxidation of hydroquinone by air to give pure quinone

Amongst the manganese complexes, Mn-**6**(Me/COOEt) shows high activity in MeCN [33], but only low activity in Py. Mn-**6**(OEt/CN) is totally inactive. Most of the cobalt complexes - with exception of the inactive Co-**6**(OEt/CN) - exhibit low to moderate activity [34]. Surprisingly, no hydrogen peroxide was formed as a side product of the reduction of dioxygen.

The most surprising differences have been observed with iron complexes. Some pairs of structurally closely related active (upper row) and inactive compounds are given in scheme 6. The ester substituted **TAA** derivative, Fe-**3**(Me/COOEt), is nearly two orders of magnitude more active than all the other complexes, and it is clearly the best catalyst - but only in MeCN. In contrast, the tetramethyl derivative, Fe-**Me$_4$TAA**, is inactive under the same conditions [32a]. Another good catalyst - but only in Py - is Fe-**4**(Me/COOEt-Etn) [35], whereas Fe-**6**(OEt/CN) is totally inactive in the same solvent. These striking differences are probably caused by quite different reasons for both pairs.

In the last pair, Fe-**6**(OEt/CN) is the only one of the about thirty investigated iron(II) complexes that does not react with dioxygen in pyridine solution as well as in solid state as adduct with Py. This solid complex is stable at air over years [24b]. In this case, the inactivity is thermodynamically determined: the oxidation potential ($\geq$ 0.6 V *vs.* NHE in Py [24a]) and the selectivity for binding of Py as axial ligand in the iron(II) state are too high for an exergonic displacement of one axial ligand by dioxygen - probably the essential first step of the catalytic *activation* of dioxygen at the metal. In contrast, Fe-**4**(Me/COOEt-Etn) reacts in Py as well as in solid state very fast with dioxygen to give the μ-oxo-iron(III) derivative - probably a key intermediate in the special case of hydroquinone oxidation. Interestingly, the molecular structures of both octahedral high-spin iron(II) complexes with two axial Py show no significant deviations in their atomic distances or angles within the first coordination sphere [24b, 34] - apart from different orientations of the pyridine planes (Figure 3).

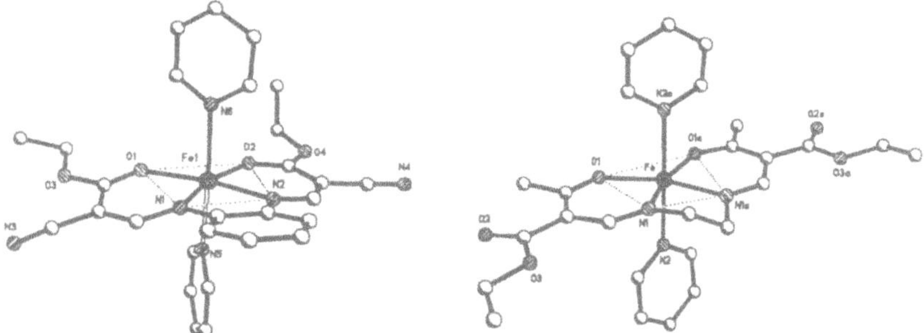

Fig. 3. Molecular structures of the inactive complex [Fe-**6**(OEt/CN)(Py)$_2$] (left) and the active complex [Fe-**4**(Me/COOEt-Etn)(Py)$_2$] (right)

The marked difference in activity of Fe-**3**(Me/COOEt) and Fe-**Me$_4$TAA** cannot be explained by similar thermodynamic arguments. Both of these iron(II) complexes react with dioxygen in MeCN containing a small content of water to give quantitatively the μ-oxo-iron(III) derivatives whose structures are, however, quite different (Figure 4). The lattice of Fe-**Me$_4$TAA** [30b] consists of isolated *dinuclear* μ-oxo units and one molecule of the solvent MeCN that shows no interactions with the complex. The active complex Fe-**3**(Me/COOEt) forms *tetranuclear* molecules wherein two μ-oxo units are linked by H-bridges of two water molecules (Figure 5). In this way, a characteristic chain of four H-bridged oxide ions is formed. The network of H-bridges is completed by weaker interactions with two of the eight peripheral ester groups (Figure 6). These tetranuclear building blocks are isolated from each other in the lattice by the nonpolar part of the ligands. Of course, their existence in the solvent MeCN is not very probably. It seems, however, that the high tendency of the carbonyl substituted complexes to form H-bridges improves the simultaneous transfer of electrons *and* protons to the iron(III) centers and the oxide ions, as it is schematically shown in Figure 6.

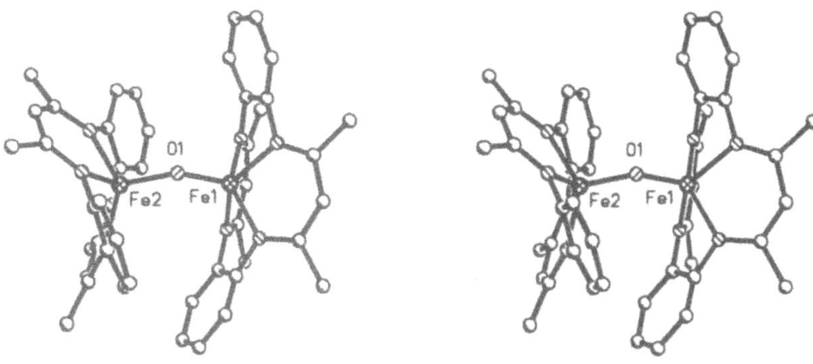

Fig. 4. Stereoview of the dinuclear μ-oxo unit [Fe-**Me_4TAA**]$_2$O [30b]

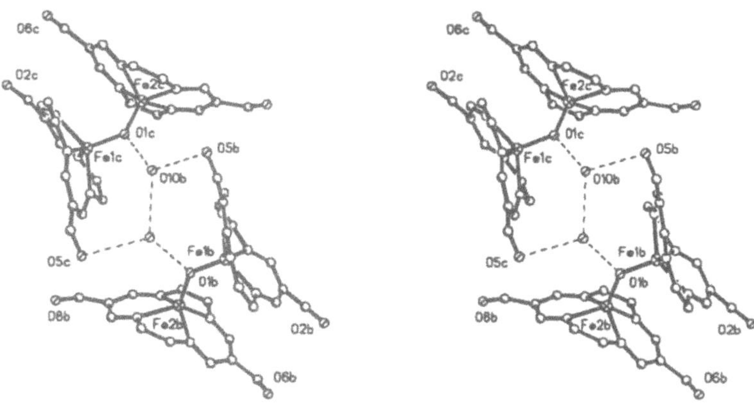

Fig. 5. Stereoview of the tetranuclear unit {[Fe-**3**(Me/COOEt)]$_2$O×(H_2O)]$_2$}; substituents Me, OEt and bridges Phn omitted for clarity.

Other significant differences between the structures of the μ-oxo derivatives of both complexes which could influence their catalytic behavior are i) the twist angle between both metal chelate units (37° in [Fe-**3**(Me/COOEt)]$_2$O; 63° in [Fe-

Me_4TAA$]_2$O), ii) the angle Fe-O-Fe of the μ-oxo group (133° in the first, 149° in the latter). Both i and ii results for [Fe-**3**(Me/COOEt)$]_2$O in a funnel-like structure wherein the "open" side is occupied by the additional oxygen of one water (Figure 7). Force-field calculations show clearly [32b] that this arrangement offers optimal conditions for the interaction with one molecule of hydroquinone. Finally, the possibility of oxidation $Fe^{III} \Rightarrow Fe^{IV}$, which is electrochemically strongly reversible for both iron in [Fe-**3**(Me/COOEt)$]_2$O at relatively low potentials (0.71 V and 0.94 V *versus* NHE in MeCN), but irreversible for the Fe-**Me_4TAA** derivative, could improve the consecutive transfer of four oxidation equivalents to *two* H_2Q molecules by *one* dinuclear unit of the ester substituted complex.

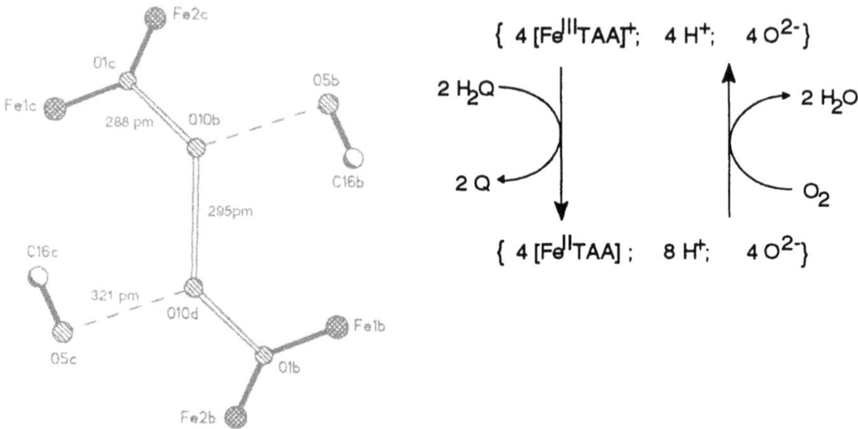

Fig. 6. Ester-carbonyl assisted chain of four H-bridged oxide ions in the core of the tetranuclear {[Fe-**3**(Me/COOEt)$]_2$Ox(H_2O)$\}_2$ and proposed mechanism of H_2Q oxidation

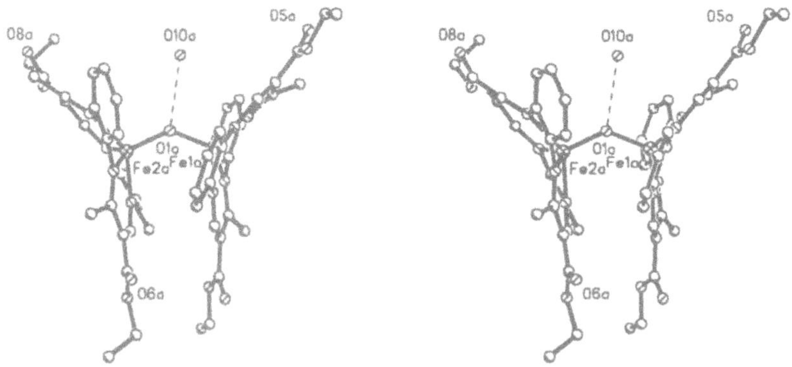

Fig. 7. Stereoview of the funnel-like structure of one half of the tetranuclear μ-oxo derivative of Fe-**3**(COOEt/Me) with the additional oxide of water on the "open" side

The structures of several derivatives of Fe-**6**(Me/COOEt) show also strong H-bridges. Their formation could be the reason for the activity of this complex which is, however, poisoned by reaction with the produced quinone.

For further experiments with dioxygen as oxidant, tri-^{t}butyl-phenol was used as an alternative model substrate. The results are contrary to those with H_2Q: all the iron complexes are inactive. The cobalt complexes show a markedly differentiated, in some cases high activity. One the other hand, the oxidation of tri-^{t}butyl-*aniline* by air is *not* catalyzed by any cobalt complex, but with good efficiency and selectivity by the macrocycle Fe-**3**(Me/COOEt) to yield the nitroso derivative and *o*-quinone as main products [36].

3.2. Electrocatalysts for reductive coupling of carbon dioxide

The electrochemical reduction of carbon dioxide in the presence of metal complexes as electrocatalysts has gained much attention over the past decades [37, 38]. In most cases, *e.g.* with cationic "cyclam" complexes of nickel, carbon monoxide and/or formic acid are formed [37]. Only a few examples are known [38] regarding the reductive formation of C-C bonds *via* $CO_2^{\cdot-}$ radicals to give oxalate, which also takes place with aromatic nitriles as catalysts [39]. Our investigations [40] have been focused on the aliphatically bridged macrocycles Ni-**1** in dry MeCN as solvent. The results show a dramatic effect of the bridges R, R' as well as the peripheral substituents R^1 and R^2. Only the carbonyl substituted complexes (R^2 = COOEt, COMe) give oxalate with good selectivity. The yield of oxalate decreases markedly in the order

Ni-**1**(Me/COOEt-Etn/Etn) >> Ni-**1**(Me/COMe-Etn/Etn)
> Ni-**1**(Ph/COOEt-Etn/Etn) > Ni-**1**(Me/COMe-Etn/Tmn)
> Ni-**1**(Me/COMe-Tmn/Tmn) > Ni-**1**(Me/H-Etn/Tmn)

Cyclic voltammetric investigations using the DigiSim® algorithm [41] for the first complex, point to the following mechanism with an inner-sphere one-electron transfer from $\mathbf{ni}^I$ to axially bound CO_2, followed by a recombination of $CO_2^{\cdot-}$ radicals in the bulk:

$$\mathbf{ni}(II) + e^- \Rightarrow \mathbf{ni}(I)^- \quad (2)$$

$$\mathbf{ni}(I)^- + CO_2 \Rightarrow [\mathbf{ni}(CO_2)]^- \quad (3)$$

$$[\mathbf{ni}(CO_2]^- \Rightarrow \mathbf{ni}(II) + CO_2^{\cdot-} \quad (4)$$

$$2\ CO_2^{\cdot-} \Rightarrow C_2O_4^{2-} \quad (5)$$

This model is not adequate, however, for the other complexes, especially those without carbonyl substituents R^2. This is probably due to side reactions between the nucleophilic CO_2 and the strongly negative charged *meso* carbon of the six-membered chelate rings in the **ni**(I) complex.

Altogether, these studies on O_2 as well as CO_2 activation demonstrate the possibility to tune the catalytic performance of our complexes by variation of the equatorial and axial (*e.g.* the solvent) ligands. It is still quite difficult, however, to rationalize the experimental results and to give convincing arguments to explain the strong differences in the behavior of individual complexes. Therefore, the aim of our studies is not superficially to develop new catalysts, but mainly to elucidate the effect of the "controlling" ligands on those general kinds of reactions and properties which should have a strong effect on catalysis: the redox-potentials, the Lewis acidity of the axial coordination sites, the spin ground state of the central atom and special geometric features. Of course, the potential catalytic performance depends on many other factors, *e.g.* the recognition of the substrate by specific interactions, the rate of all the partial steps, *etc.* The thermodynamics, however, can give information on *necessary* (but not *sufficient*) preconditions of a catalytic reaction.

4. Coordinatively unsaturated metals as probes of ligand effects

The activation of small molecules like dioxygen by a molecular metal center requires - as a rule - a coordinative binding and a more or less complete transfer of electron density. The thermodynamics of such an interaction can formally be divided - according to Hess' law - into pure redox and pure Lewis acid-base steps. These partial reactions can be more easily quantified than the complete reaction by measuring the redox potentials (if the coordination number remains constant in both oxidation steps) and the equilibrium constant for the axial addition of a suitable model ligand (if the complex molecule remains coordinatively unsaturated in the used solvent and the change of equatorial ligand solvatation by the additional ligand can be neglected).

4.1. Planar nickel complexes

The nickel complexes of scheme 5 are nearly all planar in solid state and in non or weakly coordinating solvents. Exceptions are complexes with high Lewis acidity (low electron density) of the central atom and those with donor atoms (O) or peripheral groups that can act as bridging ligands to give octahedral isomers (*e.g.* Ni-**4**(EtO/CN), Ni-**5**(EtO/COOEt) Ni-**6**(EtO/CN) and others). Most of the planar complexes retain their planarity and their diamagnetic S = 0 spin state also in weakly coordinating solvents suitable for cyclovoltammetry (MeCN, CH_2Cl_2). All investigated nickel complexes give a strongly reversible reduction

$$\textbf{ni}(II) \quad + \quad e^- \quad \Rightarrow \quad \textbf{ni}(I)^- \qquad (6)$$

without any change in the coordination number because the **ni**(I) d^9 system is clearly a weaker Lewis acid than **ni**(II). The formation of Ni^I (instead of an organic ligand-radical) could be proven by ESR spectroscopy for all the complexes [43] with exception of complexes Ni-**3** without peripheral carbonyl groups R^2. In contrast, the oxidation process

$$\mathbf{ni}(II) \Rightarrow \mathbf{ni}(III)^+ + e^- \quad (7)$$

increases the electrophilicity of the central atom, and a stronger solvatation of the formed d^7 **ni**(III) has to be considered. Also in this case strongly reversible peaks in CV can be observed if the rate of addition/elimination of solvent as an additional axial ligand is fast enough. The potentials, however, do not correctly reflect the thermodynamic data for the pure oxidation (7).

Table 1. Reduction potentials $Ni^{II/I}$ in ACN ($E_{1/2}$ *versus* SCE; $E°_H$ calculated for NHE; 293 K) and equilibrium constants for the axial addition of two pyridine in benzene ($\lg\beta_2$; 293 K).

Complex	$E_{1/2}$ [V][a)]	$E°_H$ [V][b)]	lg β_2[c)]	• in Fig.3
Ni-**4**(^{t}Bu/H-Etn)	-2.02	-1.78	d)	-
Ni-**4**(Me/H-Etn)	-1.95	-1.71	d)	-
Ni-**4**(Ph/H-Etn)	-1.82	-1.58	d)	-
Ni-**4**(Me/COOEt-Etn)	-1.70$_5$	-1.46	-3.1	4b
Ni-**4**(Ph/COOEt-Etn)	-1.58	-1.34	-1.98	4c
Ni-**4**(OEt/COOEt-Etn)	-1.58	-1.34	-0.46	4a
Ni-**4**(OEt/CN-Etn)	-1.27$_5$	-1.03	+3.68	2
*RR*Ni-**4**(OEt/CN-Chx)	-	-1.04[e)]	+2.27[f)]	-
*meso*Ni-**4**(OEt/CN-Chx)	-	-0.97[e)]	+1.24[f)]	9
*trans*Ni-**4**(Me/COOEt-Chx)	-	-1.47[f)]	d)	-
*trans*Ni-**4**(Me/COOEt-Dpn)	-	-1.31[f)]	d)	-
Ni-**5**(Me/H)	-1.69	-1.45	-0.2	-
Ni-**5**(Me/COOEt)	-1.45	-1.21	+1.46	-
Ni-**5**(Me/COMe)	-1.42	-1.18	+1.34	-
Ni-**5**(Ph/COOEt)	-1.36	-1.12	+1.92	-
Ni-**6**(^{t}Bu/H)	-1.81	-1.57	-3.9	8d
Ni-**6**(Me/H)	-1.74	-1.50	-3.48	8b
Ni-**6**(Ph/H)	-1.61	-1.37	-2.90	8c
Ni-**6**(Me/COOEt)	-1.51	-1.27	-1.48	3b
Ni-**6**(Me/COPh)	-1.44	-1.20	-0.92	7
Ni-**6**(Ph/COOEt)	-1.39	-1.15	-0.20	3c
Ni-**6**(OEt/Me)	-1.73	-1.49	-1.74	6
Ni-**6**(OEt/Ph)	-1.61	-1.37	-1.01	5
Ni-**6**(OEt/COOEt)	-1.42	-1.18	+1.52	3a
Ni-**6**(OEt/CN)	-1.15$_5$	-0.91	+4.89	1

a) From ref. [23c]. b) $E°_H = E_{1/2} + 0.24$ V. c) From ref. [23a,c]. d) Too low for precise determination by spectrophotometric titration or measurement of magnetic susceptibility in solution. e) Measured *versus* Ag/AgCl/MeCN, $E°_H = E_{1/2} - 0.16_5$; data from ref. [25d]. f) From ref. [25b].

The planar complexes with $[N_2O_2]$ donor set derived from open-chain SCHIFF bases give in most cases with strong N-bases (Py, piperidine, N-MeIm, aliphatic amines) octahedral di-adducts. The change of coordination number is accompanied by a change of spin ground state (S = 0 ⇒ S = 1) and a marked change of the d-d absorption spectra. This allows a quantitative characterization of LEWIS acidity (acceptor strength, electrophilicity) of the coordinatively unsaturated central atom by analysis of the equilibrium of axial ligand addition in non-coordinating solvents according to eq. (**8**) - (**10**) (L: additional monodentate ligand):

$$\mathbf{ni} + \mathrm{L} \Rightarrow \mathbf{niL} \quad (K_1) \tag{8}$$

$$\mathbf{niL} + \mathrm{L} \Rightarrow \mathbf{niL_2} \quad (K_2) \tag{9}$$

$$\mathbf{ni} + 2\,\mathrm{L} \Rightarrow \mathbf{niL_2} \quad (\beta_2) \tag{10}$$

The individual constants K_1 and K_2 for the consecutive steps (**8**) and (**9**) can be determined in most cases only with low precision because $K_1 << K_2$. Some exceptions observed with sterically hindered (chiral) equatorial chelate ligands are discussed in ref. [25b].

Selected data for the electrochemical reduction and the Lewis acidity of nickel complexes with tetradentate $[N_2O_2]^{2-}$ coordinated open-chain ligands are listed in Table 1. A complete thermodynamic and kinetic analysis of the coupled equilibrium of electron transfer and ligand addition has been made for Ni-**5**(Me/COMe) [42] using the DigiSim® algorithm [41].

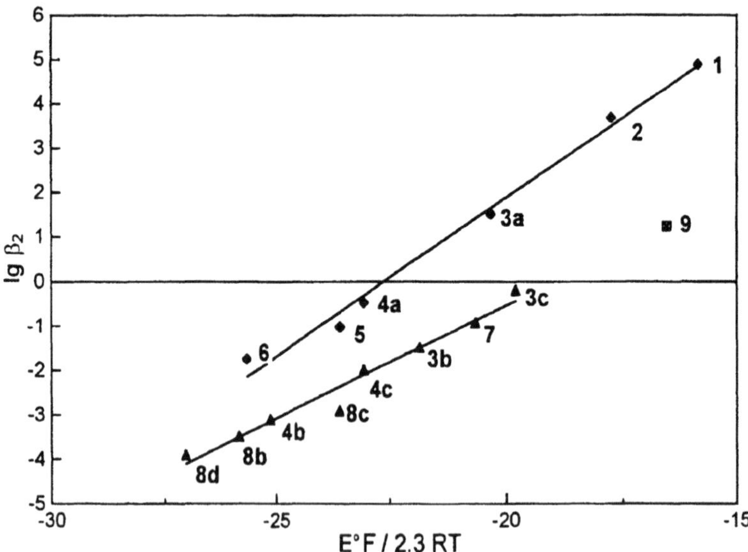

Fig. 8. Correlation between the affinity of planar nickel chelates of type **4** (R = Etn) and **6** for electrons (in MeCN; given as $E^\circ_H F / (2.3\,RT) = \lg K[\mathbf{ni} + H_2 = \mathbf{ni}^- + H^+]$; $T = 293$ K); and Py (in benzene; given as $\lg \beta_2[\mathbf{ni} + 2\,\mathrm{Py} = \mathbf{niPy_2}]$). Data from ref. [23a,c], [25a].

The electron can be interpreted as the "smallest" base. If the electron occupies in the planar nickel(I)-complex always the anti-bonding $d(x^2-y^2)$ orbital and not an empty ligand orbital, and if the addition of the axial ligands is always accompanied by a spin change from S = 0 in the planar to S = 1 in the octahedral geometry, a correlation between the free energy of reduction and that of adduct formation could be expected. Indeed, as it is shown in Figure 8, there exists a good linear relationship between $\lg K(\mathbf{Ni}^{II} + H_2 \Leftrightarrow \mathbf{ni}^{I-} + H^+)$ and $\lg\beta_2(\mathbf{ni} + 2\ \mathrm{Py} \Leftrightarrow \mathbf{ni}\mathrm{Py}_2)$ for many complexes.

Surprisingly, all complexes Ni-**4**(R^1/R^2-Etn) and Ni-**6**(R^1/R^2) with R^1 = Me and Ph fit the same line, whereas all complexes with R^1 = OEt give their own correlation with a comparatively higher Lewis acidity. This could be due to the fact that the ethoxy group causes a lower steric hindrance than the methyl or the phenyl group. The complex *meso*-Ni-**4**(OEt/CN-Chx) with the more bulky chiral bridge (point 9 in Fig. 8) does not fit the upper line for R^1 = OEt. The slope of the regression line is 0.7 for the points with R^2 = OEt and 0.5 for the others. This means that the addition of electrons is much more sensitive towards substituent effects than the addition of two axial pyridine.

In the case of nickel complexes with *macrocyclic* $[N_4^{2-}]$ ligands, the Lewis acidity of the axial coordination sites is too low for precise equilibrium measurements - even with strong N-bases such as piperidine. The reduction and oxidation potentials, $Ni^{I/II}$ and $Ni^{II/III}$, have been determined by Busch *et al.* [44] and in our laboratory [43a] for a large number of complexes Ni-**1**, Ni-**2** and Ni-**3** with a broad variety of substituents R^1 and R^2. The reduction potentials are nearly independent upon the solvent (MeCN, DMF, Py) and show clearly the electronic influence of the bridges R, R' and the peripheral substituents (selected values from ref. [20a, 43a] for MeCN calculated *versus* NHE; [V]):

Ni-**1**(Me/H-Etn/Etn):	-2.14;	Ni-**1**(Me/COOEt-Etn/Tmn):	-1.88;
Ni-**1**(Me/COOEt-Etn/Etn):	-2.03;	Ni-**1**(Me/COOEt-Tmn/Tmn):	-1.79;
Ni-**1**(Me/COMe-Etn/Etn):	-1.96;	Ni-**2**(Me/COOEt-Phn/Etn):	-1.80;
Ni-**1**(Ph/COOEt-Etn/Etn):	-1.95;	Ni-**2**(Me/COMe-Phn/Etn):	-1.76
Ni-**3**(Me/COOEt-Phn/Phn):	-1.27;	Ni-**3**(Me/COMe-Phn/Phn):	-1.25;
Ni-**3**(Me/H/Me-Phn/Phn)	-1.50 (2 e^-; ligand reduction);		
Ni-**3**(Me/H-Phn/Phn)	-1.47 (2 e^-; ligand reduction).		

A more detailed comparison of reduction and oxidation processes of macrocyclic complexes M-**1**, M-**2** and M-**3** (M = Fe, Co, Ni, Cu) is given in ref. [20a].

With respect to the free energy relationship between reduction on the one hand and addition of pyridine on the other (Fig. 8), the phenylene bridged macrocycle Ni-**3** with carbonyl substituents R^2 should give in pyridine significant concentrations of octahedral high-spin adducts. However, Ni-**3**(Me/COOEt) - like Ni-**3**(Me/H/Me) - is diamagnetic even in pure pyridine. This may be due to the fact that the "saddle-shaped" distortion of the **TAA** ligand (Fig. 1) hinders the free rotation of the axial ligand and therefore increases the positive entropy of the axial ligand binding.

4.2. Pentacoordinated organo-cobalt(III) complexes

The reaction of pentacoordinated ("base off") organo-cobalt(III) derivatives with an axial base is an important part of the catalytic cycle of coenzyme B_{12} [6]. As a model reaction it is also a good tool for evaluating the Lewis acidity of axial coordination sites in dependence on electronic effects of the equatorial ligands. In this case, also the macrocyclic complexes give equilibria in solution and allow a very precise determination of equilibrium constants [22c, 45, 46]. A special advantage of this reaction is the pure 1:1 stoichiometry. On the other hand, the high sensitivity of the complexes to light - especially of the pentacoordinated state in presence of air - complicates measurements. Some typical data are given in scheme 7.

R = Me; K [M^{-1}] =					
20	38	47	50	2600	3900
R = Et; K [M^{-1}] =					
1.3	3.5	6	-	270	620

Scheme 7. Equilibrium constants for the reaction of pentacoordinated ("base off") organo-cobalt(III) complexes with Py in benzene (293 K, data from ref. [45, 46])

The results show altogether that the following changes in the equatorial chelate ligand have an especially high effect on the increasing reactivity of the central atom with respect to the uptake of axial ligands and/or electrons:

1. Macrocyclic $[N_4]^{2-}$ $\Rightarrow$ open chain $[N_2O_2]^{2-}$ coordination;
2. R = Etn $\Rightarrow$ Phn (aliphatic 5-ring $\Rightarrow$ aromatic 5-ring; in macrocycles only valid for the exchange of the *first* Etn);
3. R = Etn $\Rightarrow$ Tmn (aliphatic 5-ring $\Rightarrow$ aliphatic 6-ring);
4. R^1 = Me $\Rightarrow$ Ph < OEt (especially regarding the axial ligand binding)
5. R^2 = H $\Rightarrow$ COOEt, COMe < CN.

The increase of the Lewis acidity of the central atom in the coordinatively unsaturated complexes caused by carbonyl or nitrile substitution of the ligand periphery can result in interesting supramolecular interactions: the more negatively polarized carbonyl oxygen or nitrile nitrogen act as axial ligands of the more acid central atom to give a two- or three-dimensional network [47, 48]. A typical example is provided by the solid state structure of the unsymmetrically bridged cobalt(II) complex **7** [47].

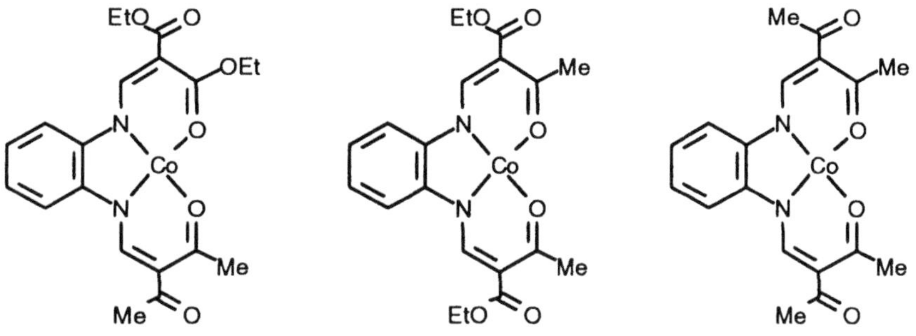

7 **8** **9**

Scheme 8. Formula units of the unsymmetrically substituted low-spin cobalt(II) complex **7**, the symmetrically substituted isomer **8** and the diacetyl derivative **9**

The substituents OEt/COOEt on the "upper" side increase efficiently the LEWIS acidity of the cobalt. The acetyl group on the lower side (Me/COMe) is a better ligand than the ester group and occupies one coordination site on the central atom. This results in stair-like chains with pentacoordinated cobalt(II) (Figure 9). It is interesting to note that the complex **7** is the only one of our cobalt complexes that is able to bind dioxygen in solid state without any additional axial ligand. In contrast, the lattice of the isomeric complex **8** (with two ester groups) and the derivative **9** (with two acetyl groups) contains planar molecules without intermolecular coordination (Figure 10).

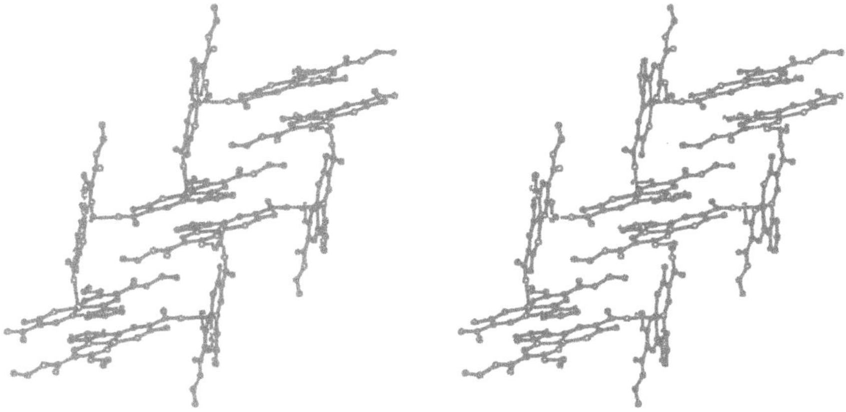

Fig. 9. Stereoview of two stair-like chains of **7** with intermolecular acetyl coordination and pentacoordinated cobalt(II)

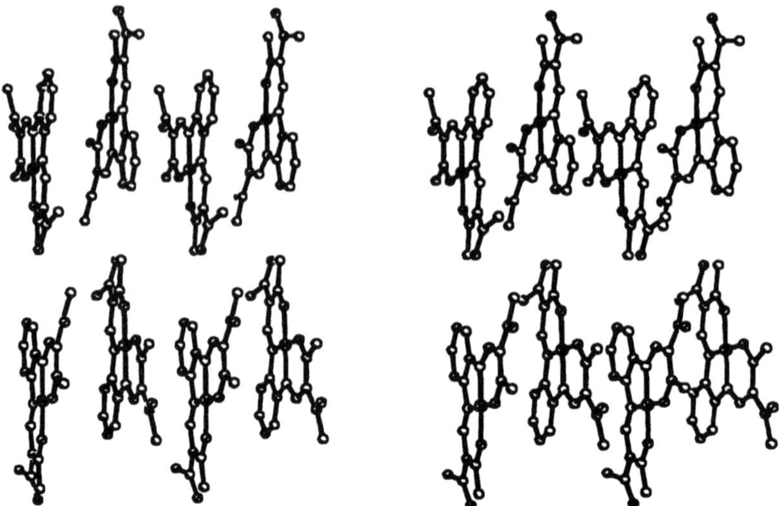

Fig. 10. Stereoview of the orientation of the molecular units in the lattice of **9** [49]

5. Chemistry of iron(III) in a hemine-like macrocyclic ligand environment

The iron complexes of the macrocyclic type **1** - **3** are of particular interest because of their structural relation to iron porphyrins. The resemblance increases with increasing extension of the π-electron system and additionally by substitution of the *meso* position of the six-membered chelate rings with electron-withdrawing groups.

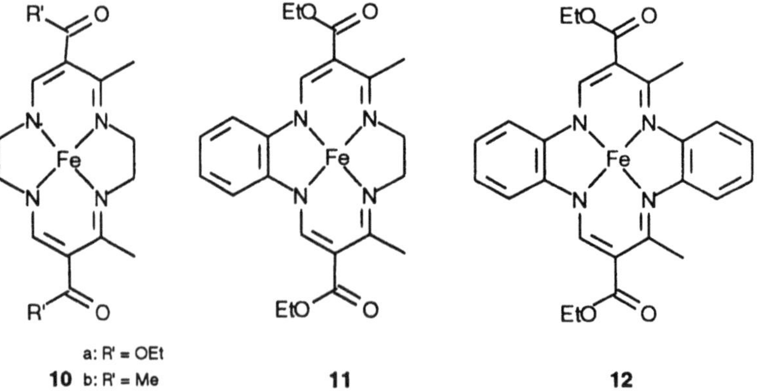

Scheme 9. Macrocyclic iron complexes with structural resemblance to heme

The complexes **10** - **12** are especially suitable for systematic studies and comparisons. The free carbonyl groups protect the organic ligand from the electrophilic attack of dioxygen and increase the stability with air. They also increase the π-acidity of the ligand and therefore the Lewis acidity of the axial coordination sites. Additionally, they seem to play a significant role as "sockets" for a fast charge transfer (thus, the electrochemical reversibility of the redox couples is - as a rule - much better with the carbonyl substituted complexes).

The effect of the increasing π-conjugation from **10** to **12** and the influence of different axial ligands on the comparative UV-/Vis- as well as the ESR spectra [20b,c] and the redox potentials [20a, 50], the differing equilibria and kinetics of axial ligand exchange [20c, 50], and special structural differences [20c] have been reported in a former review [50] and in other recent papers [20, 25]. MO calculations demonstrating some typical differences between our complexes and the porphyrins with respect to the strength and symmetry of the equatorial ligand field and the π-donor-/acceptor properties have also been reported [51]. In the following, the rich chemistry and especially the varied axial reactivity of the porphyrin-like $[N_4^{2-}]$ coordinated iron(III) will be illustrated with the cationic derivative $[Fe\text{-}\mathbf{10b}]^+$ of the complex **10b** as an example.

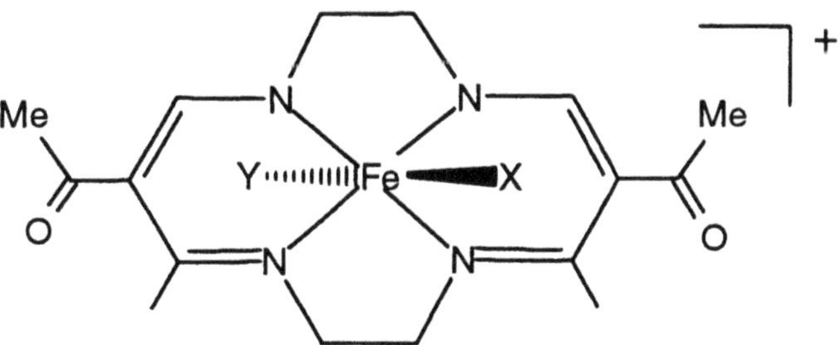

Fig. 11. The macrocyclic iron(III) complex used to illustrate the axial reactivity of iron in a hemine-like ligand environment

1. In comparison with the iron porphyrins and the complexes **11** and **12,** this aliphatically bridged macrocycle offers several advantages as a model compound to demonstrate reactions at the axial coordination sites of the central atom:
2. Because of the reduced organic part, the properties are generally more dominated by the central atom.
3. The dianion of the ligand is colorless, and the VIS-/NIR spectra are dominated by a strong and sharp (equatorial ligand to Fe^{III}) charge transfer band. The wavelength of its maximum depends strongly on the axial ligands (from 400 to 1000 nm) and is a very sensitive visual probe for changes at the axial sites. (In case of iron(III) porphyrins, including heme enzymes [52], an analogous CT band occurs only at $\lambda > 1000$ nm and requires special techniques for observation.)

4. Well resolved rhombic ESR spectra in the low spin state give further information on electronic features of the central atom.
5. The exchange of axial ligands in solution (and also in solid-gas reactions) is very fast and thermodynamically controlled.
6. The more hydrophilic acetyl substituent (instead of the ester group in **10a**) improves the solubility in water and makes possible investigations in this "biological" solvent as well as in unpolar "lipophilic" phases.

5.1. Heterogeneous ligand exchange in pentacoordinated neutral complexes

The starting material for this series of experiments (incidentally useful for teaching the rules of chemical equilibrium) is the pentacoordinated intermediate spin (S = 3/2) iodide [I-Fe-**10b**] (Y = vacancy). These iodides are the best derivatives to easily isolate all the macrocycles **10** - **12** in the iron(III) state as air stable solids. The structure of the corresponding derivative of Fe-**10a** is given in Figure 12.

Depending on the method of ligand synthesis, also the trans (or *E*-) isomer with respect to the position of the methyl groups at the ligand ring can be isolated from complexes of the general type M-**1** [53]. In case of [Fe^{III}-**10b**(X,Y)], the *trans*-derivatives are more stable in air, even in aqueous solution. The CT absorption maximum is a few nanometers shifted to shorter wavelength.

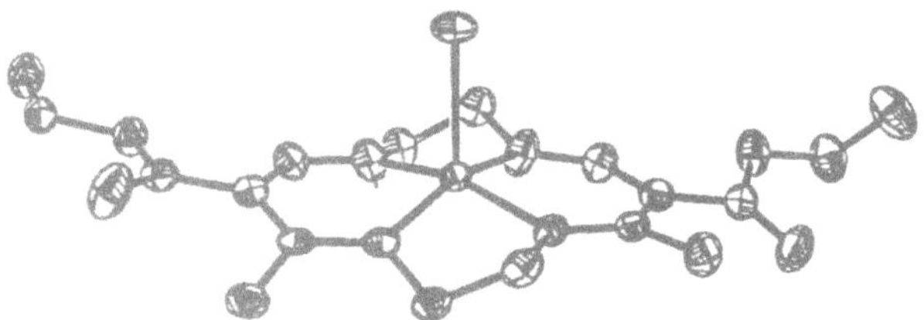

Fig. 12. Structure of the penta-coordinated S = 3/2 iron(III) complex [I-Fe-**10a**] [51a]

In water ($c_{fe}^{\circ} \approx 10^{-4}$ M; pH ≈ 3-4) the iodide dissociates nearly quantitatively to give the purple hydrated cation $[\text{Fe-}\mathbf{10b}]_{aq}^{+}$, whose spin state is not yet quite clear. In presence of a lipophilic solvent (CH_2Cl_2, $CHCl_3$, CCl_4, benzene), however, a part of the complex goes into the organic phase with the typical burgundy-red color of the undissociated iodide according to the coupled equilibrium (11) of dissociation and distribution.

With a small excess of free iodide (KI) in the aqueous phase, the equilibrium (11) is shifted to the left. The complex moves quantitatively into the organic phase,

$$[\text{Fe}\text{—X}]_{org} \rightleftharpoons [H_2O\text{—Fe—}OH_2]^{+}_{aqu} + X^{-}_{aqu} \qquad (11)$$

and the aqueous phase becomes colorless. Using a larger excess of another halide or pseudo-halide in water, the iodide bound to the complex is replaced in a heterogeneous exchange reaction, and the derivative with this other anion is formed in the organic solution. These ligand exchange reactions are accompanied by a characteristic color change. The heterogeneous metathesis is very fast and needs only a few seconds of manual shaking.

If the aqueous phase contains two salts with the potential ligand anions X^- and Y^- in a defined mole ratio c_X/c_Y, but with the constant overall concentration $c^\circ = c_X + c_Y$, the appropriate ratio of both pentacoordinated complexes is formed in the organic solvent according to the constant K of the heterogeneous equilibrium (12).

$$\text{Fe-X}_{org} + Y^-_{aqu} \rightleftharpoons \text{Fe-Y}_{org} + X^-_{aqu} \qquad (12)$$

$$K = \frac{[\text{fe-Y}][X^-]}{[\text{fe-X}][Y^-]}$$

Spectrophotometric measurements of the organic phase in dependence on the ratio c_X/c_Y in the aqueous phase lead to very precise values of K [19c, 57]. Data fitting is nearly ideal because of the constant ionic strength in the aqueous solution. Only in the case of very small total concentrations c°, more significant deviations were observed. The reason is the partial formation of the yellow dinuclear μ-oxo derivative in the organic solvent. With OH^- as one anionic ligand, the spectra show a strong dependence on the total complex concentration c_{fe}°. The data fit indicates clearly a more complicated equilibrium according to

$$\textbf{fe}\text{-X}_{org} + OH^-_{aq} \Leftrightarrow \tfrac{1}{2}\,\textbf{fe}_2\text{O}_{org} + \tfrac{1}{2}\,H_2O + X^-. \qquad (13)$$

The following data for the exchange of X = Cl by Y (lg K in water/chloroform; I = 0.5 M; from ref. [19c]) indicate a preference of the complex cation **fe**$^+$ for the higher halides. This suggests that the macrocyclic $[N_4^{2-}]$ coordinated iron(III) is rather a "soft" acid with respect to its axial coordination sites - in contrast to the free iron(III) ion. In case of Y = F and OAc the spin state at room temperature is not quite clear. These derivatives have not been isolated as solids.

Table 2. LogK values for eq. 12 in water/chloroform: X = Cl

Y =	F	AcO	Cl	Br	I	SCN	N_3	½O
lg K =	-1.16	-0.99	0	0.34	0.89	2.12	2.61	≈ 4[a)]
λ_{max} =	430	465	495	512	532	488	475	390 nm

[a)] calculated according to (13) for $c_{fe}^\circ = 1.5\times10^{-4}$ M

5.2. "Structural mimicry"

If the ions thiocyanate, azide, or hydroxide were used in an excess for heterogeneous ligand exchange experiments according to (12), quite another process takes place. The pentacoordinated uncharged complex [**fe**-X] in the organic phase takes up a second ion of the ligand to form an anionic octahedral derivative. Because of its ionic nature, this di-adduct is more hydrophilic and turns into the aqueous phase. The change of coordination number from 5 to 6 is accompanied by a change from intermediate (S = 3/2) or high spin (S = 5/2) to low spin (S = ½) and therefore by a conspicuous change of color from orange to green (X = SCN, N_3) or from yellow to raspberry-red (X = OH). In dependence on the equilibrium constant for this heterogeneous ligand addition and depending on the ligand concentration in water, the equilibrium is shifted to a greater or lesser extent to the right or the left.

$$[N_3\text{—Fe—}N_3]^{\ominus}_{aqu} \underset{H_2O}{\overset{CH_2Cl_2}{\rightleftharpoons}} [\text{Fe—}N_3]_{org} + N_3^{-}{}_{aqu} \qquad (14)$$

Low-spin (S = 1/2) Green — High-spin (S = 3/2; 5/2 ?) Orange

This experiment presents further evidence for the high mobility of the axial ligands at the hemine-like bound iron(III) and for the electronic and structural diversity of this central atom. In case of X = SCN the green anionic low-spin complex could be crystallized from the solution of Fe-**10a** as its benzyl-triethyl-ammonium salt. The X-ray analysis [54] confirms the expected octahedral structure of the anion with N-coordinated thiocyanate (Figure 13).

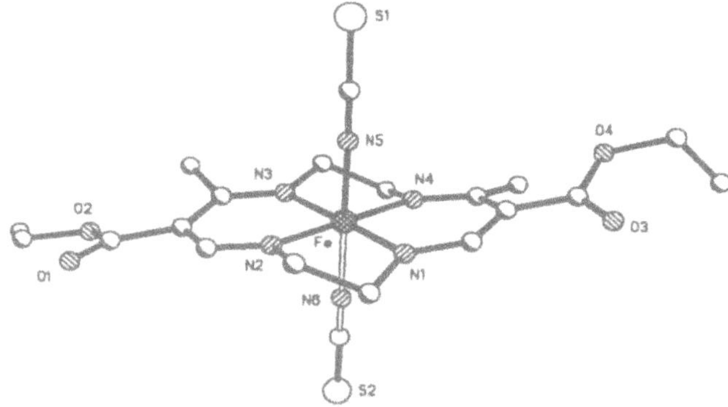

Fig. 13. Structure of the anionic low-spin complex $[Fe\text{-}\mathbf{10a}(NCS)_2]^-$

5.3. Spectrochemical CT-band series of axial ligands

The formation of octahedral low-spin di-adducts with a large variety of axial ligands is - as in the case of iron porphyrins - a typical feature of the macrocyclic complexes **10** - **12** [19c, 20b, 50, 55]. All these derivatives are characterized by an intense CT-band (lg $\varepsilon \approx 4$) that depends strongly on the type of axial ligand. In case of the complexes [Fe^{III}-**10**(X,Y)], the wavelength of the absorption maximum varies from 400 up to 1000 nm, and the colors of the solutions go from yellow over orange, red, violet, blue, turquoise, green and pale green to palish gray if the band lies above 800 nm.

DFT calculations [51b] confirm the assumption [20b] that this band is due to a charge transfer from the π-HOMO of the equatorial ligand to the t_{2g}^{5} orbital set of the low spin iron(III) which is split because of the low symmetry of the equatorial ligand. The acceptor orbital is presumably the *spin-down* d_{yz} state in a spin-polarized approximation. The strong effect of axial ligands is apparently caused by π-interactions: axial π-acceptor ligands stabilize the excited state and lead to a bathochromic shift, π-donors destabilize this state and give a hypsochromic shift. A more detailed discussion - also with respect to the ESR spectra - is given in ref. [20b].

When a sufficient excess of the ligand X is used (*e.g.* $c_X \approx 1$ M), the more hydrophilic complex Fe-**10b** forms octahedral adducts in aqueous or methanolic solution with nearly all the "classical" monodentate ligands and in aprotic solvents also with phosphanes, isonitriles *etc.* The 1:2 stoichiometry could be proven in all cases by spectrophotometric titrations (*cf.* **5.4.**) and in many cases by X-ray structure analysis [20c, 54]:

X, Y:	$MeO^{-a)}$	F^-	OH^-	$Py/OH^{-b)}$	NH_3	N_3^-	$S_2O_3^{2-}$	MeIm	NCS^-	
λ_{max}:	*424*	*469*	*513*	*620*	*655*	*656*	*673*	*699*	*730*	nm
X, Y:	Py	$NaNic^{c)}$	SO_3^{2-}	CN^-	$MeNic^{d)}$	NO_2^-	$MeCN^{e)}$	$PMe_3^{f)}$	$TMIC^{f,g)}$	
λ_{max}:	*750*	*758*	*763*	*775*	*780*	*785*	*813*	*831*	*968*	nm

a) In MeOH; presumably high-spin at room temperature [20b]; b) mixed adduct; c) sodium nicotinate; d) methyl nicotinate; e) in MeCN; f) in THF; g) tosyl-methyl-isocyanide.

A significant gradation is also found with the mono-anions of amino acids (pH $\approx$ 9), with proteins and with amino acids or proteins in dependence on pH. The different coordination in case of His^- and Cys^- is indicated by a stronger bathochromic shift of the maximum:

X, Y:	Ala^-	Met^-	Ser^-	Tyr^-	Phe^-	Trp^-	His^-	Cys^-	
λ_{max}:	*670*	*672*	*675*	*678*	*680*	*680*	*695*	*707*	nm

X, Y = His; pH:	6.5	9	12	14 (X = His^{2-}, Y = OH^-)
λ_{max}:	*707*	*695*	*660*	*585* nm

For proteins (at pH 6.5/9: human albumin 710/673 nm; human γ-globulin 707/676 nm), the values point to a histidine coordination at pH 6.5, whereas at higher pH possibly another coordination mode occurs.

Octahedral derivatives are also formed with several "exotic" ligands such as complex cyanides. The bathochromic shift of the maximum is also in this case in agreement with an increasing π-acceptor strength of the ligand. (A sensitive photo reaction was observed with Fe^{III}-**10b** in presence of a large excess of $K_4[Fe(CN)_6]$: In two clearly separated steps with the mixed adduct as intermediate, the extremely stable di-cyano-derivative [Fe-**10b**$(CN)_2]^-$ is formed [56]).

X, Y:	$[Fe(CN)_6]^{4-}$	$[Mo(CN)_8]^{4-}$	$[Ni(CN)_4]^{2-}$	$[Mn(CN)_5NO]^{3-}$	
λ_{max}:	*675*	*702*	*707*	*715*	nm
X, Y:	$[Fe(CN)_6]^{3-}$			$[Fe(CN)_5NO]^{2-}$	
λ_{max}:	*727*			*752*	nm

The complex Fe^{III}-**10a** could be used to compare the π-acceptor ability of several acetylides, R-C≡C⁻, by formation of octahedral derivatives under strongly anaerob and aprotic conditions in THF [19c]:

R:	^tBu	Ph	4F-Ph	4Cl-Ph	Me_3Si	$^iPr_3Si$	Ph_3Si	
λ_{max}:	*664*	*692*	*692*	*696*	*698*	*698*	*716*	nm

5.4. Substitution of axial ligands in octahedral derivatives - quantification of "push-pull"-effects

The strong and sharp CT-band in the VIS-/NIR region is a very suitable tool to examine quantitatively the consecutive substitution of axial ligands in the octahedral di-adducts by spectrophotometric titration. In general, the two steps according to eqn. (15) and (16) are more or less well separated in the spectra (Figure 14) and can be determined with high precision by use of non-linear fit procedures. The results of extended investigations are summarized in [20c].

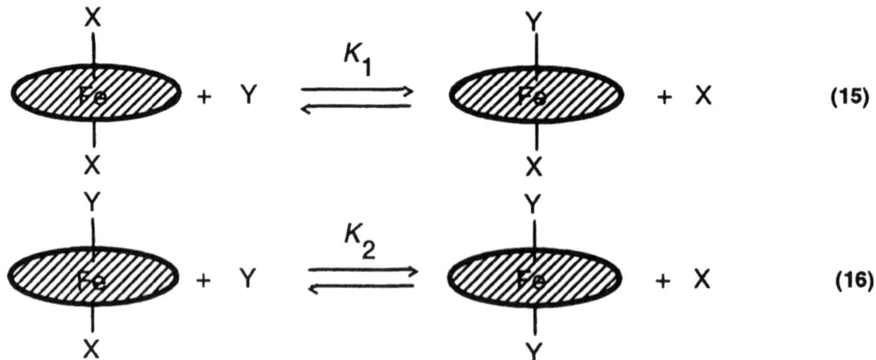

The overall constant $\beta_2 = K_1K_2$ describes the relative affinity of the axial coordination sites for different axial ligands. This overall stability increases for selected adducts of Fe^{III}-**10b** in water in the following order (lg β_2 with X = OH^- [57]). The exceptional high affinity to cyanide is particularly noteworthy; it correlates with the high toxicity of this ion.

NO_2^- (*-3.6*) < Py (*-2.6*) < NH_3 (*-0.2*) < OH^- (*0*) < MeIm (*3.3*) << CN^- (> *10*).

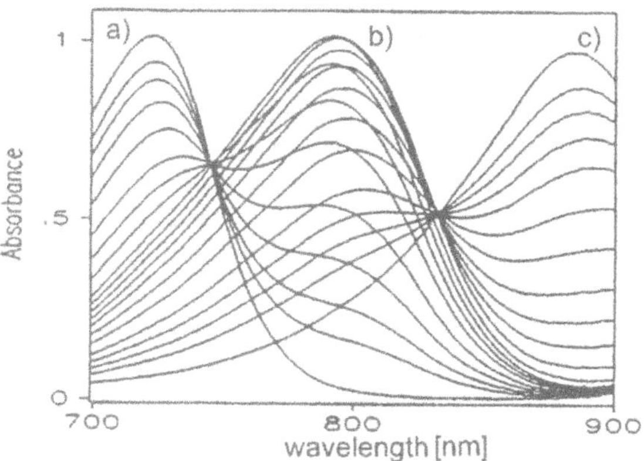

Fig. 14. Spectrophotometric titration of $[Fe\text{-}\mathbf{10a}(Py)_2]^+$ (a) with $[Fe\mathbf{10a}(P(OEt)_3)_2]^+$ (c). The well resolved spectra of the mixed adduct (b) indicate a strong extra-stabilization

The quotient K_1/K_2 is a measure of the relative stability of the mixed adduct according to the comproportionation (17). The statistic value is expected to be 4 in the case that no other enthalpy contributions occur than in the pure homo-ligand di-adducts. Much higher values indicate a stabilizing interaction between the ligands

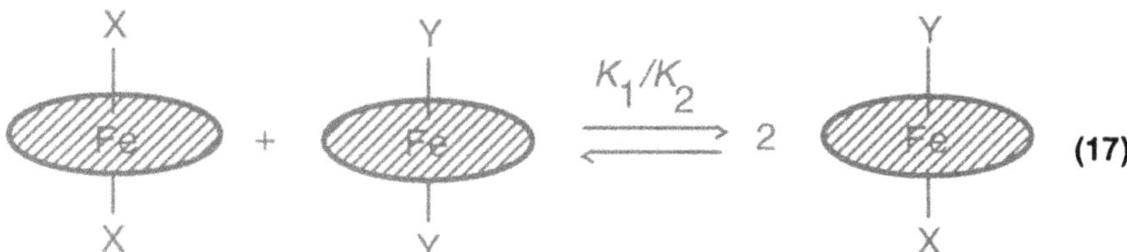

(17)

X and Y, *e.g.* π-donor-/π-acceptor interactions that are mediated through d-orbitals of the central atom. Such "*push-pull*" effects play an important role with respect to the action of special axial ligands (*e.g.* as part of a protein) in trans position to the active site of a heme enzyme. Typical combinations X/Y with significant extra stabilization of the mixed adduct are (for the complex Fe^{III}-**10a** [20c]; Fe^{III}-**10b** shows a similar behavior [57]; *cf.* ref. [20c] for a detailed discussion):

X/Y:	Py/His⁻	Py/NO_2^-	Py/OH^-	$P(OEt)_3$/MeIm	$P(OEt)_3$/Py	$P(OEt)_3$/CN-Py
K_1/K_2:	*30*	*50*	*180*	*190*	*3800*	*16000*

Some of these values could be explained by π-donor-/π-acceptor interactions between X and Y and/or by effects of charge compensation, *e.g.* for Py/OH

$$[Fe\text{-}\mathbf{10}(Py)_2]^+ + [Fe\text{-}\mathbf{10}(OH)_2]^- \Leftrightarrow 2\,[Fe\text{-}\mathbf{10}(Py)(OH)]^0. \quad \mathbf{(18)}$$

The exceptionally high stability of the combination of Py with the strong π-acceptor ligand $P(OEt)_3$ and the increasing stabilization with 4-CN-substitution of Py are difficult to understand in this way. Possibly, this surprising stabilization is caused by interactions between parts of the equatorial chelate ligand and the axial $P(OEt)_3$. Such interactions are suggested by the molecular structure of the cation $[Fe\text{-}\mathbf{10a}(CN\text{-}Py)(P(OEt)_3]^+$ that shows a striking distortion of the six-membered chelate rings of the macrocycle in direction to the axial $P(OEt)_3$ (Figure 15).

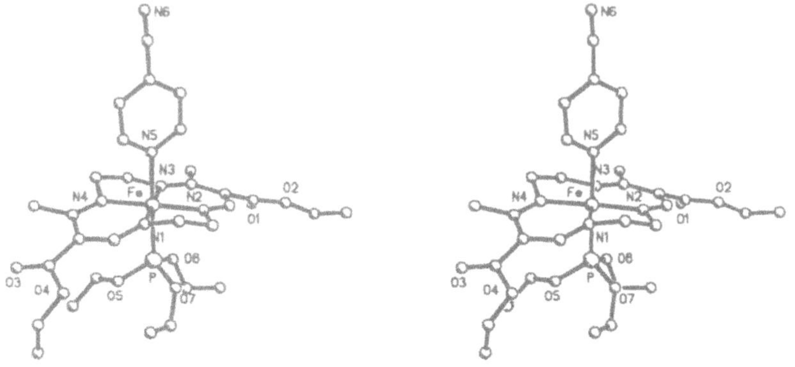

Fig. 15. Stereoview of the structure of the complex cation $[Fe\text{-}\mathbf{10a}(CN\text{-}Py)(P(OEt)_3]^+$

Another interesting and biologically more relevant example of a stabilized mixed adduct is the neutral iron(III) complex $[Fe\text{-}\mathbf{11}(NO_2)(H_2O)]$. It is derived from the unsymmetrically bridged macrocycle **2**. The axial sites are occupied by one nitrite ion and one molecule of water (Figure 15). Like all the other octahedral coordinated iron(III) macrocycles, it is also a low-spin complex - even with the rather weak ligand H_2O. This coordination mode is of interest with respect to the very recently published structure of cytochrome *c* nitrite reductase [58]. In this case, the nitrite is bound to a heme with the aliphatic amino group of lysine as an unusual trans ligand. Because the di-aquo complex $[Fe\text{-}\mathbf{11}(H_2O)_2]^+$ could not be realized, it was not possible to quantify the extra stabilization of the mixed adduct in this case.

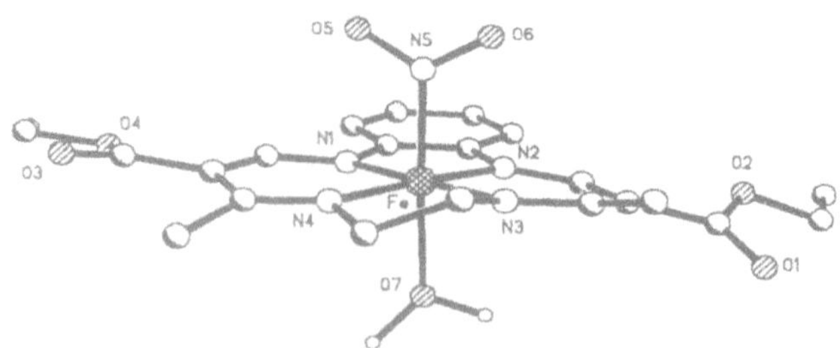

Fig. 16. Structure of the neutral mixed adduct $[Fe\text{-}\mathbf{11}(NO_2)(H_2O)]$

6. Structure and magnetic properties of the iron complexes

The reversible change of spin state within the functional cycle is a typical feature of heme proteins, especially those which are involved in oxygen metabolism. Examples are hemoglobin, myoglobin and cytochrome P 450. Our $[N_4^{2-}]$ coordinated macrocycles are stronger π- and σ-donors than porphyrins and favor more effecively the low spin state. The $[N_2O_2^{2-}]$ coordinated open chain ligands have a weaker ligand field strength and a higher tendency to form high-spin complexes. In this respect they are more related to iron porphyrins. For comparison, the complexes **13** and **14**, having the same peripheral substituents and the same bridges R as the macrocycles **10** - **12**, and the more electron-withdrawing substituted complex **15** were both included in these studies.

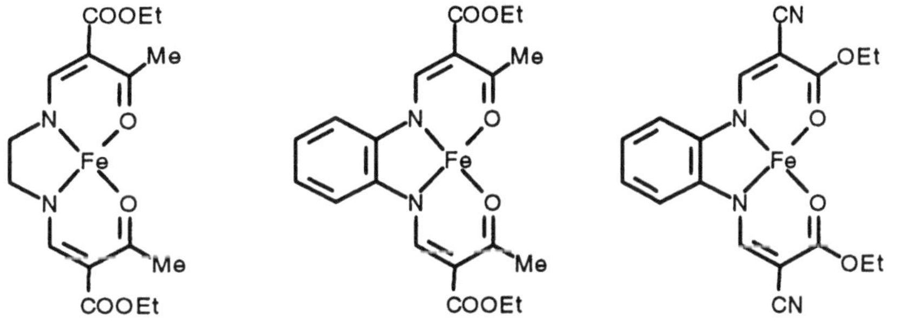

13 **14** **15**

Scheme 10. Iron complexes of open chain ligands used for magnetic investigations

6.1. Molecular structure and spin state

All the possible spin states of mononuclear iron(II/III) compounds can be realized with the complexes **10** to **15**. The structure of many prototypes with S = 0, ½, 1, 3/2, and 2 could be characterized by X-ray analysis. No attempts have been made up to now to crystallize the complexes with high-spin iron(III) (S = 5/2). Many such structures are known from porphyrins and salicylaldimines.

The low-spin states S = 0 and S = ½ are typical for all the octahedral iron(II/III) complexes with macrocyclic ligands. Surprisingly, the average $Fe\text{-}N_{eq}$-distances are independent of the oxidation step: They amount 191 pm for iron(II) (avg. distance of [Fe-**10a**(HIm_2)] [59], [Fe-**11**$(HIm)_2$], [Fe-**11**$(Py)_2$] [20c]) and 190 pm for iron(III) (avg. distance of 14 structures [20c], including those with different axial ligands). The deviations within one individual complex are in many cases larger than those between different complexes (*e.g.* in [Fe-**10a**$(CN\text{-}Py)_2$]PF_6 [20c]: 192.0(3)/191.0(3) for $Fe\text{-}N_{eq}$ on the methylated side of the macrocycle and 188.3(3)/188.2(3) pm for $Fe\text{-}N_{eq}$ on the non-methylated side). The average $Fe\text{-}N_{ax}$ distances for axial N-ligands are also the same for both oxidation steps (201 pm).

The spin state S = 1 of iron(II) could be proven by temperature-dependent measurements for [Fe-**10a**] (with a strongly distorted 4 + 1 coordination by intermolecular coordination of one neighboring CQOEt; Fe-O = 278 pm [60]), for the mono-adduct [Fe-**10aPy**] [19a], for [Fe-**11**] (planar molecules; structure only for mix-crystals with the free ligand [54]), and for [Fe-**12**] (no X-ray-analysis). The average Fe-N_{eq} distance of 190 pm is not different from those of the low-spin complexes.

The intermediate S = 3/2 state of iron(III) is relatively rare. It has been proven by (T-dependent) magnetic, ESR and Mössbauer measurements as well as by MO (DFT) calculations for the pentacoordinated complex [I-Fe-**10a**] [51a]. The ∅ Fe-N_{eq} distance is 191 pm again. A characteristic feature of this structure is the strong deviation of the central atom from the N_4-plane (Fe-Ct_{N4} = 34 pm; *cf.* Fig. 12). Presumably, all the halides [X-Fe-**10**, **11**, **12**] (perhaps with exception of X = F^-) represent this type. In contrast, the penta-coordinated halides of salicylaldimines and porphyrins are in most cases high-spin complexes.

The S = 2 state could not doubtless be obtained with our macrocyclic iron(II) complexes although such porphyrin derivatives bearing axial O-ligands exist. A very unstable di-hydrate [Fe-**10a**$(H_2O)_2$] that seems to be high-spin [19a] could not be isolated as pure solid. This high-spin state is, however, characteristic (at least at room temperature) for the iron(II) complexes [Fe-**13**, **14**, **15**] [24a] with MeOH, dioxan, Py as axial ligands or as solvent-free solids. The distances within the first coordination sphere are significantly longer than in the macrocyclic low- or intermediate-spin complexes (average of [Fe-**13**$(Py)_2$], [Fe-**15**$(Py)_2$] [24b], *cf.* Fig. 3, [Fe-**14**$(Py)_2$] at room temperature, [Fe-**13**$(MeOH)_2$], [Fe-**14**$(MeOH)_2$] [61]: Fe-O_{eq} = 203 pm, Fe-N_{eq} = 210 pm, Fe-N_{Py} = 225 pm, Fe-O_{MeOH} = 224 pm). Nearly the same values have also been found for [Fe-**4**$(CF_3$/H/Me-Etn)$(Py)_2$], the only structurally characterized pure iron(II) complex of a "classical" Schiff base [62]. A typical feature of high-spin iron(II) in an open-chain $[N_2O_2^{2-}]$-environment is the large "bite", *i.e.* the angle O_{eq}-Fe-O_{eq}. The average of 5 compounds is 110 °, whereas the average value is about 90° for low spin iron(II) complexes with the same ligands. This is obviously a consequence of the larger ion radius of the central atom in the high-spin state.

The high-spin iron(III) state (S = 5/2) is observed in the most mononuclear iron(III) complexes with $[N_2O_2^{2-}]$-ligands and with not too strong axial ligands [24a]. With strong *N*-donors such as imidazole or cyanide, also low-spin (S = ½) complexes are formed. Several derivatives of both types with classical Schiff bases have been structurally characterized [63a,b].

The spin state of the iron(II) complexes with open-chain ligands is much more susceptible to changes in the equatorial and/or the axial ligands than that of the macrocycles. For instance, the di-adduct [Fe-**15**$(Py)_2$] is high spin from room temperature down to 4 K. In contrast, the similar compound [Fe-**14**$(Py)_2$] shows a spin crossover from high spin at room temperature to low-spin below about 180 K. We were able to characterize the molecular structure of both isomers by temperature dependent X-ray measurements using the same crystal [61] (Figure 17). The distances and the "bite" of the high-spin (high-temperature) phase agree with the above-mentioned values (Fe-O_{eq} = 202/200 pm, Fe-N_{eq} = 206/205 pm, Fe-N_{Py} = 226/220 pm, ∠ O_{eq}-Fe-O_{eq} = 106°). The low-spin (low-temperature) phase fits very well the values characteristic for octahedral low spin iron(II) (Fe-O_{eq} = 196/195 pm, Fe-N_{eq} = 192/192 pm, Fe-N_{Py} = 203/202 pm, ∠ O_{eq}-Fe-O_{eq} = 92°).

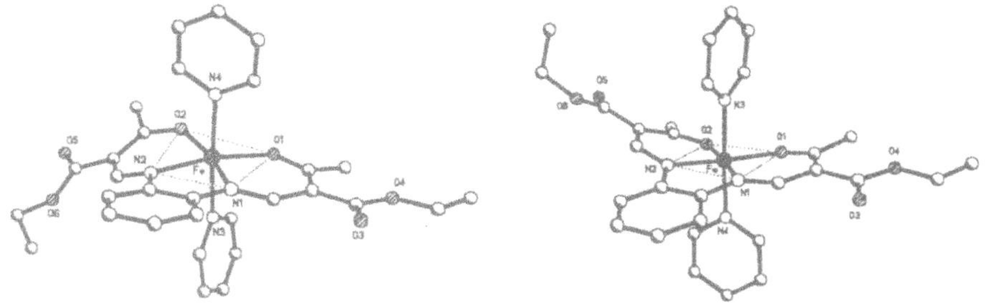

Fig. 17. High-spin (left) and low-spin (right) isomers of [Fe-**14**$(Py)_2$]

Perhaps the magnetically most interesting pair of complexes was gained by reaction of [Fe-**14**] with imidazole. Two products with different composition were obtained [61] (Figure 18): The first one, a mono-adduct [Fe-**14**(HIm)], is a penta-coordinated *high-spin* complex. With respect to its spin state, the distances of the "equatorial" donors (200/200/207/210 pm), the *single* strongly bound axial imidazole (Fe-N_{Him} = 212 pm) and the displacement of the central atom from the plane of the equatorial donors (Fe-Ct_{N2O2} = 38 pm), it resembles well the coordination sphere of iron(II) in hemoglobin. The "bite" is in the range of other high spin complexes with open-chain SCHIFF base ligands (101.5°).

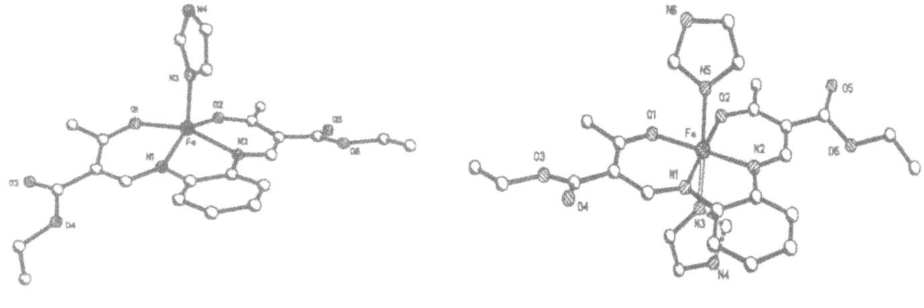

Fig. 18. High-spin mono-adduct (left) and low-spin di-adduct (right) of [Fe-**14**] with HIm

The other product, the di-adduct [Fe-**14**$(HIm)_2$], is diamagnetic from room temperature down to 4 K (apart from a small remaining paramagnetism of about 1 μ_b). The structural parameters agree with a normal octahedral low-spin iron(II) complex (Fe-O_{eq} = 196/194 pm, Fe-N_{eq} = 190/190 pm, Fe-N_{HIm} = 202/201 pm, ∠ O_{eq}-Fe-O_{eq} = 88.2°). At increasing temperature, however, the solid substance shows a spin crossover to high-spin with an unusual steep jump in magnetism and a narrow but significant thermal hysteresis (Figure 19). This surprising behavior is a consequence of strong intermolecular H-bridges. These bridges determine the three-dimensional cross-linkage of the molecular building blocks in the network of the lattice and lead to cooperative magnetic properties.

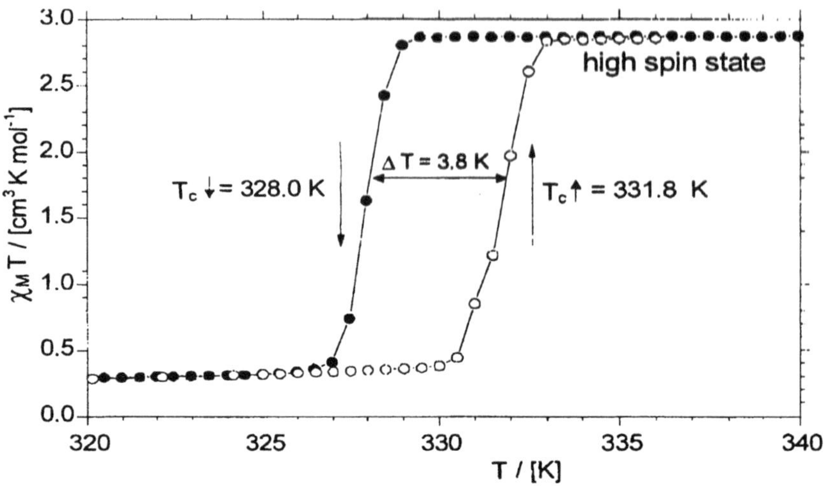

Fig. 19. High-temperature spin crossover and thermal hysteresis of [Fe-**14**$(HIm)_2$]

6.2. Intermolecular interactions and cooperative properties

The high tendency to give strong intermolecular connections - caused by the presence of free carbonyl substituents R^2 - is a typical feature of our complexes. They can be formed either by a direct intermolecular coordination of the carbonyl (or nitrile) groups or by H-bridges from axial ligands (HIm, MeOH) or incorporated solvent, *e.g.* water. An example for the first case is the cobalt(II) complex of an unsymmetrically substituted ligand shown in Figure 9. Similar interactions have been found in many other solid complexes. The strong affinity of free carbonyl groups to form H-bridges is clearly demonstrated by the fact that the solvent free planar complexes [Ni-**2**(Me/COOEt)] and [Cu-**2**(Me/COOEt)] take water from alumina during their chromatographic purification in benzene, to give chains linked by H-bridges between water and the ester groups [20a].

Especially diverse possibilities of two- or three-dimensional networks have been found in presence of axial ligands able to act as donors for H-bridges, such as imidazole or methanol. Typical examples of imidazole-linked chain-like and three-dimensional structures were described for [Fe^{III}-**10a**$(HIm)_2]^+$ and [Fe^{II}-**10a**$(HIm)_2$], respectively [59]. A complicated network of H-bridges [64] is obviously also responsible for the cooperative formation of the high-spin state during the spin crossover of [Fe-**14**$(HIm)_2$] . An example for molecular units with axial bound methanol as H-donor is given in Figure 20 for the high-spin iron(II) complex [Fe-

14$(MeOH)_2$]. In this case, the three-dimensional cross-linked structure results in a weakly ferromagnetic behavior at temperatures below 10 K [65].

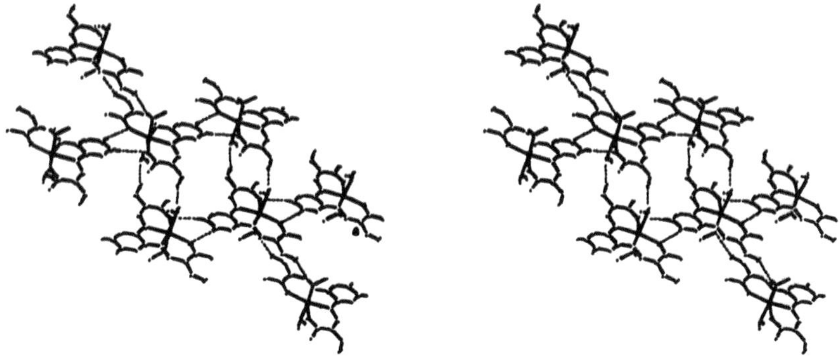

Fig. 20. Stereoview of the three-dimensional network of H-bridges in [Fe-**14**$(MeOH)_2$]

These examples show that the special structural features of our complexes can markedly influence not only their reactivity (as the catalytic performance of Fe-**3**(Me/COOEt), *cf.* **3.1.**), but also their solid state behavior. This could be useful for the development of nano-scale devices and other applications.

In conclusion, the high structural diversity and the versatile reactivity of these complexes containing porphyrin-like macrocyclic and "salen"-like open chain SCHIFF base ligands bear an interesting potential to assist in future developments in chemistry.

Acknowledgements

I am sincerely grateful to my former and present coworkers and colleagues who have contributed to the results presented here: *S. Dautz, M. Friedrich, H. Görls, F. Gräfe, I. Käpplinger, H. Keutel, B. Kirchhof, J. Knaudt, G. Leibeling, B. Müller, K. Müller, E. Poppitz, P. Renner, M. Rudolph, A. Schneider, K. Schuhmann, D. Seidel, B. Weber.* Many thanks are extended to Professors *Elias, Grodzicki, Klein, Krebs, Trautwein* for their fruitful collaboration. Financial support by the *Deutsche Forschungsgemeinschaft*, the *Fonds der Chemischen Industrie*, and the *Freistaat Thüringen* is also acknowledged.

References

[1] a) Eichhorn, G. L. (Ed.): *Inorganic Biochemistry*, Volumes 1 and 2, Elsevier, Amsterdam **1973**; b) Cowan, J. A.: *Inorganic Biochemistry - An Introduction*, VCH Publishers, New York **1993**; c) Lippard, S. J.; Berg, J. M.: *Principles of Bioinorganic Chemistry*, University Science Books, Mill Valley, Ca., **1994**; d) Kaim, W.; Schwederski, B.: *Bioinorganic Chemistry: Inorganic Elements in the Chemistry of Life*,

Chemistry, University Science Books, Mill Valley, Ca., **1994**; d) Kaim, W.; Schwederski, B.: *Bioinorganic Chemistry: Inorganic Elements in the Chemistry of Life*, John Wiley & Sons, Chichester, **1994**; e) Reedijk, J. (Ed.): *Bioinorganic Catalysis*, Marcel Dekker, Inc., New York, **1998**.

[2] a) Haag, E.; Irrgang, E. J.; Boekma, E. J.; Renger, G.: *Eur. J. Biochem.* **1990**, *189*, 47; b) Debus, R.J.: *Biophys. et Biochim. Acta* **1992**, *1102*, 269; c) Pecoraro, V. L. (Ed.): *Manganese Redox Enzymes*, VCH Publishers, New York, **1992**; d) Renger, G.: *Photosynth. Res.* **1993**, *38*, 229.

[3] a) Karlin, K. D.; Tyeklar, Z.: *Bioinorganic Chemistry of Copper*, Chapman & Hall, New York, **1993**. b) Solomon, E. I.; Sundaram, U. M.; Machonkin, T. E.: *Chem. Rev.* **1996**, *96*, 2563; c) Messerschmidt, A.: *Metal Sites in Proteins and Models,* **1998**, *90*, 37-68; d) Abolmaali, B.; Taylor, H. V.; Weser, U.: *Structure and Bonding* **1998**, *91*, 91.

[4] a) Sigel, H. (Ed.): *Metal Ions in Biological Systems*, Vol. 7, Marcel Dekker, Basel, **1978**; b) Dunford, H. B. (Ed.): *The Biological Chemistry of Iron*, D. Reidel Publishing Company, **1982**.

[5] a) Fischer, H.; Orth, H.: *Die Chemie des Pyrrols*, Akad. Verlagsgesellschaft, Leipzig, **1934**; b) Dolphin, D. (Ed.): *The Porphyrins*, Vol. I, Academic Press, New York, **1978**.

[6] a) Dolphin, D.: (Ed.) B_{12}, Wiley, New York **1982**; b) Schneider, Z.; Stroinski, A.: *Comprehensive* B_{12}, de Gruyter, Berlin, **1987**; c) Eschenmoser, A.: *Angew. Chem.* **1988**, *100*, 6; d) B. T. Golding: *Chem. Brit.* **1990**, *26*, 950.

[7] Bonacker, L. G.; Baudner, S.; Mörschel, E.; Linder, D.; Thauer, R.K.: *Eur. J. Biochem.* **1993**, *217*, 587.

[8] a) Sono, M.; Roach, M. P.; Coulter, E. D.; Dawson, J. H.: *Chem. Rev.* **1996**, *96*, 2841; b) Crane, B. R.; Arvai, A.S.; Ghosh, D. K.; Wu, C.; Getoff, E. D.; Stuehr, D. J.; Tainer, J. A.: *Science* **1998**, *279*, 2121.

[9] Langenbeck, W.: *Die organischen Katalysatoren und ihre Beziehungen zu den Fermenten*, 2nd edn., Springer-Verlag, Berlin **1949**.

[10] a) Meunier, B.: *Chem. Rev.* **1992**, *92*, 1411; b) Sheldon R. (Ed.): *Metalloporphyrins in Catalytic Oxidations*, Marcel Dekker, New York **1994**; c) Montanari, F.; Casella, L. (Eds.): *Metalloporphyrins Catalyzed Oxidations*, Kluwer Acad. Publ., Dordrecht, **1994**; d) Woggon, W.-D.: *Topics in Current Chem.* **1996**, *184*, 40; d) Song, R.; Bernadou, J.; Meunier, B.: *J. Org. Chem.* **1997**, *62*, 673.

[11] Tsumaki, T.: *Bull. Chem. Soc. Jpn.* **1938**, *13*, 252.

[12] a) Jacobsen, E. N.: *Comprehensive Organometallic Chemistry II* (Eds.: Wilkinson, G.; Stone, F. G. A.; Abel, E. W.; Hegedus, L. S.), Vol. 12, Chapter 11.1, Pergamon, New York, **1995**; b) Katsuki, T.: *Coord. Chem. Rev.* **1995**, *140*, 189; c) Younkin, T. R.; Connor, E. F.; Henderson, J. I.; Friedrich, S. K.; Grubbs, R. H.; Bansleben, D. A.: *Science* **2000**, *287*, 460.

[13] Schrauzer, G. N.: *Angew. Chem.* **1976**, *88*, 465; **1977**, *89*, 239.

[14] Uhlig, E.; Friedrich, M.: *Z. Anorg. Allg. Chem.* **1966**, *343*, 299.

[15] Costa, G.: *Coord. Chem. Rev.* **1972**, *8*, 63.

[16] Jäger, E.-G.: *Z. Chem.* **1964**, *4*, 437; **1968**, *8*, 30, 392, 470; *Z. Anorg. Allg. Chem.* **1969**, *364*, 177

[17] a) Jäger, E.-G.; Seidel, D.: *Z. Chem.* **1983**, *23*, 261; b) Müller, K.; Jäger, E.-G.: *Z. Chem.* **1985**, *25*, 377; c) Görls, H.; Reck, G.; Jäger, E.-G.; Müller, K.; Seidel, D.: *Cryst. Res. Technol.* **1990**, *25*, 1277.

[18] Müller, K.; Jäger, E.-G.: *Z. Anorg. Allg. Chem.* **1989, 577**, 195; Schade, W.; Jäger, E.-G.; Müller, K.; Seidel, D.: *J. Prakt. Chem.* **1989**, *331*, 559.

[19] a) Jäger, E.-G.; Stein, E.; Gräfe, F.; Schade, W.: *Z. Anorg. Allg. Chem.* **1985**, *526*, 15; b) Jäger, E.-G.; Gräfe, F.: *ibid.* **1988**, *561*, 25; c) Jäger, E.-G.; Hähnel, H.; Klein, H.-F.; Schmidt, A.: *J. Prakt. Chem.* **1991**, *333*, 423.
[20] a) Jäger, E.-G.; Keutel, H.; Rudolph, M.; Krebs, B.; Wiesemann, F.: *Chem. Ber.* **1995**, *128*, 503; b) Jäger, E.-G.; Keutel, H.: *Inorg. Chem.* **1997**, *36*, 3512; c) Käpplinger, I.; Keutel, H.; Jäger, E.-G.: *Inorg. Chim. Acta* **1999**, *291*, 190.
[21] Jäger, E-G.: *Z. Anorg. Allg. Chem.* **1965**, *337*, 80; **1967**, *349*, 139; **1968**, *359*, 147; Wolf, L.; Jäger, E.-G.: *Z. Chem.* **1965**, *5*, 392; *Z. Anorg. Allg. Chem.* **1966**, *346*, 76.
[22] Jäger, E.-G.; Renner, P.; Schmidt, R.: a) *Z. Chem.* **1977**, *17*, 189; b) *ibid.* **1977**, *17*, 307; c) Renner, P.: *PhD-Thesis*, University of Jena, **1977**; d) Jäger, E.-G.; Müller, R.; Renner, P.: *Z. Chem.* **1982**, *22*, 65.
[23] a) Jäger, E.-G.; E.-G.; Kirchhof, B.; Schmidt, E.; Remde, B.; Kipke A.; Müller, R.: *Z. Anorg. Allg. Chem.* **1982**, *485*, 141; b) Jäger, E.-G.; Schlenvoigt, G.; Kirchhof, B.; Rudolph, M.; Müller, R.: *ibid.* **1982**, *485*, 173; c) Jäger, E.-G.; Rudolph, M.; Müller, R.: *Z. Chem.* **1978**, *18*, 229.
[24] a) Jäger, E.-G.; Häussler, E.; Rudolph, M.; Schneider, A.: *Z. Anorg. Allg. Chem.* **1985**, *525*, 67; b) Görls, H.; Jäger, E.-G.: *Cryst. Res. Technol.* **1991**, *26*, 349.
[25] Jäger, E.-G.; Schuhmann, K.; Görls, H.: a) *Inorg. Chim. Acta* **1997**, *255*, 295; b) *Chem. Ber./Recueil* **1997**, *130*, 1643; c) Schuhmann, K.; Jäger, E.-G.: *Eur. J. Inorg. Chem.* **1998**, 2051; d) Schuhmann, K.: *PhD thesis*, University of Jena, **1998**.
[26] Cotton, F.A.; Czuchajowska, J.: *Polyhedron* **1990**, *9*, 2553.
[27] Mountford, P.: *Chem. Soc. Rev.* **1998**, *27*, 105.
[28] Weiss, M. C.; Gordon, G.; Goedken, V.L.: *Inorg. Chem.* **1977**, *16*, 305.
[29] Hanic, F.; Handlovic, M.; Lindgren, O.: *Collect. Czech. Chem. Commun.* **1972**, *37*, 2119.
[30] a) Wang, Y.; Peng, S.-M.; Lee, Y.-L.; Chuang, M.-C.; Tang, C.-P.; Wang, C.-J.: *J. Chin. Chem.* **1982**, *29*, 217; b) Weiss, M. C.; Goedken, V. L.: *Inorg. Chem.* **1979**, *18*, 819.
[31] Schneider, A. *PhD thesis*, University of Jena, **1991**.
[32] a) Knaudt, J. *PhD thesis*, University of Jena, **1998**; b) Knaudt, J.; Imhof, W.; Sternberg, U.; Jäger, E.-G.: To be published.
[33] Jäger, E.-G.; Knaudt, J.; Rudolph, M.; Rost, M. *Chem. Ber.* **1996**, *129*, 1041.
[34] Jäger, E.-G.; Knaudt, J.; Schuhmann, K.; Guba, A.: In *Peroxide Chemistry - Final Report of the DFG Priority Program*, Adam, W. (Ed.), Chapter C.4., Wiley-VCH, Weinheim **2000**, in press.
[35] Jäger, E.-G.; Rudolph, M.: Schneider, A.; Gräfe, F.: *Z. Chem.* **1985**, *25*, 445.
[36] Knaudt, J.; Förster, St.; Bartsch, U.; Rieker, A.; Jäger, E.-G.: *Z. Naturforsch.* **2000**, *55b*, in press.
[37] a) Behar, D.; Dhanasekaran, T.; Neta, P.; Hosten, C. M.; Ejeh, D.; Hambright, P.; Fujita, E.: *J. Phys. Chem.* **1998**, *102*, 2870; b) Sutin, N.; Creutz, C.; Fujita, E.: *Comments Inorg. Chem.* **1997**, *19*, 67; c) Bhugun, I.; Lexa, D.; Saveant, J.-M.: *J. Am. Chem. Soc.* **1996**, *118*, 1769; d) Nallas;. G. N. A.; Brewer; K.: *Inorg. Chim. Acta* **1996**,*253*, 7; e) Hawecker, J.; Lehn, J.-M.; Ziessel, R.: *Chem. Commun.* **1984**, 328.
[38] Ali, M. M.; Sato, H.; Mizukawa, T.; Tsuge, K.; Haga, M.; Tanaka, K.: *Chem. Commun.* **1998**, 249.
[39] Gennaro, A.; Isse, A. A. Saveant, J.-M.; Severin, M.-G.; Vianello,E.: *J. Am. Chem. Soc.* **1996**, *118*, 7190.
[40] Rudolph, M.; Dautz, S.; Jäger, E.-G.: submitted.

[41] Rudolph, M.; Feldberg, S. W.: *DigiSim® 2.0 Software*, Bioanalytical Systems Inc., West Lafayette, IN 47906, USA, **1995**.

[42] a) Rudolph, M.; Jäger, E.-G.: *Z. Chem.* **1989**, *29*, 418; b) Jäger, E.-G.; Rudolph M.: *J. Electroanal. Chem.* **1997**, *434*, 1.

[43] a) Rudolph, M.: *PhD-Thesis*, University of Jena, **1982**; b) Jäger, E.-G.; Rudolph, M.: *Z. Chem.* **1981**, *21*, 371.

[44] Streeky, J. A.; Pillsbury, D. G.; Busch, D. H.: *Inorg. Chem.* **1980**, *19*, 3148.

[45] Jäger, E.-G.; Renner, P.: *Z. Chem.* **1978**, *18*, 193.

[46] Seidel, D.: *PhD-Thesis*, University of Jena, **1985**.

[47] Wiesemann, F.; Krebs, B.; Görls, H.; Jäger, E.-G.: *Z. Anorg. Allg. Chem.* **1995**, *621*, 1883.

[48] a) Saalfrank, R. W.; Struck, O.; Toupet, L.; v. Schnering, H.-G.: *Chem. Ber.* **1993**, *126*, 837; b) Saalfrank, R. W.; Struck, O.; Danion, D.; Hassa, J.; Roupet, L.: *Chem. Mater.* **1994**, *6*, 1432.

[49] Krebs, B.: *University of Münster*, private communication.

[50] Jäger, E.-G.; Barth, St.; Keutel, H. In *Bioinorganic Chemistry: Transition Metals in Biology and their Coordination Chemistry*, Trautwein, A. X. (Ed.), WILEY-VCH, Weinheim, **1997**, 584.

[51] a) Keutel, H.; Käpplinger, I.; Jäger, E.-G.; Grodzicki, M.; Schünemann, V.; Trautwein, A. X.: *Inorg. Chem.* **1999**, *38*, 2320; b) Käpplinger, I.; Grodzicki, M.; Schünemann, V.; Görls, H.; Trautwein, A. X.; Jäger, E.-G.: *Inorg. Chem.* submitted.

[52] Gadsby, P. M. A.; Thomson, A.: *J. Am Chem. Soc.* **1990**, *112*, 5003.

[53] Jäger, E.-G.; Leibeling, G.; Friedrich, M.: unpublished results.

[54] Käpplinger, I.; Görls, H.; Jäger, E.-G.: *Manuscript in preparation.*

[55] Jäger, E.-G.; Schweder, B.; Radzuweit, *Z. Chem.* **1988**, *28*, 152.

[56] Jäger, E.-G.: *Z. Chem.* **1985**, *25*, 446.

[57] Jäger, E.-G.; Liehr, J.; Morich, E.; Dix, A.: *Proc. 12th Conf. Coord. Chem.* Smolenice/Bratislava, **1989**, 123.

[58] Einsle, O.; Messerschmidt, A.; Stach, P.; Bourenkov, G. P.; Bartunik, H. D.; Huber, R.; P. M. H. Kroneck: *Nature* **1999**, *400*, 474.

[59] Wiesemann, F.; Wonnemann, R.; Krebs, B.; Keutel, H.; Jäger, E.-G.: *Angew. Chem. Int. Ed. Engl.* **1994**, *33*, 1363.

[60] Weber, B.; Jäger, E.-G.: *to be published.*

[61] Leibeling, G.; Görls, H.; Jäger, E.-G.: *manuscript in preparation.*

[62] Liu, H. Y.; Scharbet, B.; Holm, R.H.: *J. Am. Chem. Soc.* **1991**, *113*, 9529.

[63] a) Kennedy, B. J.; McGrath, A. C.; Murray, K. S. Skelton, B. W.; Whrite, A. H.: *Inorg. Chem.* **1987**, *26*, 483; b) Nishida, Y.; Kino, K.; Kida, S.: *J. Chem. Soc. Dalton Trans.* **1987**, 1157.

[64] Leibeling, G.; Görls, H.; Müller, B.; Jäger, E.-G.: Submitted.

[65] Müller, B.; Leibeling, G.; Jäger, E.-G.: *Mol. Cryst. and Liq. Cryst.* **1999**, *334*, 389.

Temperature and Solvent effects on Facial Diastereoselectivity

Gianfranco Cainelli, Daria Giacomini, Paola Galletti, Paolo Orioli

Dipartimento di Chimica "G. Ciamician", Università degli Studi di Bologna, Via Selmi 2, I-40126, Bologna, Italy.
cainelli@ciam.unibo.it; giacomin@ciam.unibo.it

Abstract. Commonly observed, but often neglected, is the possibility of modifying the facial diastereoselectivity of nucleophilic addition to carbonyls by means of temperature and solvent. Temperature dependent measurements according to the modified Eyring equation allow the evaluation of stereoselectivity in terms of differential enthalpy and entropy contributions, and demonstrate the great importance of the entropic contribution in determining the facial diastereoselectivity. Even the reaction solvent proved to be important in directing the diastereoisomeric ratio. The inversion temperature (T_{inv}) that appears in non-linear Eyring plots mainly depends on the substrate solvation and corresponds to a transition between two differently ordered solvation states.

Stereoselective processes play a predominant role in organic chemistry and represent the main goal for researchers involved either in total synthesis or in new methodologies. The synthetic process for a new drug or for a biologically active compound cannot avoid a correct and accurate control of the final target stereochemistry and many methods are available for stereocontrolled synthesis.[1]

The nucleophilic addition to carbonyls is one of the most important strategies used in building up the molecular skeleton of organic compounds. The stereochemical aspect of this reaction is discussed in terms of face selectivity (Fig.1), the process will be stereotopically divergent whenever between the two competing attacks to either *re* or *si* face, one is preferred [2].

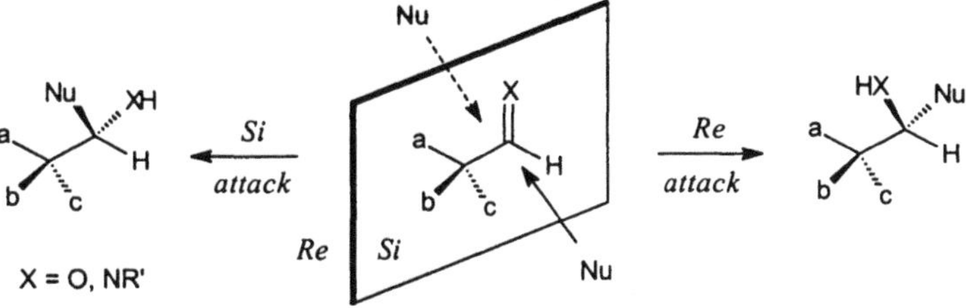

Fig. 1.

Such stereoselection is commonly observed in asymmetric induction on α-chiral π-compounds and in this case an addition reaction can afford two diastereoisomers (*anti* and *syn*). The attempts to explain the π-facial diastereoselectivity of the nucleophilic addition generated more hypotheses and models than any other subject in the field of stereoselective synthesis. Some of the earliest attempts to understand the addition of organometallic compounds were the rationalizations and predictive models made in the early 1950's by Curtin, Cram, and Prelog [3]. Nevertheless, theoretical investigations on the origin of open-chain and cyclic models continues, mainly focused on sterical and/or electronical features of the substrate. However, selectivity is a kinetic phenomenon governed by the ratio of the product formation rate for one stereoisomer with respect to the other and the possibility to influence the stereoselectivity exists not only by variations of reactants, but also by changing the reaction conditions. In this perspective, reaction temperature and solvent are readily available parameters for a control of the diastereoselectivity but they have been less well investigated.

For what concerns the dependence of the diastereomeric excess on temperature, it is generally thought that the best diastereoselectivity should be reached at low temperature, whereas at higher temperature the *de* should decrease. However, an analysis of the data reported in the literature clearly shows that the influence of temperature on diastereoselection is not uniform. Depending on the circumstances, on increasing the reaction temperature the diastereoselection can decrease, but it may be constant or even increase. In the latter case, we can observe the best diastereomeric excess (*de%*) at high temperature, a behavior that is generally not expected.

The solvent in which chemical processes take place is really a noninert medium and it plays a prominent role in solution chemistry strongly influencing the reaction. Despite that all diastereoselective reactions occur in solution, almost all transition state calculations consider the reaction only in vacuum. Until very recently, the solvent role has been considered in a relatively few number of cases and mostly just for its polarity via the unreliable continuum dielectric approach, but solvents with the same polarity can differently behave concerning the diastereoselection.

In the case of kinetically controlled diastereoisomer formation, whose kinetics are the same as those of parallel independent reactions with two different products, the selectivity (S) is expressed as:

$$S = \ln (k/k') \qquad \textbf{(1)}$$

where k *and* k' are the overall rate constants leading to the two stereoisomers.

The temperature dependence of the diastereofacial selectivity can be analyzed according to the Arrhenius equation for the two reaction paths:

$$\ln (k/k') = -(E_a - E'_a)/RT + \ln A - \ln A' \qquad \textbf{(2)}$$

which in the framework of the Eyring transition-state theory [4] becomes

$$\ln (k/k') = -(\Delta G^{\ddagger} - \Delta G'^{\ddagger})/RT = -\Delta\Delta G^{\ddagger}/RT. \qquad \textbf{(3)}$$

Since $\Delta G^{\ddagger} = \Delta H^{\ddagger} - T\,\Delta S^{\ddagger}$ **(4)**

it follows $ln\ (k/k') = -(\Delta\Delta H^{\ddagger}/RT) + (\Delta\Delta S^{\ddagger}/R),$ **(5)**

where *k* and *k'* are the observed overall rate constants, and $\Delta\Delta G^{\ddagger}$ is the difference in free activation energies for *re* and *si* face attack. In diastereoselective reactions, *k/k'* can be expressed as the final concentration ratio of the two isomers *anti* and *syn*. Temperature dependent measurements on the basis of eqn.5, allow the evaluation of stereoselectivity in terms of differential enthalpy and entropy of activation. It is important to underline that in diastereoselective processes, where only a small differential enthalpic contribution exists, temperature values in the experimental accessible range often render the differential entropy of activation a determining contribution especially at high temperature[5]. Under these conditions, any prediction of the stereoselectivity based on classical models, as for example Cram's chelated or Felkin-Anh' which are based only on enthalpic differences, are not effective because of entropy underestimation.
Entropy and enthalpy can play in favor of opposite isomers leading to an inversion of selectivity by changing the temperature[6]. In this case, we get a predominance of one diastereoisomer at low and the other one at high temperature. In the linear Eyring plot of *ln (k/k') versus* 1/T (figure 2), reversal of *de%* by reaction temperature[7] gives rise to an x-axis crossing [*ln (k/k')* = 0, *de%* = 0], thus identifying an equiselective temperature $T_0 = \Delta\Delta H^{\ddagger} / \Delta\Delta S^{\ddagger}$. Because of the positive value of absolute temperature, reversal of *de%* occurs when the differential enthalpic and entropic terms have the same sign.

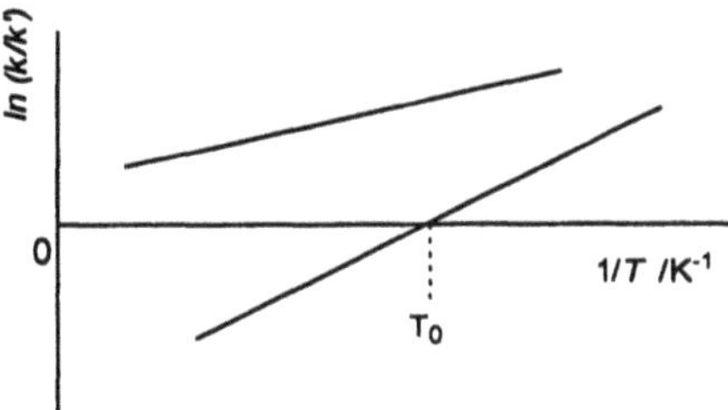

Fig. 2.

In this case, there exists a temperature range where $T\Delta\Delta S^{\ddagger} < \Delta\Delta H^{\ddagger}$ and another where $T\Delta\Delta S^{\ddagger} > \Delta\Delta H^{\ddagger}$. For example, in the formation of *anti* and *syn* isomers, whenever the former prevails, a negative value of the differential activation enthalpy ($\Delta H^{\ddagger}_{anti} - \Delta H^{\ddagger}_{syn} = \Delta\Delta H^{\ddagger} < 0$) results. Assuming that an addition reaction is accompanied by a loss of activation entropy, $\Delta S^{\ddagger}_{anti}$ and $\Delta S^{\ddagger}_{syn}$ are both negative and the condition for the *de %* inversion, $\Delta\Delta S^{\ddagger} < 0$, requires $|\Delta S^{\ddagger}_{anti}| > |\Delta S^{\ddagger}_{syn}|$: this means that the entropic loss in the formation of the *anti* is larger than those of the *syn* isomer. As a consequence, the *anti* isomer is enthalpically driven whereas the *syn* entropically: at low temperature the former prevails, while at high temperature the latter. It is important to emphasize that, if both enthalpy and entropy cooperatively work in favor of the same isomer (this condition results when $\Delta\Delta S^{\ddagger}$ and $\Delta\Delta H^{\ddagger}$ have opposite signs in eqn. 5), an inversion in the diastereomeric excess can never be obtained by solely controlling the temperature. Equation 5 shows a linear correlation between 1/T and selectivity. However, there are experimental data that show a non-linear behavior (figure 3). In these cases, the corresponding Eyring plots generally consist of two linear regions intersecting at a point defining a temperature called inversion temperature (T_{inv})[8]. This break point leads to two sets of activation parameters, one for $T < T_{inv}$ and the other for $T > T_{inv}$. Several examples of this phenomenon have been reported in the literature.

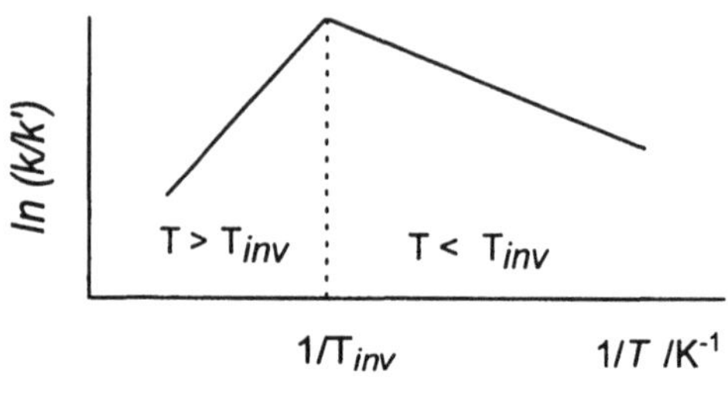

Fig. 3.

This phenomenon has been nicely reviewed by *Scharf*[9], who rationalized the condition for the existence of a T_{inv} in an Eyring plot considering a generalized kinetic scheme where the selectivity is generated at two distinct levels. He describes the T_{inv} as the temperature where an abrupt change occurs due to a change in the rate-determining step or, in other words, in the dominance of enthalpy and entropy in the partial selective steps. However, according to *Ridd*[10], which considered the same kinetic scheme, a sudden change in the dominance of selection level is unlikely. His alternative interpretation for non-linear Eyring plots is an evidence of two potential rate-determining steps in the reaction process, with only one of the two linear regions providing valid evidence on the enthalpic and entropic discrimination involved, the other coming from the transitional region in which no single stage of the reaction can be considered as rate determining. Anyway, the debate on the very nature of inversion temperature is still open and we want to show experimental results that suggest a different interpretation.

For what concerns the influence of the solvent on the stereochemical outcome of reactions[11], the complex interaction between solvent and solute molecules results in a gross modification of their free energies, and consequently, reactivities and stereoselectivities. There are many examples of solvent effects on selectivity and some of them are impressive. In many cases a variation in the diastereomeric excess is observed: as an example in the case of the Staudinger reaction between the phthalimidoketene and the α-methylphenylimine of chloroacetaldehyde[12], on going from $CHCl_3$ to CH_2Cl_2, the *de* between the two *cis*-azetidinones drops from 80% to 30%. Other examples are known where a complete reversal of diastereoselectivity occurs because of a change in the reaction solvent: in the dimethyldioxirane epoxidation of cyclohexenol[13], the *trans/cis* mixture of the products ranges from 54/46 in pure acetone to 6/94 in acetone/CCl_4. Some years ago, we reported a *de%* reversal in methylmagnesiumbromide addition to the *O*-

triisopropylsilyloxy lactal[14]. The reaction leads to an *anti/syn* ratio in THF of 64:36 and 16:84 in *t*butyl methyl ether. Very recently, *Luh*[15] reported a reversal in diastereoselectivity of MeLi and *n*BuLi addition to hydrazones of 1,4-di-*O-tert*-alkoxy-L-threitols on going from THF to diethyl ether, while *Ito*[16] reported another reversal in Diels-Alder reaction catalyzed by the aluminium complex of a chiral menthol derivative. These results show how the solvent plays a role in determining the prevailing isomer. The solvent effect on face-selectivity reflects its different influence on the two diastereomeric reaction paths through a differential contribution to k and k' (eqn.1). A change in the reaction medium corresponds to a change in the microscopic solute-solvent interactions. These interactions can differ in number or strength in different solvents, and they contribute to free activation energies $\Delta G^{\ddagger}$ and $\Delta G'^{\ddagger}$ in such a way that they can enable to a face-selectivity reversal. In the case of reactions in solution, it is hard to describe the system formed by reagents with their solvation shell or the solvated transition state, but some hypotheses can be considered. Two cases can be analyzed: *i)* the solvent enthalpically favors and entropically disfavors an isomer whenever the solute-solvent interactions are high in strength or in number, thus stabilizing conformers with the lowest intramolecular steric interactions; *ii)* the solvent entropically favors and enthalpically disfavors an isomer whenever the solute-solvent interactions are mild or low in number and the resulting solvation is less ordered, thus making the system conformationally less rigid[17]. It is obvious that any theoretical model for the analysis of facial diastereoselectivity completely looses its predictive value by ignoring solvent effects. A detailed modelling of solvation[18], and information about its effects on the dynamics of stereoselective processes, is necessary to get a deeper insight into the diastereofacial selectivity, but at the moment, research in this field is still at the beginning.

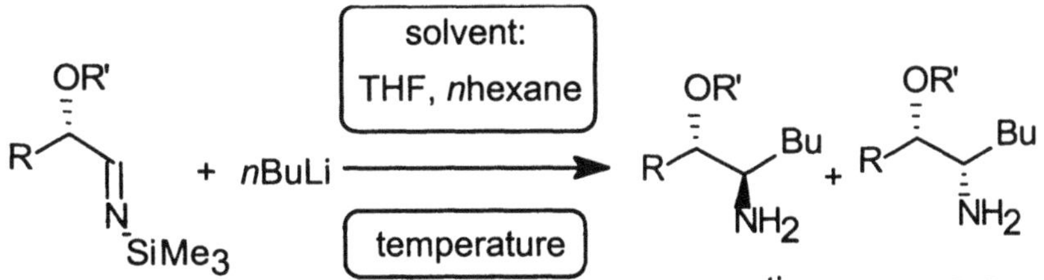

solvent: THF, *n*hexane

OR' / R / N / $SiMe_3$ + *n*BuLi → (temperature) OR' / R / Bu / NH_2 (anti) + OR' / R / Bu / NH_2 (syn)

2a-4a **2b-4b**

1: R' = Si*t*BuMe$_2$, R = Me

2: R' = Si*i*Pr$_3$, R = Me

3: R' = *t*Bu, R = Me

4: R' = Si*t*BuMe$_2$, R = *n*C$_4$H$_9$

Fig. 4.

During our studies on the stereocontrolled synthesis of 1,2-aminoethers by means of nucleophilic addition, we observed an impressive influence of solvent and temperature on the diastereofacial selectivity in the *n*BuLi addition to *N*-trimethylsilylimines of (2*S*)-lactal[19].

*n*BuLi cleanly reacted with *O*-protected *N*-trimethylsilyl imines of (2*S*)-lactal either in ether (THF) or hydrocarbon (*n*-hexane) affording optically active *anti* and *syn* 1,2-aminoethers (figure 4). The stereochemical results are strongly dependent on solvent and temperature (figure 5).

We observed a complete reversal in the *de* by changing the solvent.

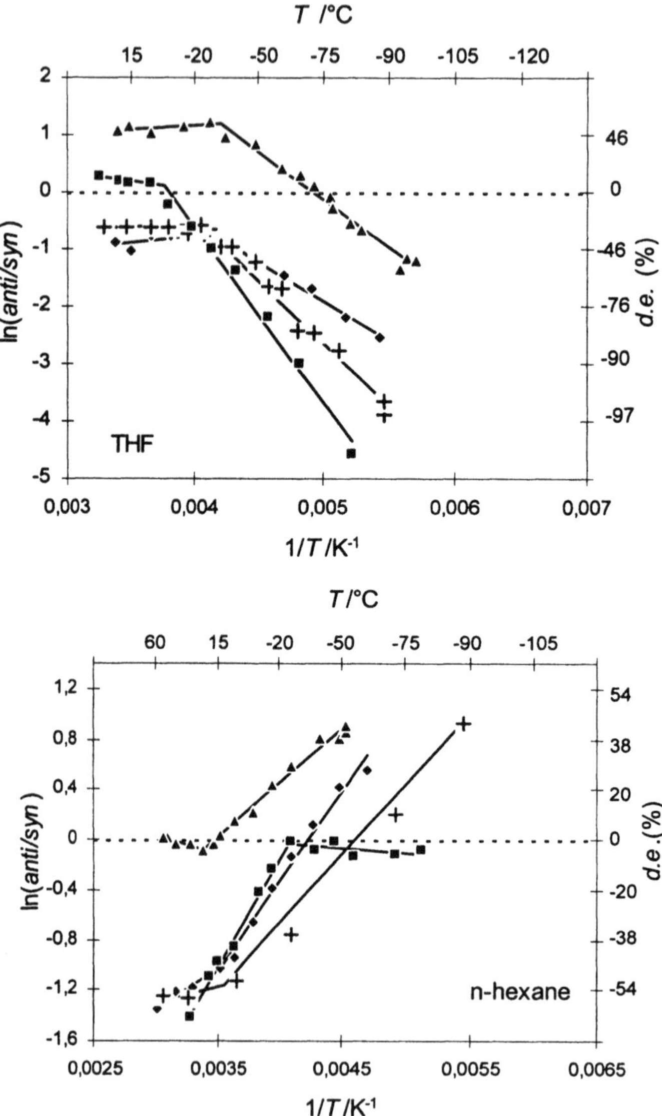

Fig. 5. Eyring plots for the diastereomeric excesses obtained in the nucleophilic additon of *n*-butyllithium to **1**(+), **2**(▲), **3**(■), **4**(◆) in THF and *n*-hexane at various temperatures.

For example, in the case of (2*S*)-*N*-trimethylsilyl-*O*-*t*butyldimethylsilyloxy lactalimine **1** at -90 °C in THF, we obtained an *anti/syn* ratio of 2:98, while in *n*-hexane *anti/syn* = 72:28. Moreover, a screening in a range of 150 °C showed a reversal diastereoselection even with the reaction temperature (e.g. in *n*-hexane at T = -90 °C *anti/syn* = 72:28, at T = 54 °C *anti/syn* = 22:78). In figure 5, the Eyring plots for all the imines **1-4** in THF and *n*-hexane are reported. The diastereomeric excess is defined as the difference *anti% - syn%*. Each imine-solvent pair showed a non-linear temperature dependence and a characteristic inversion temperature (T_{inv}) was observed in all cases.

The diastereomeric excess for imines **2** and **3** in THF crosses the x-axis (*de*% = 0) at the equiselective temperatures T_o = -71.2 °C and T_o = -8.6 °C, respectively, so that a neat inversion in stereoselectivity occurred. In *n*-hexane, the imines **1**, **2**, and **4** showed equiselective temperatures at T_o = -54.9 °C, 15.4 °C, and -36.5 °C.

In THF for T > T_{inv}., all the imines show a flatten trend of the *de%*, meaning that in this range there is no temperature control on the stereoselectivity. According to eqn.5, this result requires that the face-selection that still exists is entirely determined by $\Delta\Delta S^{\ddagger}$.

In this entropy driven region, the differentiating parameter between the four substrates is the protecting group. The *O*-TIPS- and the *O*-TBDMS-*N*-TMS imines show opposite diastereofacial selectivity, for **2** the differential entropy $\Delta\Delta S^{\ddagger}$ is positive, for **1** it is negative. As a matter of fact, in one case the *anti* isomer is favored, in the other the *syn* is preferred. It is remarkable to note that for imine **1** the negative value of $\Delta\Delta S^{\ddagger}$ implies that the transition state leading to the *anti* isomer is more ordered than that which leads to the *syn* one. This result excludes the hypothesis that the prevailing *syn* isomer derives from a chelated transition state[20]. At low temperature, all substrates give *syn* isomers. However, slopes are not markedly different and all four plots can be reproduced by translation of each other. Once again, in front of small differential enthalpic contributions, an entropic factor determines the face-selectivity.

In *n*-hexane, imines **1**, **2**, and **4** have a positive slope and exhibit a reversal of de% on going from a predominance in *syn* isomers at high temperature to a predominance of *anti* isomers at lower temperature. Even the *O*-*t*Bu-*N*-TMS imine has a positive slope, and should behave in the same way as the others, but, in this case, the T_{inv} exerts a breakdown and starts up a new linear trend with a different intercept and slope, so that the *de*% inversion is prevented.

All the activation parameters for T > T_{inv}. and for T < T_{inv}. are listed in Table 1. It is significant to note that in all cases differential entropies and enthalpies have equal signs leading to an inversion of diastereoselectivity by the temperature, as discussed in the previous section. However, for imines **1**, **2**, and **4**, but not **3**, the activation parameters for T< T_{inv} showed a further switch in sign from THF to *n*-hexane, thus showing the overwhelming role of the solvent in determining the thermodynamic parameters.

Table 1. Differential activation parameters and inversion temperatures for *n*BuLi addition to imines **1**- **4**

			$T>T_{inv}$		$T<T_{inv}$		
imine	**solvent**	T_{inv} [°C]	$\Delta\Delta H^{\ddagger}$ (kcal/mol)	$\Delta\Delta S^{\ddagger}$ (cal/mol K)	$\Delta\Delta H^{\ddagger}$ (kcal/mol)	$\Delta\Delta S^{\ddagger}$ (cal/mol K)	$\Delta\Delta G^{\ddagger}$ (298 K) (kcal/mol)
1	THF	-26.8	-0.10 ± 0.01	-1.59 ± 0.02	4.3 ± 0.3	16.4 ± 1.4	0.37
1	*n*-hexane	7.9	-0.58 ± 0.22	-4.3 ± 0.7	-2.2 ± 0.4	-10.4 ± 1.7	0.71
2	THF	-35.8	-0.28 ± 0.22	1.2 ± 0.8	3.3 ± 0.2	16.4 ± 0.9	-0.63
2	*n*-hexane	22.4	0.65 ± 0.11	2.0 ± 0.4	-1.7 ± 0.1	-5.9 ± 0.3	0.05
3	THF	-7	0.59 ± 0.18	2.4 ± 0.6	5.9 ± 0.3	22.7 ± 1.5	-0.14
3	*n*-hexane	-28.4	-3.36 ± 0.13	-13.8 ± 0.5	0.14 ± 0.1	0.52 ± 0.5	0.75
4	THF	-31.8	-0.52 ± 0.37	-3.6 ± 1.4	2.7 ± 1.5	9.8 ± 0.7	0.05
4	*n*-hexane	8.6	-1.17 ± 0.07	-6.2 ± 0.3	-2.9 ± 0.2	-12.3 ± 0.7	0.67

We have also studied the solvent effect in the stereoselective addition of *n*-BuLi to 2-phenylpropanal[21] in a series of linear and cyclic aliphatic hydrocarbons (figure 6).

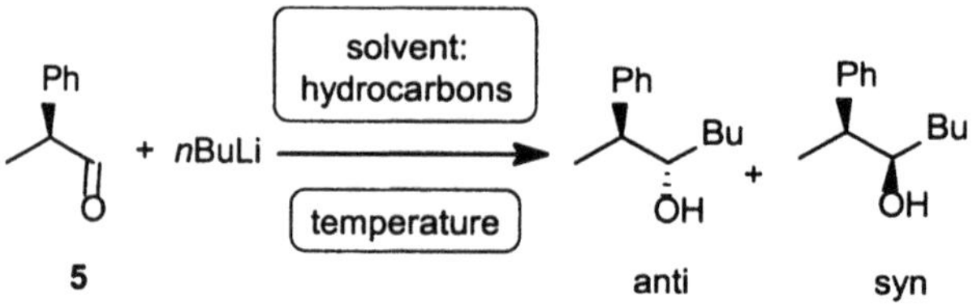

Fig. 6.

The results obtained are quite peculiar. We found inversion temperatures in aliphatic hydrocarbons and Eyring plots for a series of linear and cyclic hydrocarbons are shown in figure 7.

Interestingly, for cyclic and linear solvents plot concavities are opposite. For linear hydrocarbons the pattern is consistent with the chain length of the solvent: the longer the chain, the lower the *de%*; for example the diastereofacial selectivity obtained with propane is doubled respect to that of *n*-dodecane (propane *de%* = 70, *n*-dodecane *de%* = 30).

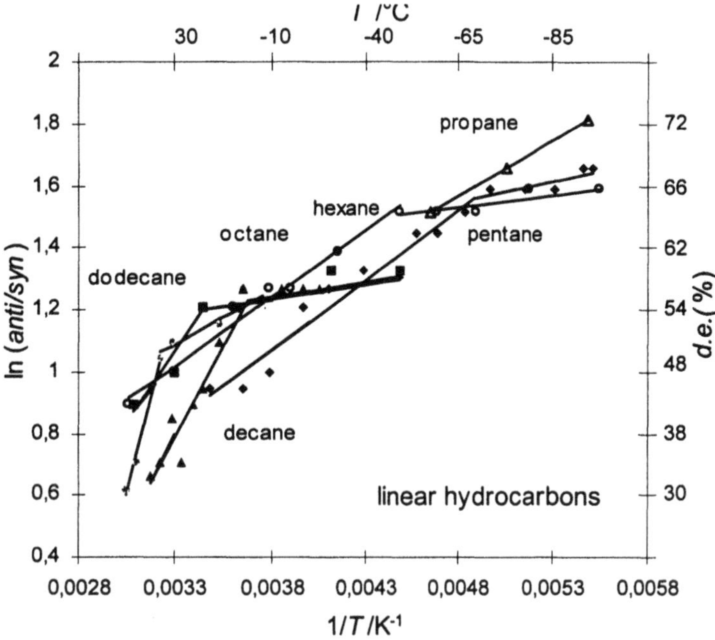

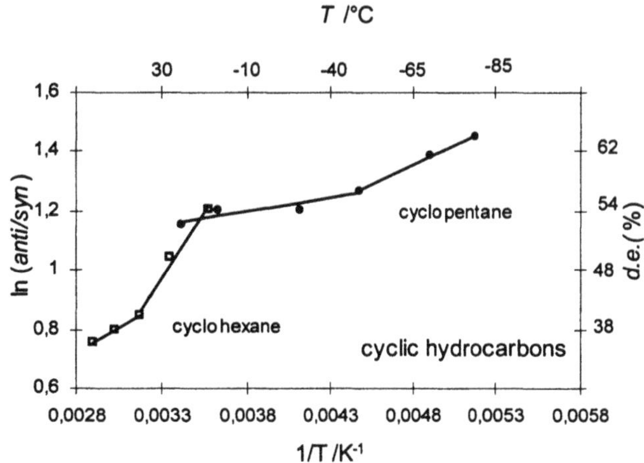

Fig. 7. Eyring plots for the diastereomeric excess obtained in the nucleophilic additon of *n*-butyllithium to **5** in propane (Δ), *n*-pentane (◆), *n*-hexane (O), *n*-octane (■), *n*-decane (▲), *n*-dodecane (●), cyclopentane (●), and cyclohexane ( ).

This result reveals the strong influence of solvation forces, although weak and non-specific, on the face-diastereoselectivity. The slopes of regression lines, which reflect the enthalpic contribution, flatten at low temperatures. Once more, differential entropy modulates the selectivity because only differences in the intercepts still remain when differential enthalpies vanish and, therefore, the effect of solvent chain-length on face-selectivity might be ascribed to an entropic control. On the other hand, at high temperature, lengthening the solvent chain increases the slope, thus increasing the enthalpic contribution to *de%*. The inversion temperatures for all linear and cyclic hydrocarbon solvents examined range between 205 and 317 K, increasing on going from *n*-pentane to *n*-dodecane.

Assuming that the inversion temperature is peculiar for each substrate-solvent pair, all efforts to correlate inversion temperatures with some classical solvent parameters, such as dielectric constant and viscosity, failed. Interestingly, analyzing T_{inv} and melting points of linear and cyclic hydrocarbons, it results a straight line with a correlation coefficient r = 0.96 (figure 8).

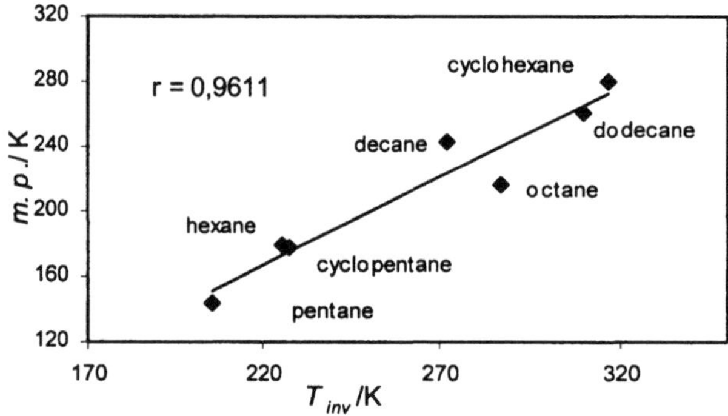

Fig. 8. Correlation between hydrocarbon melting points and inversion temperatures found for aldehyde **5**.

The dependence of facial selectivity upon hydrocarbon solvents becomes more strictly in the case of *n*butyllithium addition to *O*-*t*butyldimethylsilyloxy- mandelic aldehyde **6** (figure 9). With this substrate the reaction with *n*butyllithium was performed in a series of linear hydrocarbons with odd and even numbers of carbon atoms[22].

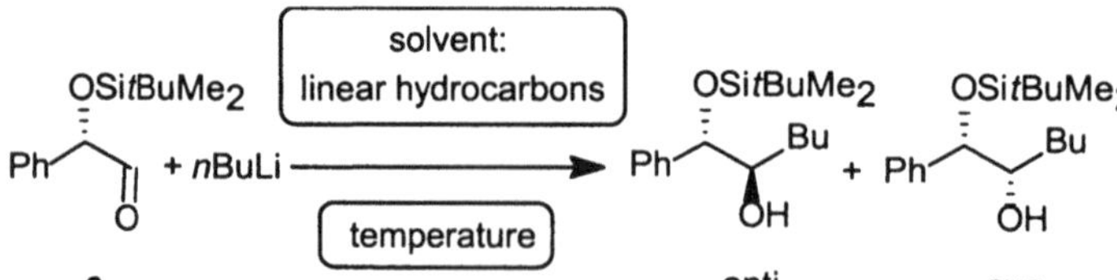

OSi*t*BuMe$_2$ Ph (**6**) + *n*BuLi → (solvent: linear hydrocarbons; temperature) → anti **6a** + syn **6b**

Fig. 9.

Although this aldehyde produced diastereomeric excesses in a low range (from 20 to -10 *de%*), in all plots an inversion temperature was recognized (figure 10).

Noteworthy, this substrate exhibits the best diastereoselectivity at high temperature (T = 60.5 °C, *de%* = 19).

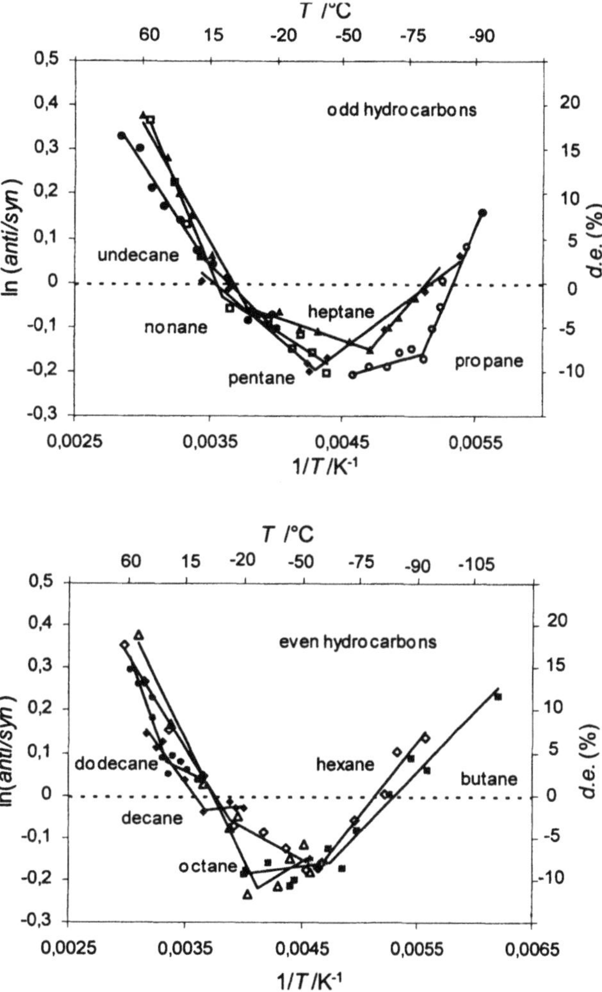

Fig. 10. Eyring plots of *n*BuLi addition to **6** in linear odd and even hydrocarbons: propane (o), pentane (■), heptane (▲), nonane ( ), undecane (●),butane (■), hexane (✧), octane (Δ), decane (◆) and dodecane (●).

The majority of these plots crosses the x-axis because of an opposition between enthalpic and entropic contribution in the two isomers as discussed above. Some plots cross the x-axis twice, once in the high and the other in the low temperature region. It is interesting to note that in the case of *n*-hexane and *n*-heptane two inversion temperatures were obtained. It is hard to intend this experimental result as a consequence of a double change in the reaction mechanism in the light of the above mentioned interpretation of the T_{inv}. As in the case of phenyl propanal **5**, the T_{inv} increases with lengthening the solvent alkyl chain. For mandelic aldehyde **6** each series shows a concavity opposite to **5**.

Plot of the inversion temperatures *versus* the number of solvent carbon atoms evidences the same swinging feature as corresponding melting points (figure 11). Even in this case, the correlation of T_{inv} and melting points exists and it results better if we separately consider the odd and even series of hydrocarbons.

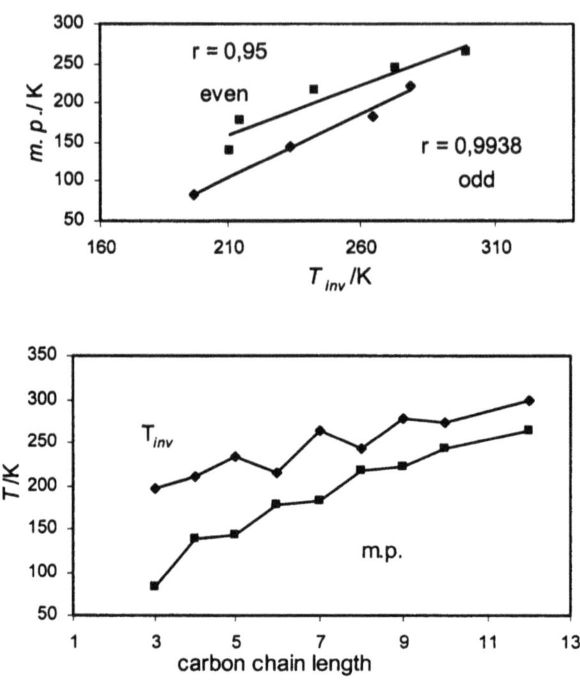

Fig. 11. Correlation between hydrocarbon melting points and inversion temperatures found for aldehyde **6**

It is really surprising how the simplicity of a linear relationship with its two parameters come over the hard complexity of the physical events involved.

It is known that even and odd linear hydrocarbons differ in the crystalline form[23], and this fact causes the melting points swinging. The same behavior has been observed for the inversion temperatures and this, along with strict linear correlationships found for both aldehydes **5** and **6**, suggests a relationship with the phase modification that happens on melting. An attractive possible explanation can be formulated: the inversion temperature could constitute a sort of transition between two "phases" which, in case of solutions, could be represented by two different solute-solvent clusters with a different order[24]. This "phase transition" could be interpreted as the interconversion of two solute-solvent clusters having a more defined three-dimensional structure than generally supposed. These supramolecular structures behave like different molecules producing a measurable change in the thermodynamic properties and therefore in diastereoselectivity. In this hypothesis, the T_{inv} represents the interconversion temperature between two supramolecules and it does not imply any change either in the rate-determining step or in the reaction mechanism. At temperatures lower than T_{inv}, one supramolecule with a given diastereoselectivity is present in solution , at temperatures higher than T_{inv} the other supramolecule is present with a different diastereoselectivity and therefore different slope and intercept (figure 12).

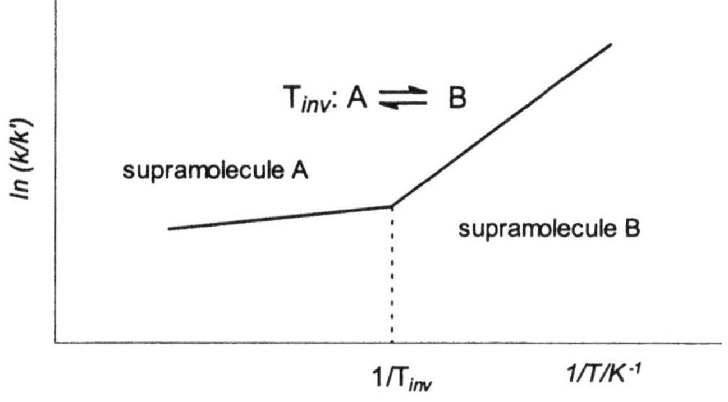

Fig. 12.

Because of the complexity of solvation processes, it is difficult to formulate a detailed microscopic model of these supramolecules. However, their interconversion could involve either a solvent reorganization phenomenon due to transfer of solvent molecules between the bulk and supramolecules or an internal rearrangement of the solvation cluster.

If this interpretation is correct, the existence of a T_{inv} and the shape of the Eyring plot depends on solvation. In this perspective, mixing two solvents should have an important impact on the solvation and, therefore, should influence the Eyring plot. Some preliminary results in this field show that this is just what happens[25].

In the case of 2-phenylpropanal (figure 13), the Eyring plot in pure THF does not show any T_{inv} whereas it is present in *n*-hexane. Following our interpretation, two molecular clusters exist at different temperatures in the hydrocarbon, whereas only one is present in the explored range of THF. By adding 5% of THF in *n*-hexane, a straight line results, meaning that just one kind of solvation exists even in the mixture of solvents.

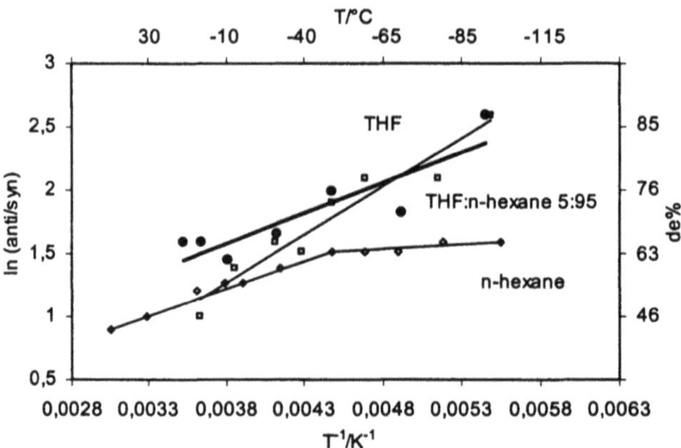

Fig. 13. Eyring plots for the diastereomeric excess obtained for the additon of *n*-butyllithium to **5** in THF ( ), *n*-hexane (◇), and a mixture of 5 mol% of THF in *n*-hexane (●).

As we reported in figure 10, in mandelic aldehyde two inversion temperatures and two isoselective points (T_0) are present in *n*-hexane. In pure THF just one T_{inv} and one T_0 is present. In the mixture THF : *n*-hexane 5 : 95, all T_{inv} and T_0 disappear and the Eyring plot becomes a straight line completely lying in the upper quadrant (figure 14).

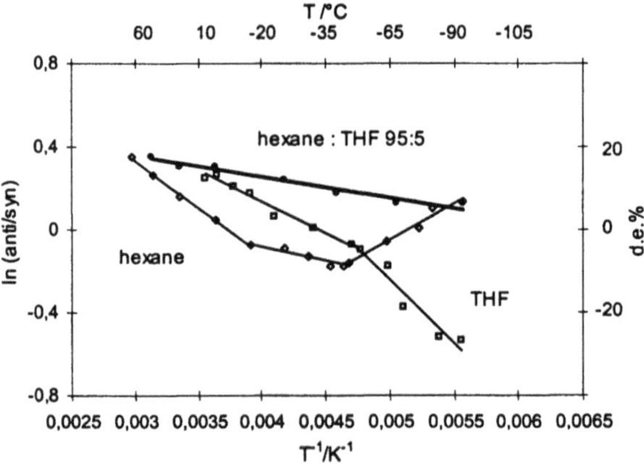

Fig. 14. Eyring plots for the diastereomeric excess obtained for the additon of *n*-butyllithium to **6** in THF ( ), *n*-hexane (◇), and a mixture of 5 mol% of THF in *n*-hexane (•).

More intriguing is the result obtained when a mixture of *n*-hexane and *n*-decane were used in a molar ratio of 95 : 5 (figure 15). In this case, a T_{inv} is found at a higher temperature than those presented by the two pure solvents.

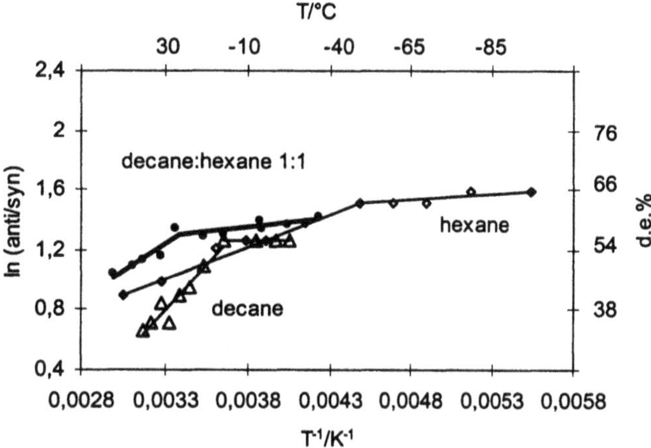

Fig. 15. Eyring plots for the diastereomeric excess obtained for the additon of *n*-butyllithium to **5** in hexane (✧), decane (Δ), and a mixture of hexane : decane 95 : 5(●).

The behaviour of the solvent mixtures is really complicated and certainly deserves much more research for a complete comprehension. However, what results clear is that the inversion temperature and therefore the diastereoselectivity exclusively depends on each substrate-solvent pair or in other words on the stereospecific solvation of the carbonyl compounds.

A common objection to our interpretation of the T_{inv} phenomenon is that temperature and solvent effects on *de%* might be attributed to different organometallic species involved. In fact, it is known that the aggregation state of *n*BuLi changes with these parameters[26]. However, the dynamic process that involves the aggregation state of *n*BuLi acts before the reacting event, so that their rate constants are identical for both diastereomers and vanish in equation 5. Thus, a mere ground state effect would be irrelevant but the *n*BuLi tendency to aggregate may in some extent influence the structure of transition states. However, the *de%* reversal with temperature and the existence of T_{inv} are observed in several different reactions where no organometallic species are involved, so that the phenomena cannot be tied to the reaction of different aggregated species[27].

In order to strengthen our proposal on substrate solvation as the predominant factor in the non linear effect of temperature on diastereofacial selectivity, we have compared the Eyring plots of different nucleophiles for a given substrate in the same solvent. We report evidences that T_{inv} does not depend on nucleophiles and its value can be obtained from a ^{13}C NMR experiment recording the evolution of C=O chemical shift *vs* temperature[28].

(2*S*)-*O*-(*tert*butyldimethylsilyl) lactal (**7**), (2*S*)-*O*-(*tert*butyldimethylsilyl) mandelic aldehyde (**6**), and (2*S*)-*O*-(tri*iso*propylsilyl)lactal (**8**) were reacted with *tert*butyllithium, *n*butyllithium, in THF and *n*hexane, and *n*butylmagnesium bromide in THF giving the corresponding *anti* and *syn* monoprotected diols (figure 16). The results are elaborated in the corresponding Eyring plots (figure 17).

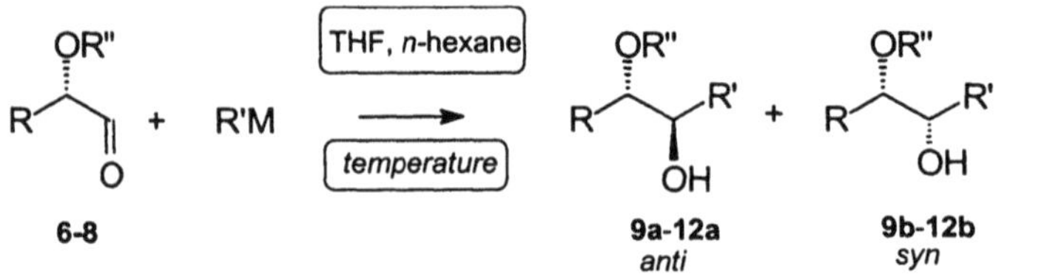

6-8 + R'M —(THF, *n*-hexane; *temperature*)→ **9a-12a** *anti* + **9b-12b** *syn*

R'M = *n*BuLi, *t*BuLi, *n*BuMgBr

7: R = Me, R'' = tBuMe$_2$Si

6: R = Ph, R'' = tBuMe$_2$Si

8: R = Me, R'' = iPr$_3$Si

9: R = Me, R' = *n*Bu, R'' = tBuMe$_2$Si

10: R = Ph, R' = *n*Bu, R'' = tBuMe$_2$Si

11: R = Me, R' = *n*Bu, R'' = iPr$_3$Si

12: R = Me, R' = *t*Bu, R'' = tBuMe$_2$Si

Fig. 16.

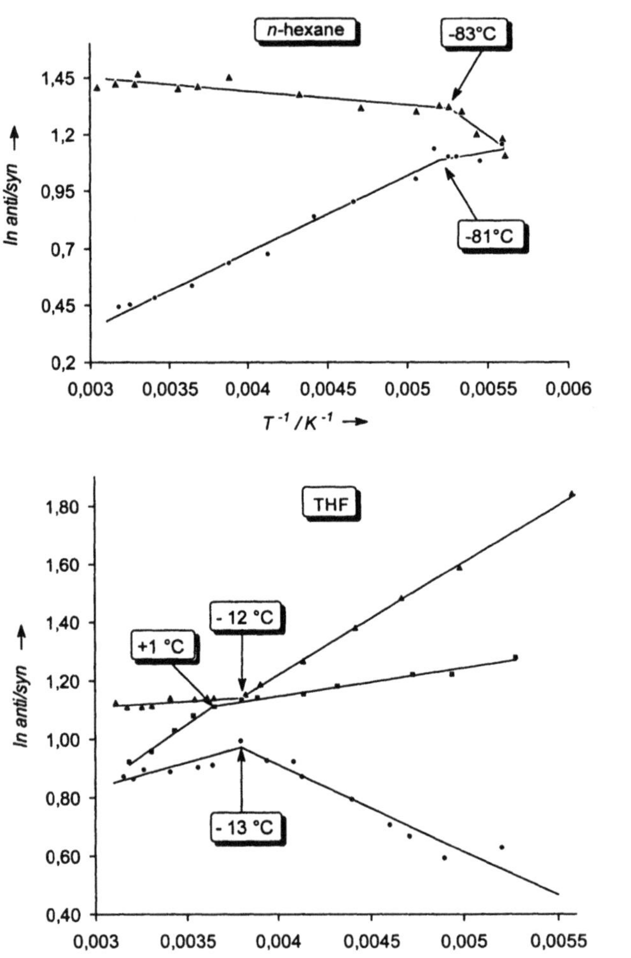

Fig. 17. Eyring plots for the nucleophilic addition of *n*-butyllithium (▲), *t*-butyllithium (●), and *n*-butylmagnesium bromide (■) to **7** at various temperature in *n*-hexane and THF

Even in these cases, in the explored temperature range, each aldehyde-solvent combination shows two linear regions and in all cases we determined the characteristic T_{inv}.

In *n*-hexane, the diastereofacial selectivity of *n*BuLi addition to lactal is only slightly temperature-sensitive, this notwithstanding, a T_{inv} is well recognised at -83 °C. With *t*BuLi the *de*% is notably affected and increases with reducing the temperature, until the *t*BuLi plot becomes equiselective to the *n*BuLi one at -95 °C. Interestingly, both organolithium plots exhibit quite similar inversion temperatures (figure 17).

In THF at low T, *t*BuLi and *n*BuLi give specular plots on account of an opposite temperature effect on *de*%. The diastereofacial selectivity in the addition of *n*BuMgBr to **7** is less affected by temperature than the above discussed organolithium reactions. In this case, the T_{inv} is +1 °C. With *t*BuLi, *n*BuLi, and *n*BuMgBr, plots are more differentiated at low T. Remarkably, all three plots in figure 17 present T_{inv} in a narrow temperature range.

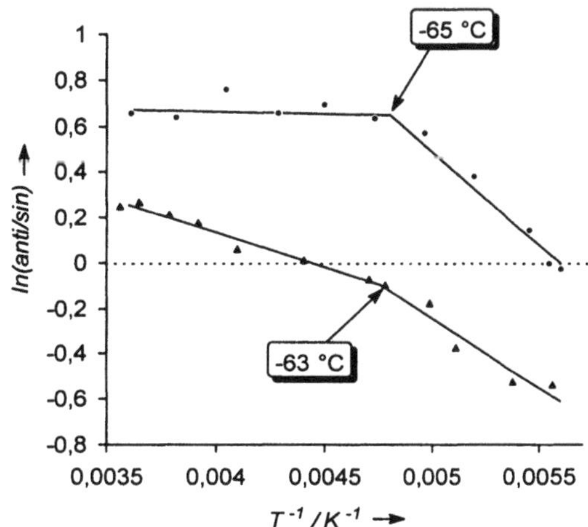

Fig. 18. Eyring plots for the nucleophilic addition of *n*-butyllithium (▲), and *n*-butylmagnesium bromide (●) to **6** at various temperature in THF

*n*BuMgBr and *n*BuLi were added to mandelic aldehyde **6** (figure 18). As it clearly appears in this figure, both plots reach the x axis thus allowing the possibility to invert the diastereofacial selectivity by changing the reaction temperature. Notwithstanding the slight change in *de*% with T, these two plots have a very similar T_{inv} (-65 and -63 °C respectively).

We obtained that different organometallic reagents with different molecular structures in solution have similar inversion temperatures. So that, the influence of aggregation and solvation of the organometallic compound on stereoselectivity is limited. All the above results clearly show that the T_{inv} is greatly dependent on the couple aldehyde-reaction solvent and less on the nucleophile. This fact emphasises our idea that the solvation of the starting π compound accounts for the T_{inv}.

NMR techniques provide a powerful method to investigate the local electronic environment due to molecular structure. Local variations in angles and bond lengths can affect chemical shieldings through change in electronic structure. More distant groups in the environment can contribute to chemical shifts as well, by affecting the magnetic field or electron density at nuclei of interest *via* anisotropic magnetic susceptibility, electrostatic, and close contact interactions[29]. Therefore, the nuclear shielding is a sensitive probe of intermolecular interactions and solvent effects.

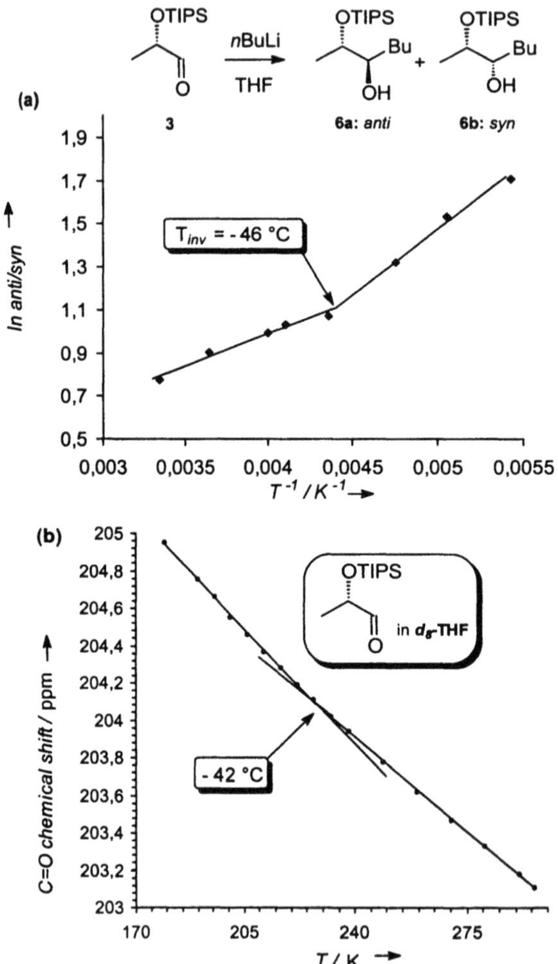

Fig. 19. a) Eyring plot for the addition of *n*-butyllithium (▲) to **8** at various temperatures inTHF. b) Plot of the ^{13}C chemical shifts for the C=O *vs* temperature for **8** in deuterated THF. Solid lines are the respective fitting results. See text about fitting details.

We performed a serie of ^{13}C NMR experiments in deuterated THF and *n*hexane with aldehydes **6**, **7**, and **8** to evidence any modification in chemical shift that should occur on changing the temperature. Figure 19 displays the evolution of the C=O ^{13}C chemical shift as a function of T in d_8-THF for the aldehyde **8**, that can be directly compared with the Eyring plot of the same aldehyde with *n*BuLi in THF. In figure 20 are reported plots of C=O chemical shift *versus* T for **6** and **7** in d_8-THF and d_{14}-*n*hexane.

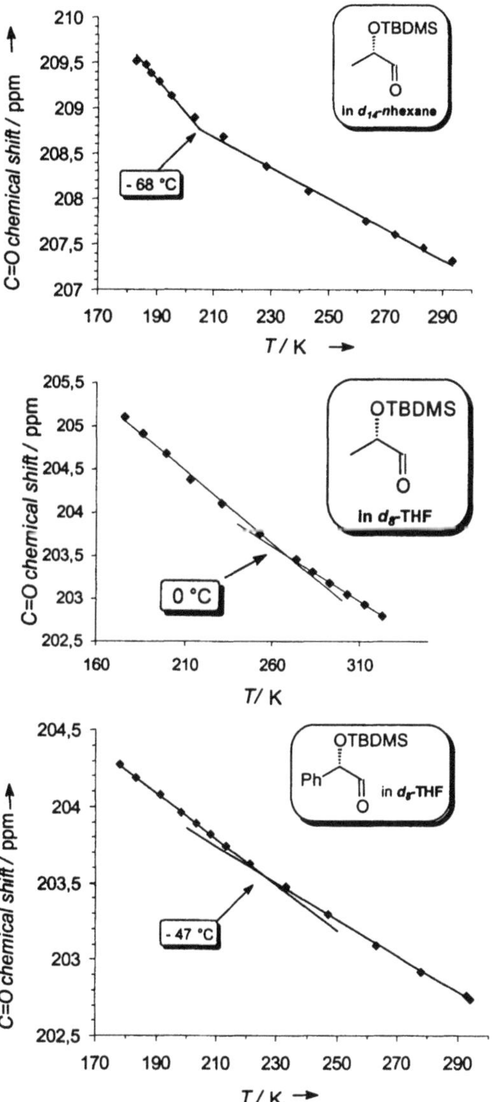

Fig. 20. Plots of the ^{13}C chemical shifts for the C=O *vs* temperature for **7** in deuterated THF and deuterated *n*-hexane, for **6** in deuterated THF.

All substrates show a ^{13}C downfield shift on warming. This could be the result of a loss of electron density near the observed nucleus. At a further check of the experimental data points, we recognised a break in the linear variation of chemical shift *vs* temperature. To determine the T value of these breaks more precisely we applied the least median squares (LMS) regression[30]. Use of the LMS regression improves the reliability and the ability to objectively define the segments of the straight line given by experimental data. In figure 19 and 20 are reported all the break temperatures for all studied substrates. These temperature values change with the solvent and the substrate and, surprisingly, they correspond to or are in a narrow range with the inversion temperatures obtained with the same substrate/solvent pair in the nucleophilic addition reaction.

This result reinforces our hypothesis that the T_{inv} corresponds to a transition between two differently ordered solvation states. As we pointed out, the inversion temperature reflects a feature of the solvation state of the carbonyl compound. Change in the solvation with the temperature can be not continuous and makes an abrupt change corresponding to the T_{inv} .

Because of its very nature, this phenomenon should not be tied to the presence of a stereocenter on the π compound. In fact, Figure 21 reports the evolution of C=O ^{13}C chemical shift for the 2-*O*-*t*butyldimethylsilyloxy ethanal in deuterated THF which shows a break at T = -24 °C. This achiral aldehyde has a change in its solvation shell that could have consequences on its reactivity simply because two different solvation clusters exist in the two temperature regions and acting as two different molecules. This latter result constitutes a challenge to find out a reaction that will show a non-linear behaviour against temperature with a T_{inv} near to the break temperature revealed in the ^{13}C experiment.

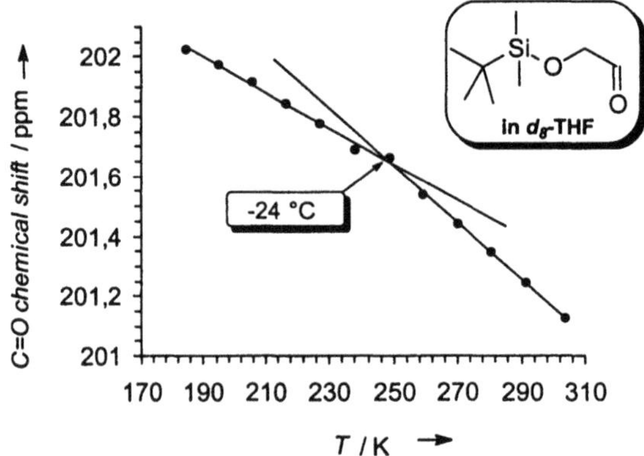

Fig. 21. Plot of the ^{13}C chemical shifts for the C=O *vs* temperature for 2-*O*-*t*butyldimethylsilyloxy-ethanal in deuterated THF.

Conclusion

Our results confirm that not only enthalpic contributions act in determining diastereo control, but even entropic factors can acquire importance with reaction temperature. Indeed, whenever enthalpy and entropy oppositely favor an isomer, a *de%* reversal can be obtained.

Solvent effects are determinant on face-selectivity because solute-solvent interactions modulate the free activation energies acting on both enthalpic and entropic terms. In this way different solvents can lead to an opposite diastereoselectivity.

The correlation with melting points suggests an alternative interpretation of the very nature of inversion temperatures. We propose that T_{inv} is the temperature value for the interconversion between two different solvation clusters represented by two different supramolecules.

We also demonstrated that the inversion temperature observed in Eyring plots obtained in diastereoselective nucleophilic addition to α-chiral aldehydes does not depend on nucleophiles. ^{13}C NMR results clearly connect the phenomenon of T_{inv} with the solvation of the starting π compound confirming our hypothesis of a solvent-dependent nature of T_{inv}.

All these arguments support the hypothesis that the main factor in the diastereofacial control is a stereospecific solvation of the π-system[31].

References

1. R. E. Gawley, J. Aubè, in "*Principles of Asymmetric Synthesis*", Tetrahedron Organic Chemistry Series, vol.14, ed. J. E. Baldwin, F. R. S. Magnus, P. D. Magnus, Pergamon, Oxford, 1996; D. J. Ager, M. B. East in "*Asymmetric Synthetic Methodology*", CRC Press, 1996; M. T. Reetz, *Angew. Chem., Int. Ed. Engl.,* **1984**, *23*, 556.
2. M. T. Reetz, *Angew. Chem., Int. Ed. Engl.,* **1984**, *23*, 556; N. C. DeMello, D. P. Curran, *J. Am. Chem. Soc.* **1998**, *120*, 329.
3. E. L. Eliel, S. H. Wilen, L. N. Mander, in "*Stereochemistry of Organic Compounds*", Wiley: New York, 1994, p. 875; D. Y. Curtin, E. E. Harris, E. K. Meislich, *J. Am. Chem. Soc.* **1952**, *74*, 2901; D. J. Cram, F. A. A. Elhafez, *J. Am. Chem. Soc.* **1952**, *74*, 5828; V. Prelog, *Helv. Chim. Acta* **1953**, *36*, 308.
4. H. Eyring, *J. Phys. Chem.* **1935**, *3*, 107; S. Glasstone, K.J. Laidler, H. Eyring, *The Theory of Rate Processes.* McGraw-Hill, New York, 1941, Chap.4
5. B. Giese, *Acc. Chem. Res.*, **1984**, *17*, 438, and references cited therein; R. E. Rosenberg, J. S. Vilardo, *Tetrahedron Lett.*, **1996**, *37*, 2185.
6. See for instance: N. Hoffmann, H. Buschmann, G. Raabe, H-D. Scharf, *Tetrahedron,* **1994**, *50*, 11167; D. Awandi, F. Henin, J. Muzart, J-P. Pete, *Tetrahedron: Asymmetry,* **1991**, *2*, 1101; C. Zioudrou, P. Chrysochou, *Tetrahedron*, **1977**, *33*, 2103.
7. Y. Inoue, N.Yamasaki, T. Yokoyama, A. Tai, *J. Org. Chem.*, **1992**, *57*, 1332.
8. H. Buschmann, H-D. Scharf, N. Hoffmann, M. W. Plath, J. Runsink, *J. Am. Chem. Soc.*, **1989**, *11*, 5367; M. Palucki, P. J. Pospisil, W. Zhang, E.N. Jacobsen, *J. Am. Chem. Soc.,* **1994**, *116*, 9333; J. Brunne, N. Hoffmann, H.-D. Scharf, *Tetrahedron,* **1994**, *50*, 6819; T. Göbel, K.B. Sharpless, *Angew. Chem.,* **1993**, *105*, 1417, *Angew. Chem., Int. Ed. Engl.,* **1993**, *32*, 1329; J. Muzart, F. Hénin, J.-P. Père, A. M'boungou-M'Passi, *Tetrahedron: Asymmetry,* **1993**, *4*, 2531; I. Tóth, I. Guo, B. E. Hanson, *Organometallics,* **1993**, *12*, 477; I. E. Markò, A. Chesney, D. M. Hollinshead, *Tetrahedron: Asymmetry*, **1994**, *5*, 569.
9. H. Buschmann, H.-D. Scharf, N. Hoffmann, P. Esser, *Angew. Chem.,* **1991**, *103*, 480; *Angew. Chem., Int. Ed. Engl.*, **1991**, *30*, 477.
10. K. J. Hale, J. H. Ridd, *J. Chem. Soc., Perkin Trans. 2,* **1995**, 1601; K. J. Hale, J. H. Ridd, *J. Chem. Soc., Chem. Commun.*, **1995**, 357.
11. C. Reichardt, in "*Solvents and Solvent Effects in Organic Chemistry,*" 2nd ed., VCH, Weinheim, 1990.

12. Tschaen, D. M.; Fuentes, L. M.; Lynch, J.E.; Laswell, W. L.; Volante, R. P.; Shinkai, I.; *Tetrahedron Lett.* **1988**, *29*, 2779.
13. R. W. Murray, M. Singh, B .L Williams, H. M. Moncrieff, *J. Org. Chem.* **1996**, *61*, 1830.
14. G. Cainelli, D. Giacomini, F. Perciaccante, *Tetrahedron: Asymmetry,* **1994**, *5*, 1913
15. See for instance: Y-T. Hsieh, G-H. Lee, Y. Wang, T-Y. Luh, *J. Org. Chem.*, **1998**, *63*, 1484.
16. G. Naraku, K. Hoti, Y. N. Ito, T. Katsuki, *Tetrahedron Lett.*, **1997**, *38*, 8231.
17. B. Lecea, A. Arrieta, F. P. Cossío, *J. Org. Chem.*, **1997**, *62*, 6485; C. L. Perrin, M. A. Fabian, I. A. Rivero, *J .Am. Chem. Soc.*, **1998**, *120*, 1044.
18. D. J. Giesen, C. J. Cramer, D. G. Truhlar, *J. Phys. Chem.*, **1995**, *99*, 7137; G. D. Hawkins, D. A. Liotard, C. J. Cramer, D. G. Truhlar, *J. Org. Chem.*, **1998**, *63*, 4305; M. F. Ruiz-Lopez, X. Assfeld, J. I. Garcia, J. A Mayoral, L. Salvatella, *J .Am. Chem. Soc.*, **1993**, *115*, 8780; M. Sola, A. Lledos, M. Duran, J. Bertran, J.-L. M. Abboud, *J. Am. Chem. Soc.*, **1991**, *113*, 2873.
19. G. Cainelli, D. Giacomini, M. Walzl; *Angew. Chem., Int. Ed. Engl.*, **1995**, *34*, 2150; G. Cainelli, D. Giacomini, P. Galletti *Eur. J. Org. Chem.* **1999**, 61.
20. The preponderant formation of *syn* isomers is generally attributed to a chelated transition state (chelation control), while the *anti* isomer should arouse from an open-chain transition state (non-chelation control). See for instance M. T. Reetz, *Acc. Chem. Res.* **1993**, *26*, 462. M. T. Reetz *Angew. Chem., Int. Ed. Engl.* **1991**, *30*, 556. However, the chelated transition state should be more rigid and more ordered than the open-chain one.
21. G. Cainelli, D. Giacomini, P. Galletti, A. Marini, *Angew. Chem., Int. Ed. Engl.*, **1996**, *35*, 2849.
22. G. Cainelli, D. Giacomini, P. Galletti *Chem. Commun., Feature Article* **1999**, 567.
23. M. S. Searle, D. H. Williams, *J .Am. Chem. Soc.*, **1992**, *114*, 10690.
24. The possibility of partial ordering of molecules in solution exists, and a temperature dependent change in this order is already manifested in a non-linear behaviour of some spectroscopic properties. See for instance: J. B. Robert, *Mol. Phys.*, **1997**, *90*, 399; M. A. Wendt, J. Meiler, F. Weinhold, T. C. Farrar, *Mol. Phys.*, **1998**, *93*, 145.
25. G. Cainelli, D. Giacomini, P. Galletti, P. Orioli, submitted for pubblication.
26. M. Schlosser, in " *Organometallics in Synthesis* ", Wiley, New York, 1994, 11.
27. Preliminary experiments using different organometallic reagents with surely different aggregations, such as *t*BuLi and *n*-butyl magnesium bromide, show with the same carbonyl compound the same $T_{inv.}$
28. G. Cainelli, D. Giacomini, P. Galletti, P. Orioli, in press, *Angew. Chem., Int. Ed. Engl.* **2000**, 000.
29. For solvent effects on chemical shift see for instance: T. Helgaker, M. Jaszunski, K. Ruud. *Chem. Rev.* **1999**, *99*, 293; E.Y. Lau, J.T. Gerig, *J. Am. Chem. Soc* **1996**, *118*, 1194; E.Y. Lau, J.T. Gerig, *J. Chem. Phys.* **1995**, *103*, 3341 and references cited therein.
30. P. J. Rousseeuw, *J. Am. Stat. Ass.* **1984**, *79*, 871; M. Ortiz, A.Herrero-Gutierrez, *Chemometrics and Intelligent Laboratory Systems* **1995**, *27*, 231.
31. H. Pracejus, A. Tille, *Chem.Ber.*, **1963**, 854.

Biological Performance of Materials

Rolando Barbucci, Stefania Lamponi, Agnese Magnani

Dipartimento di Scienze e Tecnologie Chimiche e dei Biosistemi, Università di Siena, via E. Bastianini 12, I-53100 Siena, Italy
e mail: barbucci@unisi.it

Abstract.

The biomaterial science, the study of the application of materials to biological and biomedical problems is a field characterised by medical needs, basic research, advanced technological development, industrial involvement, ethical considerations and regulations. The biological performance of materials largely depends on their bulk and surface properties.

Some general concepts necessary to define the biological performance of materials are discussed. Current approaches of surface modification of materials and methods of synthesis for creating surfaces with specific biofuncionality are reviewed.

An approach to biocompatibility: general considerations

Biomaterials and the devices made from them occupy an increasingly important role in clinical medicine. Seldom these devices cause significant clinical problems; however complications arise with the use of both implantable and temporary devices. For example, vascular catheters need to be replaced frequently because of functional failure caused by mechanical injury to the vessel wall, leading to thrombosis, inflammation and sometimes infection; in addition, the material itself activates platelet, leukocyte, blood coagulation and complement pathways. This common clinical observation illustrates the importance of understanding the nature of biocompatibility in order to design materials and devices that are more biocompatible than those currently in use.

Biocompatibility has commonly been defined in negative terms. That is, the property of materials and devices not to initiate intolerable clinical responses (such as thrombosis, thromboembolism, intimal hyperplasia, calcification, hemorrage, etc.).

Alternatively, a contemporary definition of biocompatibility is „the ability of a material to perfom with an appropriate host response in a specific application" [1]. By this definition, biocompatibility is viewed as not merely a passive state in which biomaterials are designed to have specific, desirable functions.

The design of such biomaterials is currently under way. The biomaterial community has begun to move away from complete reliance on materials already „on the shelf" and toward design of materials with specific biological functions and clinical applications.

The label „biocompatible" usually suggests that the material described displays universally „good" or harmonious behviour in contact with tissue and body fluids. It is an absolute term without any referent.

Futhermore, as already mentioned, the traditional ideas of biocompatibility refer essentially to the effect of the material on the biological system. Effects of biological processes on materials are rarely included in the meaning, unless the results of material changes elicit a change in biological response. The effects of the biological system on the material are usually lumped in the term of biodegradation; this implies „bad" behaviour again without referent.

However, the real issues in the use of biomaterials in medical and surgical devices are not absolute, any more than is the choice of a material for any other engineering application. The choice of materials for construction of a device or machine is made early in the designed process.

The properties of the candidate materials, particularly those properties which bear on the intended function of the complete mechanical assembly, then interact strongly with the design.

The ultimate test of the appropriateness of the choice of materials is the performance of the completed design. In this performance we can see the interaction between design choices (shape, size, linkage, etc.) and materials properties (strength, density, conductivity, etc.).

The real issue of biocompatibility is not whether there are adverse reactions to a biomaterial, but whether that material perform satisfactorily (that is, in the intended fashion) in the application under consideration. This should lead directly to the traditional engineering design process of considering the advantages and disadvantages inherent with the selection of a particular material for a design in a specific application. Among the factors considered must be the interaction of the material with the biological processes in its intended site of operation.

The „biological performance" concept

The „biological performance" is a description of materials and replaces the present idea of biocompatibility.

Two important aspects of this „performance" are:

- *host response*: the local and systemic response, other than the intended therapeutic response of living systems to the material;
- *material reponse*: the response of the material to living systems.

These two terms, however are not sufficient to discuss the biological performance of materials, since the interaction of materials with living systems should be considered on a relative, rather than an absolute, basis.

Other terms are thus needed to implement such a concept.

Reference material: a material that, by standard test, has been determined to elicit a reproducible, quantifiable host or material response.

A reference material might be a material with minimal host response (negative reference) or extreme host response (positive reference).

Level of host (or material) response: the nature of the host (or material) response in a standard test with respect to the response obtained with a reference material.

A standard test is simply any well defined repeatable test.

Biocompatibility: the biological performance in a specific application that is judged suitable to that situation.

Thus, when the host and material response are known and the particular device application is examined, a final value judgement can be made, which leads to the acceptance or rejection of the material.

Such a selection of adequate performance does not „qualify" a material. Rather, it increases the confidence in the use of the material and points to „possible" successful use in similar applications.

Another important point should also be taken into account when discussing the biological performance of materials.

Living systems differ most from machines in respect to the constant flux and change of their components, that is, in their physiology.

The biological performance, in particular the host response, ought not to be defined in terms of issue structure and pathology, but primarily in terms of physiology.

Deviation from normal physiological conditions may lead to changes in the structure and function of living tissues.

Extrapolation of results obtained in tissue culture and animal models rests significantly on knowledge of how both normal and abnormal physiological processes in these systems differ from those in humans.

So in addition to making „relative" rather than „absolute" determination, whe should also take care to attend to physiology (either normal or abnormal) and its interspecie variations. Biomaterials can, thus, be classified as either „biotolerant", „bioinert" or „bioactive" on the basis of respectively negative (but tolerable) local host response, absence of local host response, and positive (desired) local host response.

Considering the historical development of biomaterials, four types of biomaterials can be defined, based upon changing concepts of host response:

1. *Inert biomaterials*: implantable materials which elicit little or no host response;
2. *Interactive biomaterials*: implantable materials which are designed to elicit specif, beneficial response;
3. *Viable biomaterials*: implantable materials, possibly incorporating live cells at implantation, which are treated by the host as normal tissue matrices and are actively resorbed and/or remodelled;
4. *Replant biomaterials*. Implantable materials consisting of native tissue, cultured in vitro from cells obtained previously from the specific implant patient.

Type 1 materials are pointless.

Many biomaterials in present clinical use and ones in development are propertly called type 2 materials. Preliminary research reports reveal great interest and promises in type 3 materials, and advances in control and manipulation of the genetic code in mammals, suggest that not intellectual barrier exists to prevent the future realisation of type 4 materials at both the tissue and organ level.

Medical practice today utilises a great number of artificial devices and implants, and a wide variety of designs for any application exists with different degree of efficacy.

In contrast, only few of the many available metal, polymer and ceramic compositions have proven useful in medical devices and implants. The limiting factors is the biological performance; thus, better understanding of the biological performance and the factors influencing it, will lead to a variety of useful new material options, widening the expansion of the role that artificial devices can play in the prevention and treatment of humans deseases.

At the end of this progression, artificial devices will be called upon to serve only as „bridges" to replantation and biomaterials will emerge in their rightful place as one of the healing arts.

Approaches to improve the biological performance of materials

Surface modification

Millions of medical devices comprised of synthetic materials (biomaterials) are used in humans each year. These medical devices can be designed, synthesised and fabricated to have appropriate mechanical properties, durability and funcionality. Examples are a knee prosthesis that should withstand high localised mechanical stresses, a blood oxygenator membrane that should have a requisite permeability characteristics, and the leaflets in a heart valve that should flex for millions of cycles without failure.

The bulk structure of the materials governs these properties.

Biological response to biomaterials, on the other hand, are dominated by their surface chemistry, structure and morphology.

Thus, the rationale for the surface modification of biomaterials is straightforward: retain the key physical properties while modifying only the outermost surface to influence the biointeraction. If surface modification is propertly effected, the bulk mechanical properties and functionality of the medical device will be unchanged, but the biological performance will be improved.

As mentioned before, the terms „biocompatibility" or „biological performance" are widely used to describe appropriate or desirable interactions between a synthetic material and a living organism.

The biological reaction to synthetic materials implanted in living organism, in every case, starts with the material becoming coated with a layer of adsorbed protein of the body fluids. Substantial amounts of adsorbed protein have been seen on solid surfaces in adsorption times as short as 1 second.

Cells then interrogate the material, or more accurately, the protein layer coating the material. The cells, in response to the protein layer, can produce soluble biochemical signaling agents that instruct other cellular systems in the organism about how to react to the implant. Different reactions are noted in blood, or in other soft tissues or in bone. These ideas can be schematically illustrated as in Figure 1.

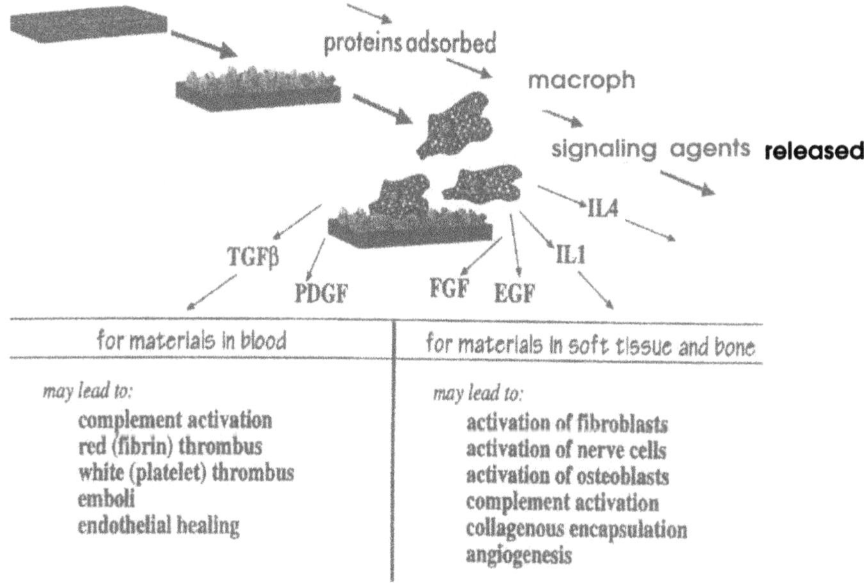

Fig. 1. For implanted biomaterials, protein adsorption, interrogation by neutrophilis and macrophages, and cell release are always observed. The soluble agents released from the cells at the implant site direct subsequent reactions. These subsequent reactions differ in the blood or in other soft tissue or bone implant sites

Two end results that might be discussed are blood compatibility (an acceptable performance in the blood stream) and healing (a process that occurs around the implant and noted in soft tissues and bone) [2,3].

Surface modification strategies have been used to engineer biomaterials to optimise specific surface properties such as: lubricity, protein resistance, degradation protection, enhanced protein retention, incorporation of chemical sites for subsequent reactions (including biomolecule immobilisation), inhibition of cell adhesion, enhancement of cell attachment and growth, and antibacterial properties.

Blood compatibility, in particular, has frequently been addressed with a range of chemical and biochemical surface treatments including heparinisation, the immobilisation of enzymes and a variety of specific chemistry and morphologies to inhibit the induction of blood clot [4-8].

Surface modifications for biomedical applications have been performed on metals, ceramics, carbons and polymers. Here, will be emphasised only the surface modifications of polymers. Five broad categories of surface modifications can be described:

1. chemically or physically alterating the atoms or molecules in the existing surface (treatment, etching, chemical modification);
2. coating over the existing surface with a material having a new composition (coating, grafting, thin film deposition);
3. implantation of ions, clusters or particles into a surface zone;
4. inducing roughness and texture;
5. immobilising molecules with specific bioactivity.

These approaches can be schematically shown as in Figure 2.

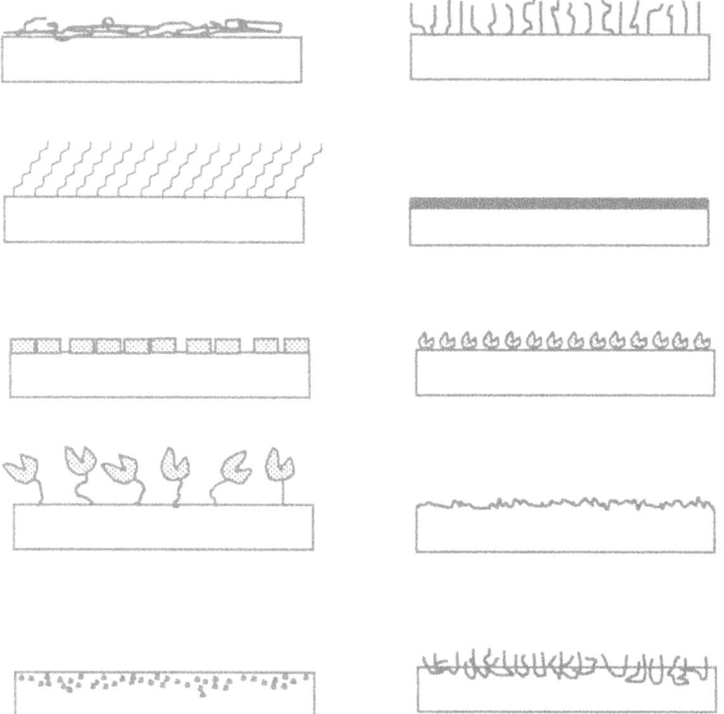

Fig. 2. The possibilities for surface modification. (a) cross-linked graft, (b) polymeric brush graft, (c) assembled monolayer, (d) chemical reaction of the surface molecules, (e) patterned overlayed on the surface, (f) biomolecule immobilised directly to a surface, (g) biomolecule immobilised to a surface via a tether, (h) surface etching, (i) ions implanted into the surface region, (j) surface interpenetrating graft.

In successfully performing any surface modification for biomedical applications there are important considerations that must be addressed.

These are, in particular, the chemical, surface analytical and biological challenges for surface modification.

Chemical challenges

A wide range of polymer surface modifications for biomedical applications have been proposed. Some of these are summarised in Table 1.

Table 1. Surface modification for polymeric materials

Chemical modification	Biological modification
Hydrogel grafts	Heparinisation
Poly(ethylene oxide)	Hyaluronic acid and sulphated
Poly(2-hydroxyethylmethacrylate)	hyaluronic acid coatings
polyacrylamide	Polysaccharide immobilisation
n-alikyl chains grafted to the surface	Enzyme immobilisation
Sulphonate groups	Tethered growth factors
Sulphate groups	
Fluoropolymer treatments	
Incorporation of biomolecules	
amine	
carboxyl	
hydroxyl	

For all surface modifications, some important generalisations can be made.

First, thin surface modifications are desirable; modified surface layers that are too thick can change the mechanical functional properties of the material. Also, thick coatings are subjected to delamination due to mismatch of physical properties.

Also extremely thin layers may be more subject to surface reversal and mechanical erosion. Some coatings have a specific thickness, for example the thickness of Langmuir-Blodgett films is related to the length of the surfactant molecules that comprise them (20-40 Å).

Other coatings, such as poly(ethylene glycol) protein-resistant layer, may require a minimum thickness (i.e. a dimension related to the molecular weight of chains) to work.

Surface-modified layers should resist to delamination in aqueous environments. Delamination resistance is achieved by covalently binding the modified region to the substrate, intermixing the substrate and the surface film at the interfacial zone, incorporating a compatibilising layer at the interface, or incorporating appropriate functional groups for strong intermolecular adhesion (i.e. Lewis acid-base interactions) between the substrate and the overlayer [9,10].

Surface rearrangement is commonly observed. Surface chemistry and structures can drift from the „as formed“ state due to rotation, translation or diffusion of surface atoms or molecules in response to the external environment.

„Reconstruction", „relaxation" and „surface segregation" are often used to describe alterations in surface structures and chemistries [11-14].

The driving forces for these changes are the minimisation of the interfacial free energy (orienting the component providing the lowest interfacial energy for the system to the surface) and the increase of the entropy of the system (the concentrated modified region at the surface will distribute itself uniformily throughout the bulk).

A major challenge in surface modification is the precise control over functional groups. Many surface modifications produce a spectrum of functional groups such as hydroxyl, ether, carbonyl, carboxyl, etc., in contrast to a single functional group, that was, in most cases, intended for the surface.

Glow-discharge plasmas, corona or chemical oxidation are examples of methods yielding non-specific reaction [15].

A greater control and precision in the depositoin of surface structure has been addressed for RF-plasmas [16,17].

Self-assembled monolayers are particularly effective for surface modifications when precise control of surface functional groups is required [18,19].

Finally, the uniform surface treatment of complex shapes and geometries is often required for the biomedical application of a material. This includes long rolls of film stocks, tubing, particulate materials, foams, and shaped parts with complex contours.

Surface analysis

The surface modified region is usually very thin and consists of only minute amounts of materials. Moreover, undesirable contamination can readily be introduced during modification reactions.

Surface reversal can also occur during surface modification. The surface reaction should be monitored to ensure that the intended surface is indeed being formed.

Since conventional analytical methods are often insufficiently sensitive to detect surface modifications, special surface analytical tools are called for.

Contact angles, electron spectroscopy for chemical analysis (ESCA), secondary ion mass spectrometry (SIMS), scanning probe microscopies and the surface infrared methods are standard tools for those concerned with high quality surface modifications [20-25].

Spatial control of surface modification: micro and nano patterned surfaces able to influence the cell behaviour

The surface treatments or coatings can be applied to specific, localised regions of the surface to generate patterns or features, often of the size of micro.or nanometers. Micro and nanotechnologies as well as micro and nanostructured materials are continuously developing in several scientific areas. Structures with micro and nanodimensions are produced on a large number of materials through micro and nanofabrication techniques previously developed by the electronic industry [26-30]. Such structured materials have shown the ability of controlling the cellular behaviour, adhesion and signaling [27-30], and influencing cell functionality and phenotype [31].

They may be extremely important in directing blood interactions, tissue reaction and healing [32,33]. Many of the basic characteristics of micro and nanostructured materials, and in particular their biofunctionality, depend in fact on the dimensions of their domains which are of the same order of magnitude of those of the biological moiety they have to interact with. That allows us to correlate the chemical and physical characteristics of a material surface with their specific functional properties and to obtain modified materials which respond to specific requirements.

The surface control of the cellular behaviour such as adhesion, diffusion, migration and proliferation is on the other hand of fundamental importance for the formation of tissues and organs as well as for the realisation of functional biomaterials. The exact manipulation of two external signals such as the cell-substrate and cell-cell interaction by a surface with well defined chemistry and

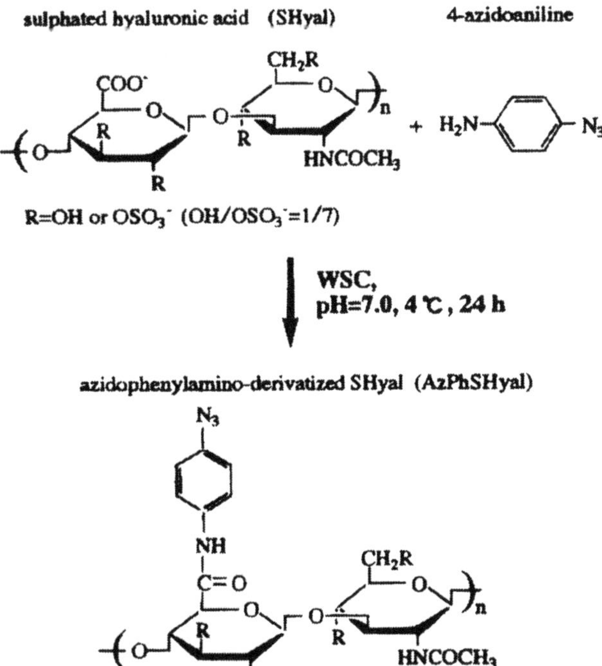

Fig. 3. Preparation of azidophenylamino-derivatised sulphated hyaluronic acid

morphology allows the realisation of a „pattern“ of cell highly oriented and differentiated. The advanced control of the cellular morphogenesis is a requisite necessary to obtain advanced tissue devices.

Micro and nanofabrication methods are usually coupled with techniques of surface chemistry modification to obtain surfaces able to control the celllular behaviour. Many methods have been applied to effect such spatial surface modifications, including photolitography, plasma and other immobilisation reactions with masking and mechanically inducing surface features with appropriate tools [34-38].

A polysaccharide with heparin-like properties such as sulphated hyaluronic acid (HyalS) has been immobilised on poly(ethylene terephthalate) (PET) substrate in a specific pattern by photolithography.

In this procedure, schematised in Figures 3 and 4, the polymer is first coupled with azidoaniline and then cast on the PET film from aqueous solution. After drying, the film is photoirradiated in the presence of a photomask to create the micro or nanodomains on the substrate surface [39].

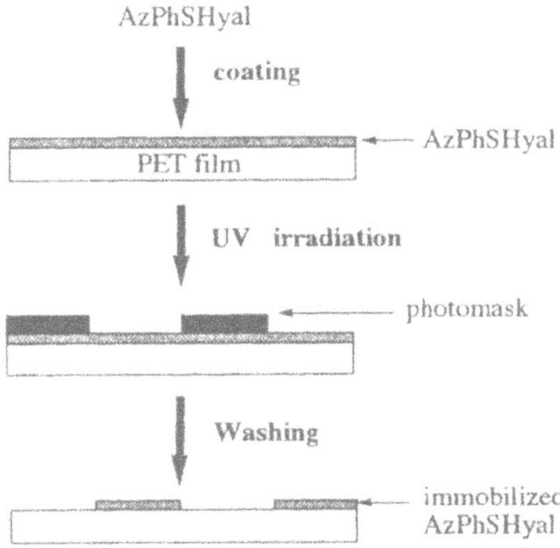

Fig. 4. Photoimmobilisation of azidophenylamino-derivatised sulphated hyaluronic acid. The photoimmobilisation was carried out with a photomask.

Platelet adhesion and thrombus formation was reduced on the HyalS-immobilised areas [39]. The micropatterning of anticoagulant polysaccharides by the photolithography technique also allowed the simultaneous visualisation of the blood interactions with the regions with or without immobilised polysaccharide.

Dynamics of cell movement in response to topographic and chemical micro and nano-features has been investigated by the analysis of primary ovine chondrocytes on planar fused silica or microfabricated grooves in silica, ranging from 0,75 μm to 9 μm in depth and 2 μm to 20 μm in width [40]. Observations showed that the cells did not spread appreciably on any groove size, and did not align to the grooves, but did move (Figures 5 and 6).

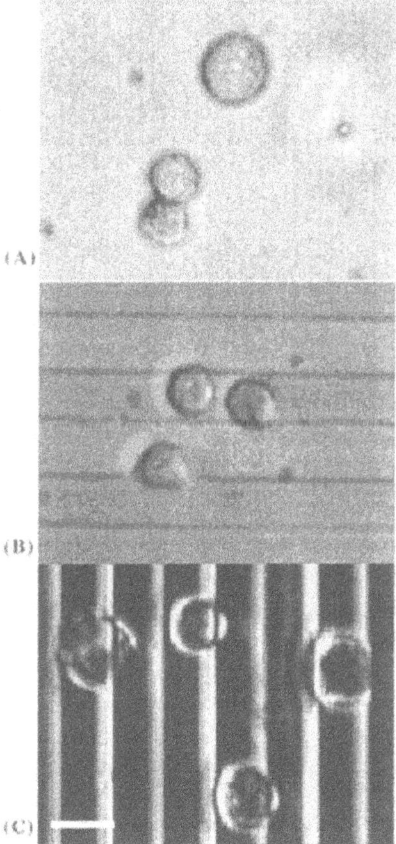

Fig. 5. Phase contrast image of primary chondrocytes on (A) plain fused silica, (B) 750 nm deep, 12,5 μm wide grooves and (C) 8 μm deep, 12.5 μm wide grooves. Bar = 20 μm

Analysis of cell movement showed that the cells did respond to grooves of 750 nm deep with longer time spent moving and further distance travelled compared with chondrocytes on a flat surface. However, groove depths of 3 μm or over inhibited cell movement. F–actin arrangement was not altered for up to 4 days after plating on flat or grooved surfaces and actin organisation was similar to that seen in site.

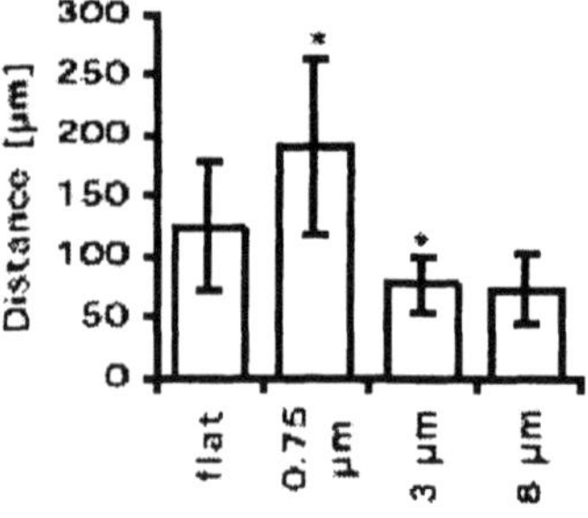

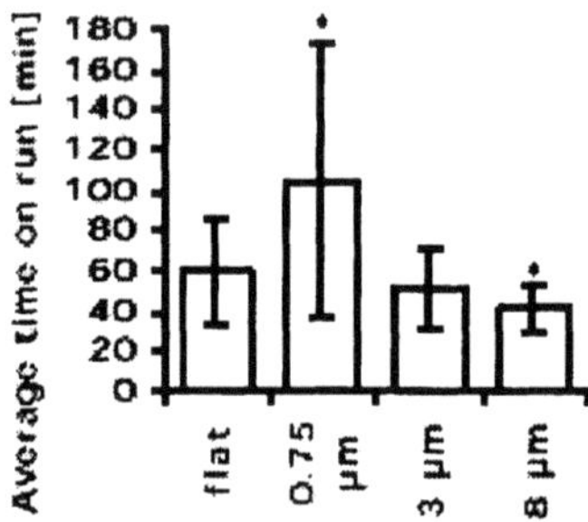

Fig. 6. Movement analysis of chondrocytes for 20 hours after plating. (A) Average distance moved, (B) average time on the run, and (C) average velocity per run * = comparison with flat surface, $P < 0.01$+ = comparison with 750 nm grooves, $P < 0.01$, student paired-t-test

When primary chondrocytes were plated on poly(ethylene terephthalate) coated with 100 µm bands of HyalS, they migrated from PET to the HyalS (Figure 7) and the cells on HyalS also showed a greater degree of spreading (Figure 8) and F-actin reorganisation [40].

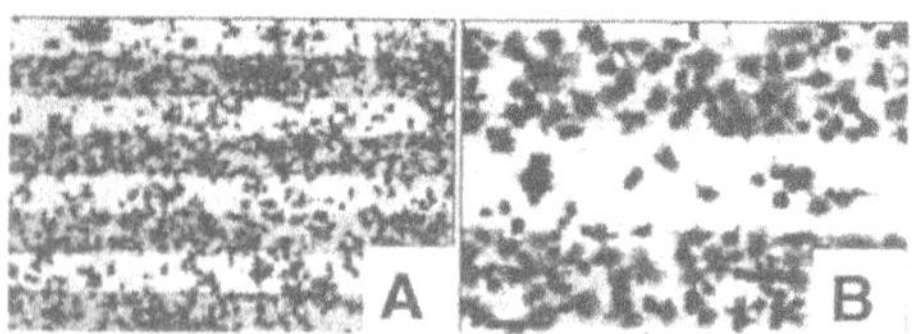

Fig. 7. (A) Coomassie blue staining of primary chondrocytes on PET and HyalS (darken stripes) after 96 hours. The cells cluster on the HyalS, but not on PET. Bar = 100 µm. (B) The same preparation at higher magnification. The cells are more spread on HyalS than on PET. Bar = 50 µm

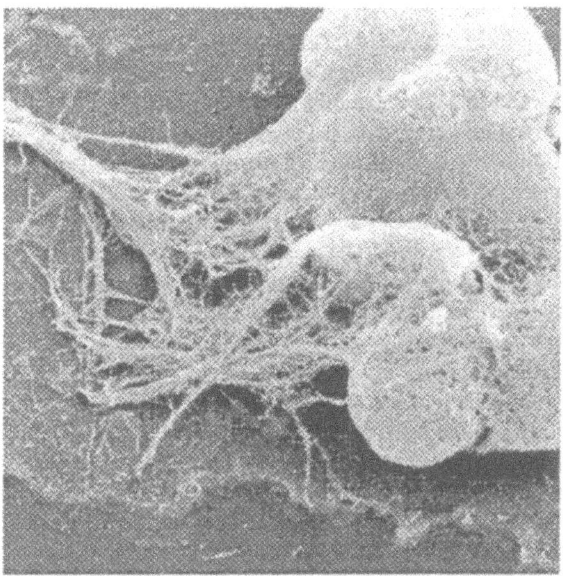

Fig. 8. Scanning electron micrograph of primary chondrocytes on HyalS. Bar = 5 μm

These data emphasise how the primary chondrocyte behaviour is affected by grooves of specific depth and different chemical patterns.

Design for biological interaction: precision surface engineering

The most difficult task for the surface engineer is to know what structure is needed to produce a given biological reaction.

The ambiguity in the definitions of such terms as blood compatibility and biocompatibility has been addressed [41,42], but little progresses have been made in altering the basic biological response of living systems to synthetic materials, thus enhancing their biological performance.

It has been often observed that different surfaces (e.g. hydrogels, teflon, polyurethane, etc.) implanted in soft tissue lead to almost indistinguishable healing reactions and encapsulation in a thin, quiescent collagenous sac. Similarly in bone implantations, surprisingly little difference is seen in biological reaction with large changes in surface properties of the implanted materials [43]. The exception to this may be for blood interaction where large and clear changes in biological response are produced with alterations in surface properties [7,6].

The future of surfaces modification for manipulating biological response will take advantage of precision of surface engineering. We must know which biochemical and biological pathways we need to activate or deactivate, and design surfaces to precisely turn on or off those pathways.

Surface heparinisation for improving blood compatibility in situations where the intrinsic coagulation system is activated, is one example of biomaterial surface modification where the rationale for the modification is based upon the understanding of well defined biochemical pathways [4]. The rationale for the application of heparinisation to platelet interactions or soft tissue implantation is less well defined, although instances where good results were obtained have been reported. Polymeric surfaces containing ionically-bound heparin [44] or sulphated hyaluronic acid [45], for example, showed anticoagulant and non-thrombogenic properties as demonstrated by the prolongation of thrombin time and whole blood clotting time. The immobilised heparin and HyalS molecules also allow the adsorbed proteins (human serum albumin and human fibrinogen) to maintain their native conformation [46,47] while reducing the platelet adhesion and aggregation and inhibiting the bacterial colonisation [45].

Other approaches to create surfaces for controlling biointeraction with precision and specificity include self-assembled monolayers [18,19], multifunctional surfaces to mimic receptor sites [48], template materials [49], immobilised biomolecules [45,46], special surfaces patterns or textures [31,33,50].

Recently great interest and work have been particularly devoted to the synthesis of highly hydrophilic chemical gels based on natural macromolecules. It has been proposed by Andrade et al. [51,52] that hydrophilic gels should be blood compatibles because of their very low interfacial energy. This hypothesis has been questioned [53,54], but the assumption that hydrogels are more natural and endothelial-like remains.

Natural macromolecule-based hydrogels have been obtained by several types of cross-linking methods such as physical treatment with U.V., γ irradiation, and dehydrothermal and chemical treatment.

In our laboratory we investigated a new approach for the synthesis of hydrogels involving the carboxylate moitey which, thus, has to be present on the macromolecule chain [55].

This approach allowed a linear polyelectrolyte to become a polymeric network. These polyelectrolyte gels had the peculiarity of absorbing up to several hundred times their own weight of water, while retaining coherence and elasticity.

Among these polyelectrolytes, some natural polysaccharides such as hyaluronan, carboxymethylcellulose and alginate appeared to be interesting.

The alginate molecule (AA) was a block copolymer composed of regions of sequential (1,4) linked β-D-mannuronic acid (M unit), regions of sequential α-L-guluronic (G unit) and regions of atactically organised M and G units.

Hyaluronan (Hyal) was a straight chain polymer consisting of alternating $\beta(1\rightarrow4)$ linked 2-acetamide-2 deoxy-D-glucose and $\beta(1\rightarrow3)$ linked D-glucuronic acid. The monomeric unit of the carboxymethylcellulose (CMC) backbone consisted of D-glucose residues linked through β-1,4 bonds. Some H atoms of the cellulose glucose hydroxyl groups were substituted by CH_2COO^- moieties. The repeating units of these polysaccharides are shown in Figure 9. Their use as gels in medical applications has recently been proposed [56,57].

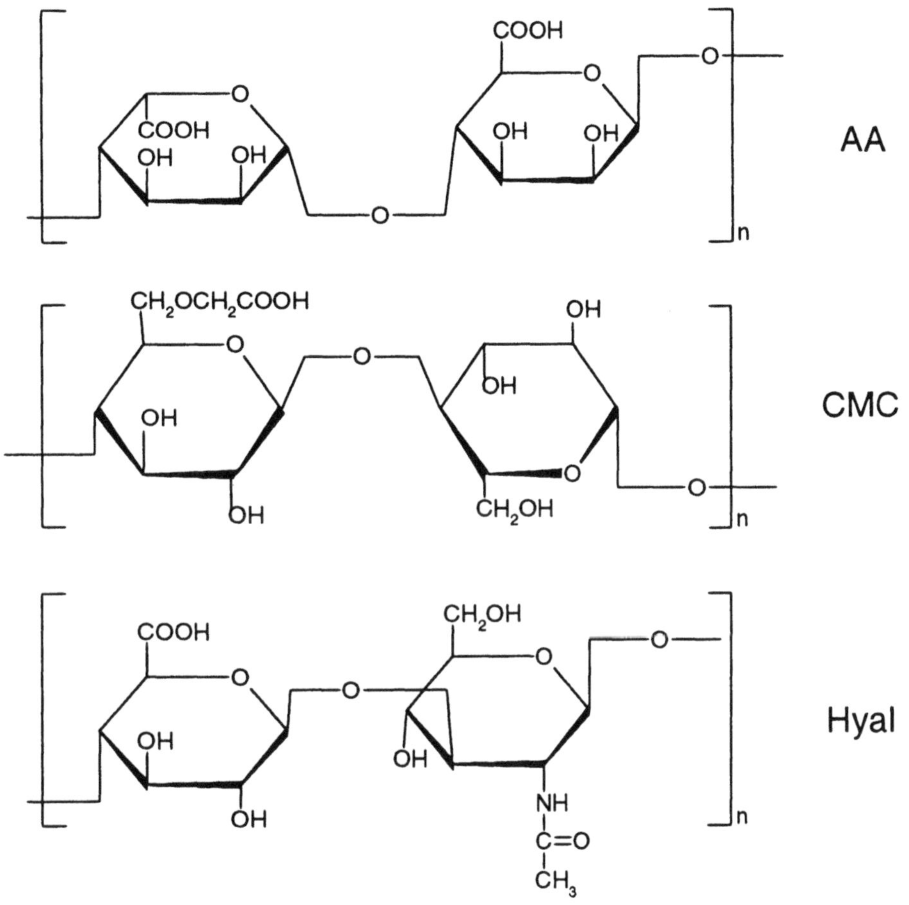

Fig. 9. Disaccharide unit for alginate (AA) carboxymethylcellulose (CMC) and hyaluronan (Hyal)

In our approach the cross-linking reaction between the polysaccharide chains occurred via 1,3-diaminopropane, using 2-chloro-1-methyl pyridinium iodide (CMPJ) as activating agent. Simply by changing the amount of CMPJ, gels with different cross-linking degrees and well-defined molecular structure could be obtained.

The reaction scheme for the synthesis of the natural-polysaccharide networks is reported in Figure 10.

In the case of AA the cross-linking reaction was not stoichiometric, as for CMC and Hyal, because of the closeness of the carboxylate groups.

Figure 11 shows, as an example, the FTIR/ATR spectra of the CMC gels with three different cross-linking degrees (5%, 50% and 100%).

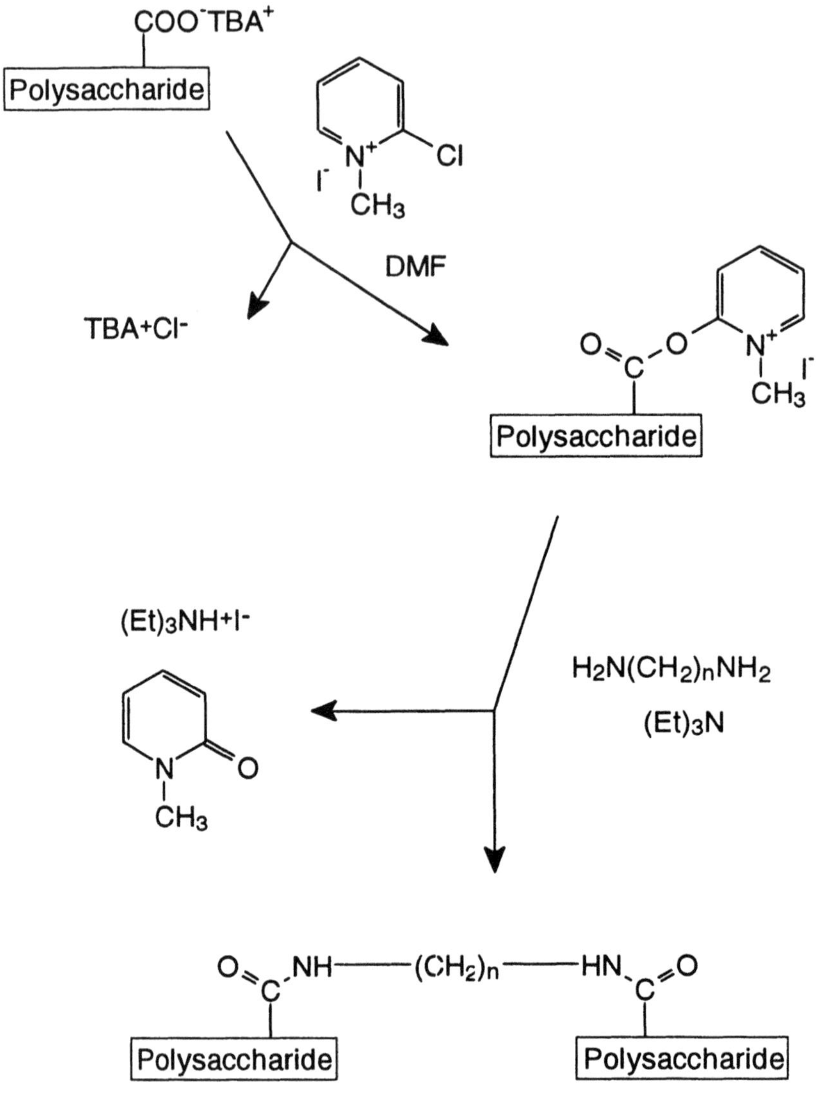

Fig. 10. Scheme of the cross-linking procedure for the synthesis of polysaccharide gels

During the cross-linking reaction the COO^- moiety was converted to a -CONH- group. The intensity of the band at about 1630 cm^{-1}, attributed to the C=O stretching of the amide moiety, should, thus, increase with increasing the cross-linking degree. This occurred for both CMC and Hyal, whereas in the case of AA, the infrared spectrum of the theoretically 100% cross-linked gel did not significantly differ, within this region, from that of the 50% cross-linked gel, emphasising that the reaction was not stoichiometric for this polysaccharide.

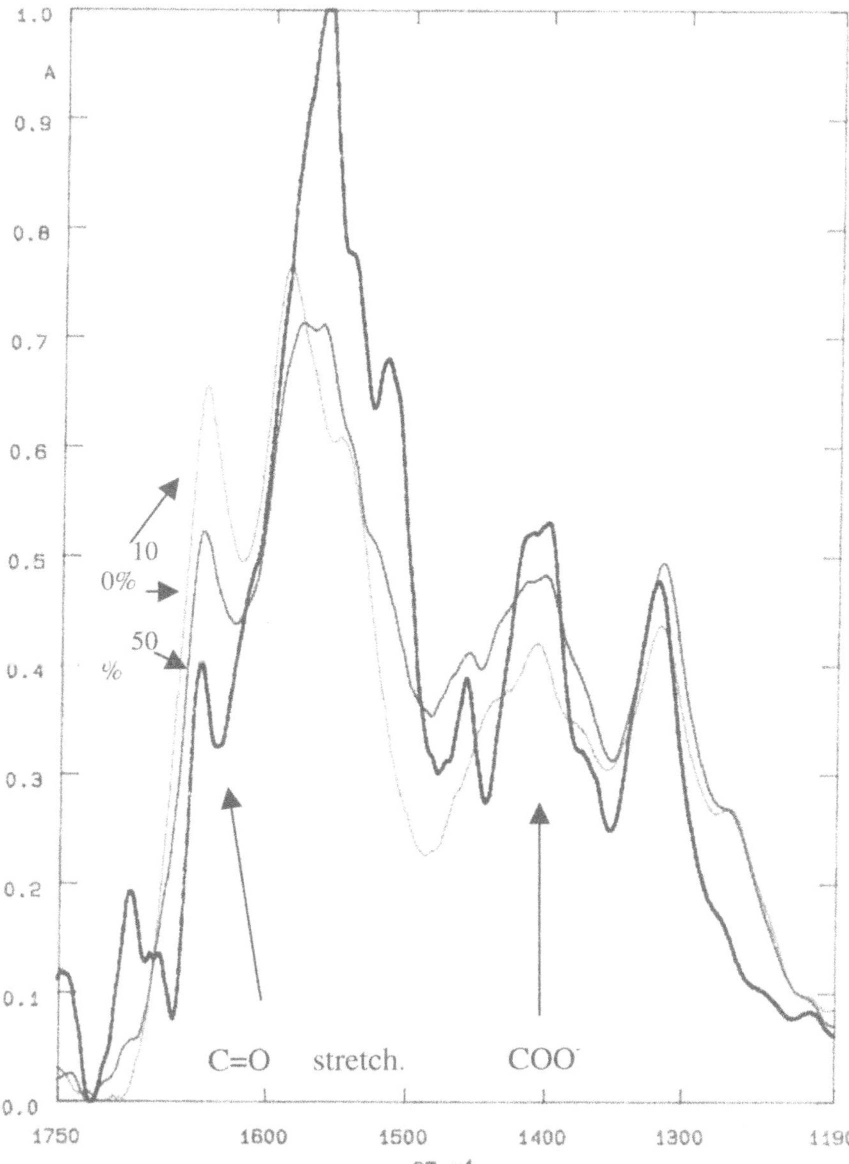

Fig. 11. FTIR/ATR spectra of CMC gel with three different cross-linking degrees

The gels swelled in water within a few minutes. Their swelling degree (S.D.) roughly depended on the cross-linking percentage and increased with decreasing the cross-linking percentage. The Hyal gels had the highest S.D. (two or three times that of AA and CMC) as is shown in Figure 12.

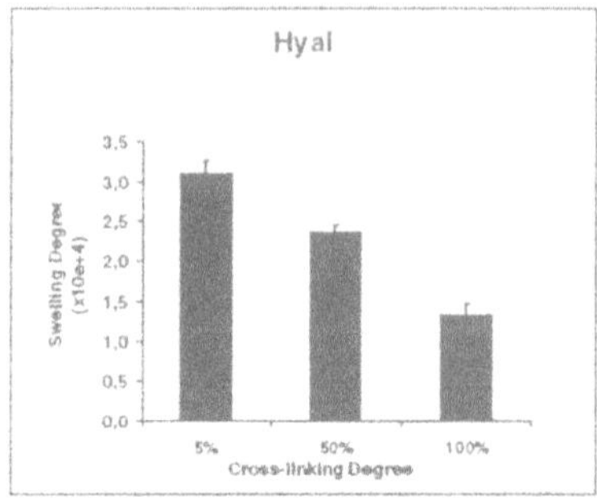

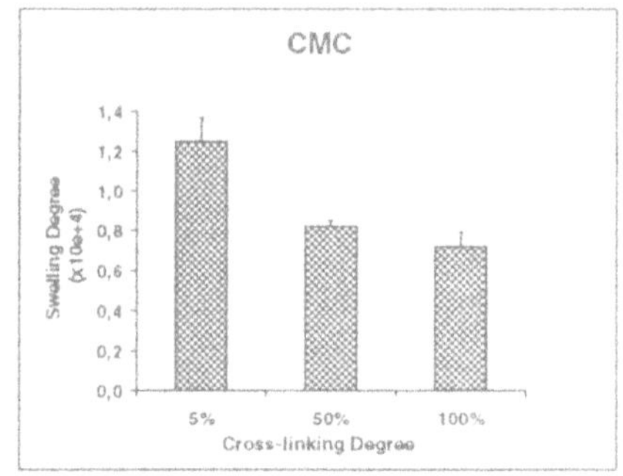

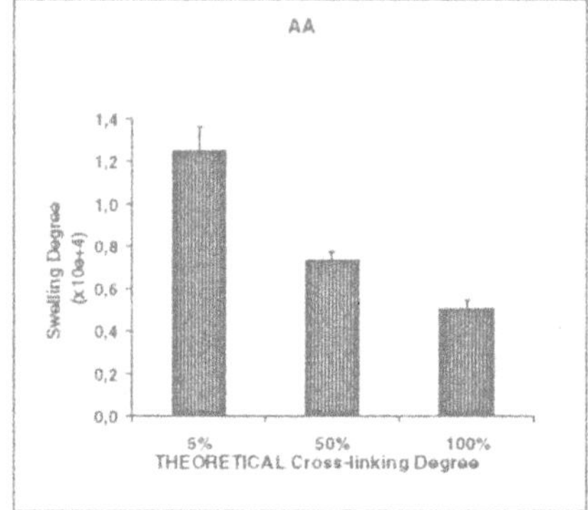

Fig. 12. Swelling degree versus cross-linking percentage for (A) Hyal, (B) CMC, (C) AA networks

Table 2 summarises the morphology and structure for these hydrogels as a function of their cross-linking degree.

The platelet adhesion experiments performed on the Hyal, CMC and AA gels with different cross-linking degree showed that for Hyal and CMC gels the cross-linking degree did not affect the gel-platelet interaction: in fact only few platelets were found to adhere without aggregating on all the samples independently of the cross-linking percentage. Unlike CMC and Hyal, AA gel with a 5% of cross-linking degree seemed instead to stimulate the platelet adhesion and aggregation.

With regard to the cell-hydrogel interaction, the results of the contact of human hepatocytes (Hep G2) with Hyal, CMC and AA gels with the three cross-linking degree showed that all the three differently cross-linked AA gels were unable to stimulate the cell adhesion and proliferation, according to the fact that mammalian cells do not specifically interact with alginate [58].

In contrast, the CMC and Hyal gels with 50% and 100% of cross-linking degree were able to stimulate the cell adhesion and proliferation, allowing the hephatocytes to maintain their round morphology. This feature makes these hydrogels suitable substrates for cell adhesion and proliferation.

On the basis of these data, the role of both the chemical structure and morphology of precisely engineered surfaces in inducing specific biological interactions can, thus, be addressed.

Table 2. Morphology and structure of hydrogels as a function of their cross-linking degree

Hydrogels and their cross-linking degree	Morphology and pore cross-section
5% CMC	Filament-like structure
50% CMC	13-25 μm
100% CMC	No pores. Presence of meshes (15-35 μm) delimiting compact structures
AA[a]	Filament-like structure
30% AA	No pores. Presence of meshes (24-35 μm) delimiting compact structures
AA[b]	Compact lamellar structures devoid of pores
5% Hyal	Filament-like alternating with porous structures (75-113 μm)
50% Hyal	Lamellar structures
100% Hyal	63-211 μm

a The cross-linking degree is less than 5%
b The cross-linking degree is more than 30%

Acknowledgements

The authors would like to thank Progetto Finalizzato MSTA II of the National Research Council for financial support.

References

1. D.F. Williams. Consensus and definition in biomaterials. In: Implant materials in biofuncion. Advances in Biomaterials, C. De Putter, G.L. deLange, K. De Groot, A.J.C. Lee (Eds.), Elsevier, Amsterdam, 8:11-71 (1988)
2. L.P. Tang, J.W. Eaton. Inflammatory response to biomaterials, Am. J. Clin. Path., 103:466-471 (1995)
3. B.D. Ratner, A.S. Hoffman, J.E. Lemons, F.J. Schoen. Biomaterials Science. An introduction to materials in medicine. Academic Press, New York, (1996).

4. N.A. Plate', J.I. Valuev. Heparin-containing polymeric materials. Advances in Polymer Science, 79:95-137 (1986).
5. A.S. Hoffman, G. Schumen, C. Harris, W.G. Kraft. Covalent binding of biomolecules to radiation. Grafted hydrogels on inert polymer surfaces. Trans. Am. Soc. Artif. Int. Organs, 18:10-17 (1972).
6. A.S. Hoffman, B.D. Ratner, A.M. Garfinkle, L.O. Raynolds, T.A. Horbett, S.R. Hanson. The small diameter vascular graft: a biomaterial challenge. In: Polymers in Medicine II. Biomedical and Pharmaceutical Applications. E. Chielllini, P. Giusti, C. Migliaresi, L. Nicolais (Eds.), Plenum Press, New York, 157-173 (1985).
7. G.H. Engbers, J. Feijen. Current techniques to improve the blood compatibility of biomaterial surfaces. Int. J. Artif. Org., 14(4):199-215 (1991).
8. C.H. Bamford, K.G. Al-Lamee. Chemical methods for improving the haemocompatibility of synthetic polymers. Clinical Materials, 10:243-261 (1991).
9. S. Wu. Polymer interface and adhesion. Marcel Dekker Inc., New York (1982).
10. F.M. Fowkes. Acid-base interactions. In: Encyclopedia of Polymer Science and Engineering Supplement. H.F. Mark, N.M. Bikales, C.G. Overberger, G. Menges, J.I. Kroschowitz (Eds.), John Wiley & Sons, New York, 1-11 (1988).
11. B.D. Ratner, S.C. Yoon. Polyurethane surfaces: solvent and temperature induced structural rearrangements". In: Polymer Surface Dunamics. J.D. Andrade (Ed.), Plenum Press, New York, 137-152 (1988).
12. F. Garbassi, M. Morra, E. Occhiello, L. Barino. R. Scordamaglia. Dynamics of macromolecules: a challenge for surface analysis. Surf. Interface. Anal., 14:585-589 (1989).
13. G.A. Somorjai. Modern concepts in surface science and heterogeneous catalysis. J. Phys. Chem., 94:1013-1023 (1990).
14. G.A. Somorjai. The flexible surface. Correlation between reactivity and restructuring ability. Langmuir, 7(12):3176-3182 (1991)
15. H.J. Griesser, R.C. Chatelier. Surface characterisation of plasma polymers from amine, amide and alcohol monomers. J. Appl. Polym. Sci. Appl. Polym. Symp., 46:361-384 (1990).
16. B.D. Ratner. Plasma deposition of organic thin films-control of film chemistry. ACS Polym. Prepr., 34(1):643-644 (1993).
17. G.P. Lopez, B.D. Ratner. Substrate temperature effects on film chemistry in plasma deposition of organics: II. Polymerizable precursors. J. Polym. Sci., Polym. Chem. Ed., 30:2415-2425 (1992).
18. G.S. Ferguson, M.K. Chandhury, H.S. Biebuyck, G.M. Whitesides. Monolayers on disordered substrates: self-assembly of alkyltrichlorosilanes on surface-modified polyethylene and poly(dimethylsiloxane). Macromolecules, 26:5870-5875 (1993).
19. F. Sum, D.W. Grainger, D.G. Castner. Ultrathin self-assembled polymeric films on solid surfaces. III. Influence of acrylate dithioalkyl side chain length on polymeric monolayer formation on gold. J. Vac. Sci. Technol. A., 12(4):2499-2506 (1994).

20.B.D. Ratner. Charactherisation of biomaterial surfaces. Cardiovasc. Pathol., 2 suppl. (3):875-1005 (1993).
21.K.B. Lewis, B.D. Ratner. Observation of surface restructuring of polymers using ESCA. J. Coll. Interf. Sci., 158:77-85 (1993).
22.A. Chilkoti, B.D. Ratner, D. Briggs. Static secondary ion mass spectrometric investigation of the surface chemistry of organic plasma-deposited films created from oxygen-containing precursors. 3. Multivariate statistical modeling. Anal. Chem., 65:1736-1745 (1993).
23.M. Nocentini, R.Barbucci. Fourier Transform Attenuated Total Reflection Infrared Spectroscopy (ATR/FT-IR).In:Test Procedures for the Blood Compatibility of Biomaterials. S. Dawids (Ed.), Kluwer Academic Publishers (NL), 151-170 (1993).
24.A. Magnani, R.Barbucci. Fourier Transform Attenuated Total Reflection Infrared Spectroscopy: application to proteins adsorption studies.In: Test Procedures for the Blood Compatibility of Biomaterials. S. Dawids (Ed.), Kluwer Academic Publishers (NL), 171-184 (1993).
25.A. Magnani, R.Barbucci, K.B. Lewis, D. Leach-Scampavia, B.D. Ratner. Surface properties and restructuring of a crosslinked polyurethane-poly(amido-amine) network. J. Mater. Chemistry, 5:1321-1330 (1995).
26.A.S.G. Curtis, S. Britland. Surface modification of biomaterials by topographic and chemical patterning. In: Advanced Biomaterials in Biomedical Engineering and Drug Delivery Sistems. N. Ogata, S.W. Kim, J. Feijen, T. Okano (Eds.), Springer-Verlag, Tokio, 158-167 (1996).
27.A.S.G. Curtis, C.D.W. Wilkinson. Topography control of cell migration. Motion analysis of living cells. D.R. Soll, D. Wessels (Eds.), 7:141-155 (1988).
28.B. Wojciak-Stothard, Z. Madeja, W. Korohoda, A.S.G. Curtis, C.D.W. Wilkinson. Activation of macrophage-like cells by multiple grooved substrata. Topographical control of cell behaviour. Cell Biology International, 19(6):485-490 (1995).
29.B. Wojciak-Stothard, A.S.G. Curtis, W. Monaghan, K. Macdonald, C.D.W. Wilkinson. Guidance and activation of murine macrophages by nanometric scale topography. Experimental Cell Research, 223:426-435 (1996).
30.B. Woiciak-Stothard. M. Denyer, M. Mishra. R.A. Brown. Adhesion, orientation, and movement of cells cultured on ultrathin fibronectin fibers. In vitro-cell. Dev. Biol. Animal, 33:110-117 (1997).
31.R. Singhvi, A. Kumar, G.P. Lopex, G.N. Stephanopoulos, D.I.C. Wang, G.M.. Whitesides, D.E. Ingber. Engineering cell shape and function. Science, 264:696-698 (1994).
32.T. Okano, K. Suzuki, N. Yui, Y. Sokurai, S. Nakahama. Prevention of changes in platelet cytoplasmic free calcium levels by interaction with 2-hydroxyethyl methacrylate/styrene block copolymer surfaces. J. Biomed. Mater. Res., 27:1519-1525 (1993).

33.A.F. von Recum, T.G. van Kooen. The influence of micro-topography on cellular response and the implications for silicone implants. J. Biomater. Sci. Polymer. Edu., 7(2):181-198 (1995).
34.A. Kumar, H.A. Biebuyck, G.M. Whitesides. Patterning self-assembled monolayers: application in material science. Langmuir, 10(5):1498-1511 (1994).
35.D.J. Prithchard, H. Morgan, J.M. Cooper. Patterning and regeneration of surfaces with antibodies. Anal. Chem. 67(19):3605-3607 (1995).
36.L.F. Rozsnyai, M.S. Wrighton. Selective deposition of conducting polymers via monolayer photopatterning. Langmuir, 11(10):3913-3920 (1995).
37.J.P. Ranieri, R. Bellamkonda, J. Jacob, T.G. Vargo, J.A. Gardella, P. Albisdrer. Selective neuronal cella attachment to a covalently patterned monoamine on fluorinated ethylene propylene films. J. Biomed. Mater. Res., 27:917-925 (1993).
38.S.K. Bhatia, J.J. Hickman, F.S. Ligler. New approach to producing patterned biomolecular assemblies. J. Am. Chem. Soc., 114:4432-4433 (1992).
39.G. Chen, Y. Ito, Y. Imanishi. A. Magnani, S. Lamponi, R. Barbucci. Photoimmobilization of sulphated hyaluronic acid for antithrombogenicity. Bioconjugate Chemistry, 8:730-734 (1997).
40.D. Hamilton, M. Riehle, R. Rappuoli, W. Monagham, A.S.G. Curtis, R. Barbucci.The dynamics of primary chondrocyte movement in response to topographical and chemical cues: Implications for cartilage. Submitted.
41.B.D. Ratner. The blood compatibility catastrophe. J. Biomed. Mater. Res., 27:283-287 (1993).
42.B.D. Ratner. New ideas in biomaterial science: a path to engineered biomaterials. J. Biomed. Mater. Res., 27:837-850 (1993).
43.J. Piglowski, I. Gangazz, J. Staniszewska-Kus, D. paluch, M. Szymonowicz, A. Konieezny. Influence modification on biological properties of polyethyleneterephthalate. Biomaterials, 15(11):909-920 (1994).
44.A. Albanese, R.Barbucci, J. Belleville, S. Bowry, R. Eloy, H.D. Lemke, L. Sabatini.In vitro biocompatibility evaluation of a heparinizable material (PUPA), based on polyurethane and poly(amido-amine) components.Biomaterials, 15(2):129-136, (1994).
45.A. Magnani, R. Barbucci, L. Montanaro, C.R. Arciola, S. Lamponi. In vitro study of haemocompatibility and effect on bacterial adhesion of a polymeric surface with immobilised heparin and sulphated hyaluronic acid. Submitted
46.R. Barbucci, A. Magnani.Conformation of human plasma proteins at polymer surfaces: the effectiveness of surface heparinisation.Biomaterials, 15(12):955-962 (1994).
47.A. Magnani, A. Albanese, S. Lamponi, R. Barbucci. Blood-interaction performance of differently sulphated hyaluronic acid.Thrombosis Research, 81(3):383-395 (1996).
48.J. Jozefowicz, M. Jozefovicz. Review: Interactions of biospecific functional polymers with blood proteins and cells. Journal of Biomaterials Science: Polymer Edition, 1(3):147-165 (1990).

49.S. Mallik, S.D. Plunkett, P.K. Dhal, R.D. Johnson. D. Pack, d. Shnek, F.H. Arnold. Towards materails for the specific recognition and separation of proteins. New J. Chem., 18:299-304 (1994).
50.J.H. Branker, V.E. Care-Brendel, L.A. Martison, J. Crudele, W.D. Johnston, R.C. Johnson. Neovascularisation of synthetic membranes directed by membrane microarchitecture. J. Biomed. Mater. Res., 29:1517-1524 (1995).
51.J.D. Andrade, H.B. Lee, M.S. John, S.W. Kim, J.R. Hibbs, Trans. Am. Soc. Artif. Inter. Organs, 19:1 (1973)
52.M.S. John, J.D. Andrade. J. Biomed. Mater. Res., 7:509 (1973).
53.S.R. Hanson, L.A. Harker, B.D. Ratner, A.S. Hoffman. J. Lab. Clin. Med., 95:289 (1980).
54.M.N. Godo, M.V. Sefton. Characterization of transient platelet contacts on a polyvinyl alcohol hydrogel by video microscopy. Biomaterials, 20:1117-1126 (1999)
55.A. Magnani, R. Rappuoli, S. Lamponi, R. Barbucci. Novel polysaccharide hydrogels: characterization and properties. PAT99 (in press).
56.J.A. Rowley, G. Madlambayan, D.J. Mooney. Alginate hydrogels as synthetic extracellular matrix materials. Biomaterials, 20:45–53 (1999).
57.The chemistry, biology of Hyaluronan and its derivatives. T.C. Laurent (Ed.), Werner-Green International Series, Portland Press Ltd, London (1998).
58.K. Smetana. Cell biology of hydrogels. Biomaterials, 14:1046-1050 (1993).

Oxide superconductors: a Chemist's View

Giorgio Spinolo, Paolo Ghigna, Umberto Anselmi Tamburini and Giorgio Flor

Dipartimento di Chimica Fisica, Università di Pavia, Viale Taramelli 16, I 27100 - Pavia (Italy).
E-mail: gs@chifis.unipv.it

Oxide superconductors are at the crossroad between physics, chemistry and engineering, but their typical language is the language of solid state physics. It is the aim of this presentation to show that point of view and language of chemistry are valuable tools (in addition to those of solid state physics) for understanding some basic aspects of the science of oxide superconductors and, conversely, that this science gives outstanding challenges to chemistry.

In its guidelines this work is deeply indebted to Roald Hoffmann's presentation of chemist's view of bonding in crystals and surfaces [1]. In this book, Hoffmann notes that *what is most interesting about many of the new solid state materials are their electrical and magnetic properties...* and emphasizes that ... *Chemist must be able to reason intelligently about the electronic structure of the compounds they make in order to understand how these properties may be tuned... This leads to the problem that learning the language necessary for addressing these problems, the language of solid state physics and band theory, is generally not part of the chemist's education.* At the same time, however, he also stresses the importance of the chemist's point of view: *I suspect that physicists don't think that chemists have much to tell them about bonding in the solid state. I would disagree. Chemists have built up a great deal of understanding... From empirical experience and simple theory, chemists have gained much intuitive knowledge of the what, how, and why of molecules bonding together. To put it as provocatively as I can, our friend physicists sometimes know better than we how to calculate the electronic structure of a molecule or a solid, but often they do not understand as well as we do, with all the epistemological complexity of the meaning that 'understanding' can involve.*

This presentation will show through examples that similar considerations can be made for other aspects of the chemical approach to the science of oxide superconductors. Two main areas will be discussed: relationship between properties and structure and synthesis.

Structure/properties relations.

The chemical language is characterized by a concise and comprehensive approach to microscopic objects (atoms, molecules, and ions). In discussing, for instance, a water solution of sodium carbonate, a great deal of chemical knowledge drives us to focus attention to the carbonate anion CO_3^{2-}. With this simple formula we do not mean the naked ion, we know very well that the solvent is somehow organized around it in a complex microscopic structure, as shown by theory and experiments. All this complexity is implicitly contained in the CO_3^{2-} symbol and we are allowed by chemical thermodynamics to write chemical reactions relating this with other microscopic objects { HCO_3^- , H_3O^+ , ...}, to assign in some way a chemical potential to each of these and to discuss the chemical equilibrium between them. Then, using more o less approximate assumptions (ideal behavior, Debye-Hückel laws, ...) on the relations between chemical potential and amount of these objects and between these amounts and some property of the solution, it is possible to *explain* and *predict* the behavior of the complex system under different external conditions.

Going on with the same example, the elementary theory of electrolyte solutions shows how to solve the chemical equilibrium equations to obtain the ion concentrations under different conditions (temperature, total quantity of dissolved carbonate, ...) and how to calculate the electrical conductivity of a sodium carbonate solution from these ion concentrations and from specific properties of the microscopic objects (the so-called infinite-dilution ionic conductivities). Making reference to these or to other microscopic objects, it is also possible to investigate the validity of the approximations, to think intelligently about the limits of the underlying theories, and to proceed with more refined assumptions and theories.

Solid state chemistry shares with fluid phase chemistry much of these ideas. We start with the geometrical arrangement of atoms or ions investigated by crystallography: this is the *reference* structure. Coupled to this, there is the electronic structure, which is addressed with the language of solid state physics and band theory. As with the simple MO pictures of bonding in atoms and molecules, many-electrons states are typically build up from sets of one-electron states and from the exclusion principle, with more or less complete neglect of electron-electron interactions. These are the aspects addressed by Hoffmann's book.

The crystallographic and electronic structures are zero - Kelvin ideal structures: real crystals are characterized by *defects*. For a large class of problems, and for the aim of this presentation, the relevant defects are those which are able to reach equilibrium with each other and with the surrounding environment and it is useful to classify them into two sets: point defects and electronic defects.

To illustrate meaning and properties of these defects, let us first consider the particular case of a binary insulating oxide with a simple crystal structure with only two sets of crystallographically equivalent sites (one for cations, the other one for anions). Without going into deeper details, the assumption of insulating behavior means that the relevant feature of the band structure is the presence of a completely filled band (let say the valence band, VB, in the language of solid state physics, but we may say the 'HOMO' *band* if the language of MOs is preferred) and a completely empty band at energies just above it (the conduction band, CB, or the 'LUMO' band). When the temperature is 'high enough' (*i.e.* when kT is of the same order of magnitude as the energy gap between these bands) the oxide behaves as a semiconductor: some electrons are thermally promoted into the CB leaving empty one-electron states in the VB. This process can be written:

$$\text{nil} \rightleftarrows \text{e}' + \text{h}^{\bullet} \tag{1}$$

where the symbols e' and $h^{\bullet}$ stand for an *additional* electron in the CB and an *additional* hole in the VB, respectively and <'> and <˙> are used instead of plus or minus signs to remind that these are charges *relative to the reference* (0 K) *electronic structure*. The reason of using a chemical equation is that a chemical potential can be assigned to these particles, that these chemical potentials follow the same rules as the chemical potentials of other microscopic objects in other chemical reactions {*i.e.* $\mu(\text{e}') + \mu(\text{h}^{\bullet}) = 0$}, that a free energy change can be assigned to this *reaction* and that it is safe to write an equilibrium constant {$K_e = [\text{e}']\cdot[\text{h}^{\bullet}]$ } because it is (here) safe to assume that both $\mu(\text{e}')$ and $\mu(\text{h}^{\bullet})$ are related to [e'] and $[\text{h}^{\bullet}]$ by a logarithmic dependence. The phase behaves as a semiconductor essentially because at higher temperatures the equilibrium constant is higher and more charge carriers become available.

The crystalline phase is also in equilibrium with the surrounding atmosphere. What is most important in an oxide material is obviously equilibrium with external oxygen. The oxygen atoms present in the reference structure as oxide anions can escape from the crystal phase into the gas phase and conversely oxygen atoms from molecular oxygen can enter the crystal structure. This gives rise to two point defects, which are respectively denoted as:

1. (fully ionized) oxygen vacancy: $V_O^{\bullet\bullet}$ and
2. (fully ionized) oxygen interstitial: O_i''

In this notation, which is known as Kröger-Vink notation [2],

1. the subscript indicates *where* the defect is (O is a site pertinent to an oxide anion according to the reference structure; i - interstitial - is a position which is not occupied according to the reference structure),
2. the main symbol indicates *what* is placed there (V: a vacancy, nothing; O: an oxygen atom), and
3. the superscript indicates the *difference charge* with respect to the reference structure (in the above symbols, a 2+ charge, <••>, is used because a 2- ion is lacking; a 2- charge, <">, because a 2- ion is added, while an <x> sign is used when necessary to denote a neutral defect).

The formation equations of these defects are then written as:

$$O_O^x \rightleftarrows \frac{1}{2} O_2(\text{gas}) + V_O^{\bullet\bullet} + 2\text{e}' \qquad \text{and} \tag{2}$$

$$\frac{1}{2} O_2(\text{gas}) \rightleftarrows O_i'' + 2\text{h}^{\bullet} \tag{3}$$

If the relations between defect amounts and chemical potentials are reasonably well described by ideal laws (broadly speaking: if point defects are diluted 'enough' and if the chemical potentials of the electronic defects do not lye within a band), one is allowed to write for each of these reactions an equilibrium constant:

$$K_V = \left[V_O^{\bullet\bullet}\right] \cdot \left[e'\right]^2 \cdot \sqrt{P(O_2)} \quad \text{and} \quad K_i = \frac{\left[O_i''\right] \cdot \left[h^\bullet\right]^2}{\sqrt{P(O_2)}} \tag{4}$$

Some notes are required here. A) The reactions between defects are *formal* equations: as with ions in water, the detailed structure of the microscopic object can be very complex and, for instance, some amount of lattice distortion is always implied by the symbol of a point defect. B) A chemical equation involving defects (a quasi-chemical equilibrium) must obey balance rules. There is a balance for each kind of atom and for charge: this closely corresponds to what is required when writing a reaction between ions in water. A less familiar rule is the site balance: the number of *regular* lattice sites of each kind must be the same on the left and right sides of the reaction sign. Site creation or annihilation is allowed, but then the difference between left side and right side *must* correspond to the content of a whole unit cell. C) The concentrations of regular lattice sites do not usually appear in the equilibrium constants: again, this closely corresponds to the usual practice of neglecting solvent concentration when writing a fluid phase equilibrium constant, and is due to the same reason. D) Some degree of non-stoichiometry arises as a consequence of point defect formation: for instance, in the oxygen partial pressures range where eq. (2) predominates, the oxygen amount in the phase is lower than the value corresponding to the stoichiometric coefficient of the reference structure. E) The reactions above have been directly written as formation of twice ionized point defects (interstitials or vacancies): this is by no means mandatory, and also neutral defects or defects with a single (relative) charge can be taken into account, if theory or experiments indicate that it is convenient or necessary to do that. In this case, the above equilibrium constants must be split into products of three distinct equilibrium constants. Again, there is a close relationship with what is usual done in fluid phase equilibria.

In a few cases, the amount of point defects is high enough to make possible a direct investigation with diffraction techniques (which amounts to measure slightly incomplete occupancy factors, for vacancies, or very small occupancy factors, for interstitials) or with a determination of oxygen stoichiometry: in the latter case much better accuracy is obviously obtained when measuring stoichiometry *changes* under different oxygen pressures.

More direct (and usually more reliable) approaches rely on spectroscopic or local structural techniques, when these are exclusively or at least predominantly due to defects. For instance, mono-ionized oxygen vacancies can be investigated with ESR. Particular insights are given at this regard by X-ray absorption techniques (XAS) both for clarifying an oxidation state and probing a local geometric structure.

However, in many areas of the chemistry of imperfect crystals [2, 3, 4], the method of choice is based on electrical conductivity measurements and will be briefly recalled here. In the case above described, if we assume that: a) the electrical conductivity is due only to electrons and holes (*i.e.* there is negligible ionic contribution), b) the contributions of electrons and holes are independent of each other, and c) each contribution is written as the product of a carrier concentration times an almost concentration-independent mobility:

$$\sigma = \sigma(e') + \sigma(h^\bullet); \qquad \sigma(e') = q_e \cdot [e'] \cdot u(e'); \qquad \sigma(h^\bullet) = q_e \cdot [h^\bullet] \cdot u(h^\bullet)$$

the defect equilibria can be investigated using electrical conductivity measurements under constant temperature and variable oxygen partial pressure conditions and solving the defect equations with the required balance conditions. For that purpose, let us assume that the relevant defect equilibria are only those already written (eq. 1-3) plus the anti-Frankel equilibrium:

$$O_O^x \rightleftarrows V_O^{\bullet\bullet} + O_i'' \qquad K_{AF} = [V_O^{\bullet\bullet}] \cdot [O_i''] = K_V \cdot K_i \cdot K_e^2 \qquad (5)$$

which corresponds to a linear combination of the previous reaction (1-3). The charge balance is: $2[V_O^{\bullet\bullet}] + [h^\bullet] = 2[O_i''] + [e']$. When oxygen partial pressure in the external atmosphere is 'low' enough, the predominant equilibrium is (2), the charge balance is well approximated by $2[V_O^{\bullet\bullet}] \cong [e']$, and the equilibrium equations give $[e'] \propto P(O_2)^{-1/6}$ and therefore $\sigma \propto P(O_2)^{-1/6}$. When oxygen partial pressure in the external atmosphere is 'high' enough, the predominant equilibrium is (3), the charge balance becomes $2[h^\bullet] \cong [O_i'']$, and the equilibrium equations give $[h^\bullet] \propto P(O_2)^{1/6}$ and $\sigma \propto P(O_2)^{1/6}$. Finally, in an appropriate intermediate range of oxygen partial pressures, either equilibrium (1) or (5) predominate and the conductivity does not (significantly) depend on this external variable, but the different nature of charge carriers makes possible to further distinguish the two cases with suitable electrochemical measurements.

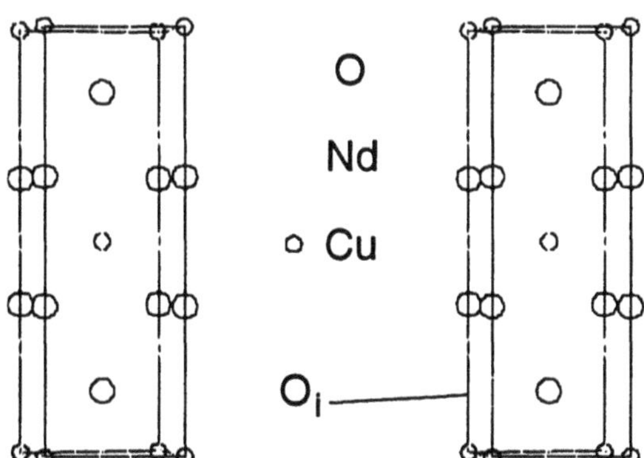

Fig. 1. Reference crystal structure of Nd_2CuO_4 (left part). The right part of the figure shows the position of the interstitial oxygen defect.

Now, the thermodynamic variable that drives the onset of superconductivity in oxide superconductors is the amount of electronic defects (holes or, in a particular family, electrons). With respect to the reference structure and stoichiometry, this amount can be modified by [*intrinsic*] doping with foreign atoms and by interaction with external oxygen [*extrinsic* doping]. In either case, point defects are involved and a quasi-chemical equilibrium is established between atomic and electronic defects.

Understanding these equilibria is an obvious prerequisite for understanding how the amount of holes (or electrons) can be tuned and superconductivity is achieved.

Oxide superconductors clearly represent a much more complex situation than a simple oxide. Cases involving, for instance, defects on cation sites, or systems with more than two independent components can be treated, however, along the same general guidelines with appropriate modifications. The pertinent details are not discussed here and interested readers are advised to refer to general treatises on the subject [2, 3, and 4]. Instead, let us discuss Nd_2CuO_4, which is the parent material of a whole family of oxide superconductors.

Pure Nd_2CuO_4 is an insulating oxide and will be discussed, for simplicity, in close analogy with the previous analysis of simple oxides in spite of its nature of ternary compound. The reference structure of Nd_2CuO_4 is shown on Fig. 1.

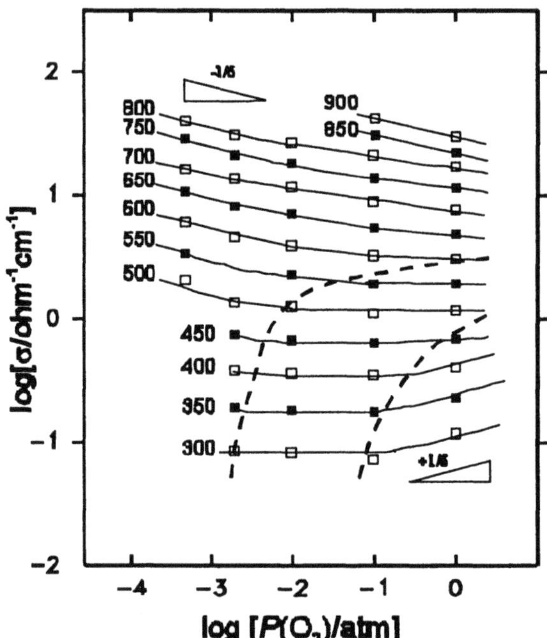

Fig. 2. Electrical conductivity of $Nd_2CuO_{4\pm y}$ as a function of oxygen partial pressure at different temperatures (°C: labels). The dashed lines roughly mark the transitions from one predominant equilibrium to another one. In the upper left region, the main point defects are oxygen vacancies [equilibrium (2)]; in the lower right region, the main point defects are oxygen interstitials [equilibrium (3)]; in the intermediate region, equilibrium (1) or equilibrium (5) predominate.

Fig. 2 clearly shows that the three main trends of the electrical conductivity of $Nd_2CuO_{4\pm y}$ *vs* $P(O_2)$ are indeed those predicted by the model based on the simple equilibria (eq. 1, 2, 3, and 5). By the way, this is what makes Nd_2CuO_4 a nice example for introducing the defect chemistry of oxide superconductors.

The above understanding of defect chemistry of the Nd_2CuO_4 phase can be further assessed with other experimental techniques. For instance, the charge of the carriers can be inferred from thermoelectric power measurements. The presence of oxygen interstitials in Nd_2CuO_4 quenched to room temperature from high $P(O_2)$ - intermediate T conditions can be directly proved [5] by EXAFS (see Fig. 3): the additional oxygen atoms occupy a normally empty apical position above copper (see right part of Fig. 1) and the EXAFS determination of their amount nicely corresponds with the value given by oxygen over-stoichiometry measurements.

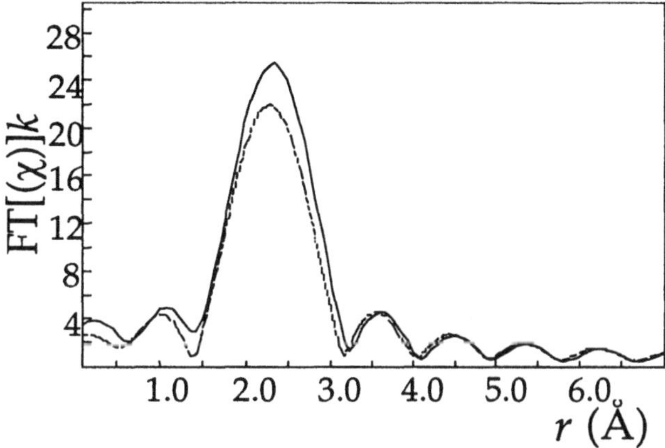

Fig. 3. Fourier Transform of the EXAFS Difference between a stoichiometric and a p-doped (oxygen over-stoichiometric) Nd_2CuO_4 sample. The peak at ~ 2 Å is due to interstitial oxygen. Dotted and continuous lines are experiment and fit.

A point of some relevance in the following discussion is the nature of the electronic defects that predominate in Nd_2CuO_4 materials quenched from high T - low $P(O_2)$ conditions. This can be inferred from their mobility, which is in turn obtained from conductivity and non-stoichiometry. Fig. 4 shows that electron-doped oxygen under-stoichiometric Nd_2CuO_4 follows an Arrhenius law in a large temperature range, with an activation energy which decreases significantly at the lowest temperatures. This is the typical behavior of a particular defect which is known as small polaron and is related to strong electron localization coupled to a local lattice distortion: following a famous statement due to Mott, a polaron is *an electron which digs its own potential well.* The contribution of a small polaron to the transport properties of the phase is due to an activated jump of the defect (both the charge and the lattice distortion associated to it).

Perhaps, the nature of this defect is better understood in chemical words: simply stated, a small polaron is an ion with a lower charge than required by the reference structure. In our case, XAS clearly shows [6] that the extra electrons are indeed localized on copper sites, formally giving rise to a Cu(I) defect where the reference structure wants a Cu(II) ion. A local lattice distortion where a Cu(I) replaces a

Cu(II) ion is precisely what must be expected on the basis of simple chemical knowledge, and the localization can be written as a chemical reaction:

$$\mathrm{Cu}_{\mathrm{Cu}}^{\mathrm{x}} + e' \rightleftarrows \mathrm{Cu}_{\mathrm{Cu}}^{'} \tag{6}$$

As a further assessment, a more detailed analysis of the transport properties of these polarons indicates that the jump distance of the defect is around 0.39 nm, a value which is in close agreement with the Cu-Cu distance in the basal planes of the crystallographic structure of Nd_2CuO_4.

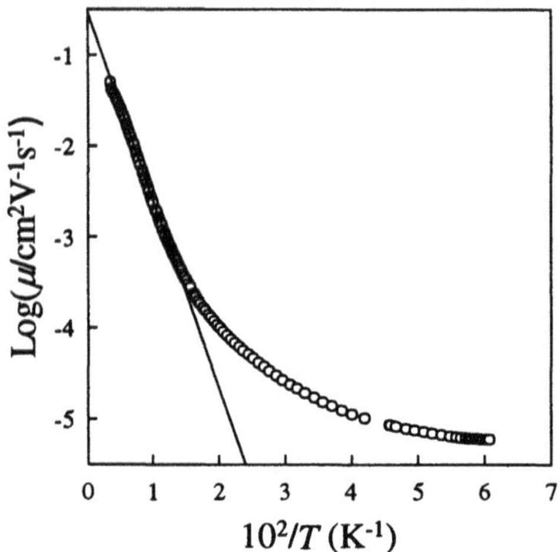

Fig. 4. Mobility (μ) of electronic carriers in (n-doped) Nd_2CuO_4 between 15 and 300 K. The solid line is an Arrhenius least squares fit of the experimental data above 80 K. At lower T the apparent activation energy vanishes. This behavior is indicative of small polaron charge transport.

Now, the most known superconductor of the Nd_2CuO_4 family is $Nd_{2-x}Ce_xCuO_4$ [7]. In the whole family, the critical transition temperature is at most 25 K, so that what makes so interesting this class of materials is not an outstanding performance as superconductor, but a number of features that simplify basic studies and approach to technologically more interesting materials. First of all is the simple crystal structure with only four crystallographically non-equivalent atoms in the primitive unit cell. A low number of orbital functions are therefore needed to grasp the basic features of the band structure. As a matter of fact, there is reasonable evidence that a single band is present at the Fermi level in doped materials. Also, the chemical synthesis is straightforward and does not show the formation of impurity phases that so frequently hinder the investigation of other high critical temperature superconductors. Moreover, this is presently the only family of oxide superconductors where the charge carriers are electrons instead of holes. Finally, the onset of superconductivity is not coupled here to a structural phase transition, as occurs in p-type (hole doped) families of superconductors.

Actually, two doping mechanisms are required for the onset of superconductivity in $Nd_{2-x}Ce_xCuO_4$. The first one is Ce doping: superconductivity is achieved when x is between 0.12 and 0.18, and the best performance corresponds to $x = 0.15$. This doping is directly performed during high temperatures synthesis by adding the appropriate amount of CeO_2 to the Nd_2O_3 and CuO main reagents. As a matter of fact, this gives an over-stoichiometric material so that a further annealing under low oxygen pressure is required to remove some of this additional oxygen before quenching to room temperature, where the diffusion coefficient of oxygen is low enough to leave its content practically unmodified.

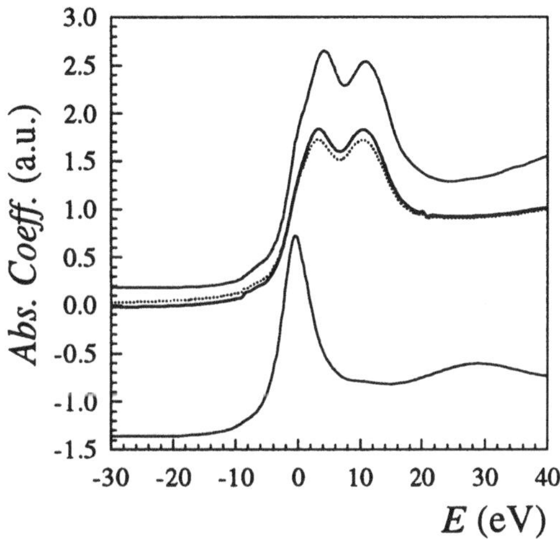

Fig. 5. Top to bottom: X-ray Absorption spectra of a CeO_2 standard for Ce(IV), a neodymium copper oxide sample with $x = 0.05$ (dots) and another one with $x = 0.15$ (continuous line), and a Ce(III) nitrate standard. The similarity with CeO_2 is clearly apparent.

A first point to be investigated on $Nd_{2-x}Ce_xCuO_4$ materials is the nature of point defects associated with Ce doping. From a chemist's point of view, there is little doubt that cerium enters the crystal structure as *substitutional* defect on neodymium sites. This has been directly proved by Ce-L and (more recently) Ce-K EXAFS determinations [8-9], which also show that Nd / Ce substitution produces a local lattice distortion in form of a collapse of the O_8 coordination cuboid. Perhaps more interesting is the valence state of the defect. On this regard, XAS spectroscopy (fig. 5) shows that Ce is present as Ce(IV), so that Ce doping formally corresponds to injecting electrons into the electronic structure of the reference material. Formally, this is described by:

$$Nd_{Nd}^{x} + Ce \rightarrow Nd + Ce_{Nd}^{\bullet} + e',$$

but, if one prefers to follow more closely the reactions sequence occurring during synthesis, annealing, and quenching, the electron injection can be described by the two reactions:

$$2CeO_2 + CuO \rightarrow Cu_{Cu}^{x} + 2Ce_{Nd}^{\bullet} + 4O_{O}^{x} + O_{i}'' \tag{7}$$

$$O_i'' \rightleftarrows \frac{1}{2}O_2(gas) + 2e' \tag{8}$$

On simple chemical grounds, one can foresee also some kind of interaction between the oppositely charged defects $Ce_{Nd}^{\bullet}$ and O_i'': Ce-K edge EXAFS indeed shows some evidence of formation of mono-ionized $[Ce_{Nd} - O_i]'$ and neutral $[O_i - Ce_{Nd} - O_i]^x$ complex defects [9].

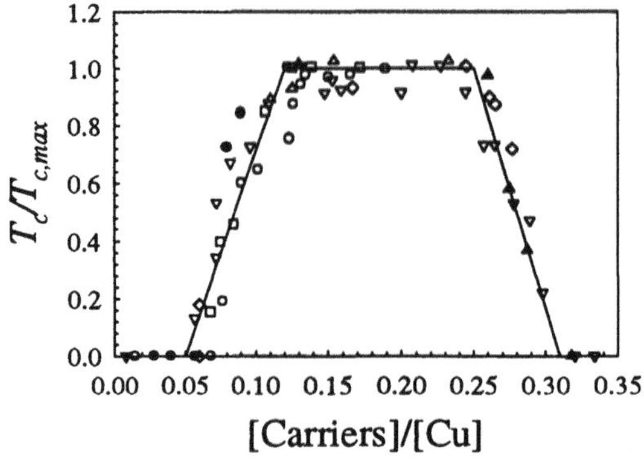

Fig. 6. 'Universal relationship' between T_c and carrier content. Different symbols indicate superconducting oxides of different families: filled circles are for electron-doped $Nd_{2-x}Ce_xCuO_{4\pm\delta}$, the other symbols are for hole-doped materials.

According to eq. (7-8), the amount of injected electrons can be obtained from the amount of Ce dopant and O overstoichiometry if other oxygen defects can be neglected. For hole-doped superconductors, Zhang and Sato [10] empirically discovered a relationship between onset of superconductivity and amount of carrier. Fig. 6 shows that the same relationship holds also for electron-doped superconductors.

The onset of superconductivity corresponds to a marked change of the nature of the predominant charge carriers (see Fig. 7). The thermally activated mobility of the injected electrons in Nd_2CuO_4 materials quenched from high T - low $P(O_2)$ conditions changes to metallic behavior in materials that are at the same time heavily doped with Ce ($x \sim 0.15$) and almost exactly oxygen stoichiometric. The small polaron model no more holds here and the transport properties of these much more concentrated carriers are better described with reference to the large polaron model [11]. In this new kind of electronic defect, the electron is not so strictly localized on a particular site and behaves in some ways as a truly free electron occupying a completely delocalized state in a band, except for a larger effective mass. A well known explanation of superconductivity in these complex oxide materials is based on the Bose-Einstein condensation of integer - spin particles produced by strong coupling of two electrons. Then, a nice support to this line of thinking is given by the good evidence, in $Nd_{2-x}Ce_xCuO_{4\pm y}$ materials, of a strong interaction between carriers well above the critical transition temperature of superconductivity [12].

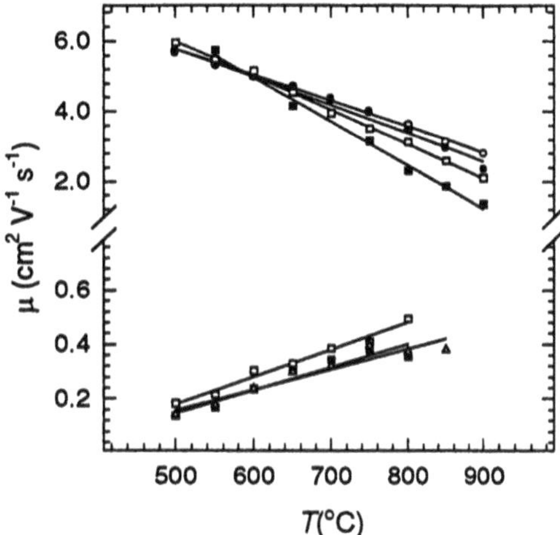

Fig. 7. Carrier mobility in Ce-undoped ($x = 0$) and Ce-doped ($x = 0.15$) neodymium-copper oxide

As said, in other families of oxide superconductors the unit cell contains a larger number of non-equivalent atoms, more complex schemes of doping with foreign cations are involved, and the presence of several electronic bands at the Fermi level makes much more difficult to reach a reliable interpretation of the experimental behavior. We think that also in these cases the chemical point of view provides useful guidelines for reaching a better understanding.

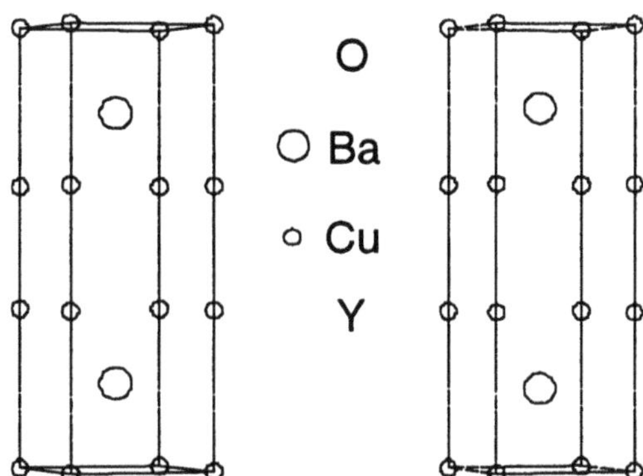

Fig. 8. Crystal structures of tetragonal (left) and orthorhombic-I YBCO ($YBa_2Cu_3O_{6+\delta}$).

In the so called YBCO-123 family, two main crystal structures are involved (see fig. 8): the first one with $YBa_2Cu_3O_6$ stoichiometry and tetragonal symmetry, the other one with $YBa_2Cu_3O_7$ stoichiometry and orthorhombic symmetry. Important building blocks of these structures are the two CuO_2 planes which are placed at ~ 1/3 and ~ 2/3 along the *c* axis and are separated from each other by 'ionic' YO and BaO layers. The additional oxygen atoms of the orthorhombic structure are arranged into other CuO chains along the *b* crystallographic axis. Tetragonal $YBa_2Cu_3O_6$ formally contains two Cu(II) and one Cu(I), while orthorhombic $YBa_2Cu_3O_7$ formally contains two Cu(II) and one Cu(III). The onset of superconductivity is typically achieved with the orthorhombic phase and with an intermediate oxygen content. Above T_c, superconducting materials show metallic behavior, and the itinerant carriers are holes. For this reason, it is easier to use here tetragonal $YBa_2Cu_3O_6$ as the reference structure. Consequently, oxygen doping is typically referred to as a hole injecting process.

Hole injection is not faithfully described by the single chemical reaction (3). Actually, many different but related processes occur along the chemical coordinate corresponding to hole injection. To make the situation even more complex, different materials of the 123 family (phases where Y is partially or totally replaced by Sm, Nd, Yb, Pr, Eu, ...) show different behavior.

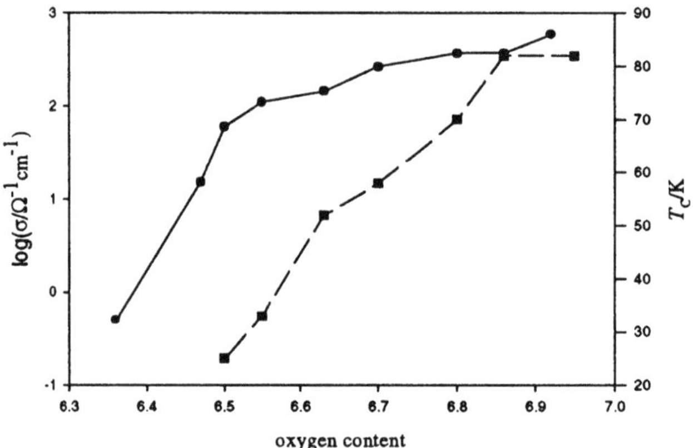

Fig. 9. Sm-123 samples with various oxygen contents: left scale and circles: conductivity (at 200 K), right scale and squares: critical transition temperature for superconductivity.

On one side, while the very first additional oxygen atoms are introduced into the crystal structure in a seemingly random way, some ordering mechanism quickly appears. The arrangement of the right part of Fig. 6 is only one of these, and other ordered structures have been reported. On the other side, the newly injected holes occupy different states as oxygen doping proceeds [13]. A first difference is between states which are strongly localized on particular sites of the reference structure and more or less delocalized (itinerant) states. Another difference is the 'character' of the hole, that is atom and orbital most contributing to the state actually occupied by the hole. A clear indication of the strongly varying transport properties of the injected holes is given for instance by Fig. 9 where the conductivity (at 200 K) of a Sm-123 sample is plotted as a function of oxygen content (*x* in $SmBa_2Cu_3O_{6+x}$). The same figure also reports (right scale) the critical transition temperatures, and provides a nice indication of the relation between onset of superconductivity and injection of mobile holes.

As the figure shows, the onset of superconductivity at $x \sim 6.5$ is related to the presence of holes that are much more mobile (note the logarithmic scale) than those injected at lower oxygen contents.

Actually, more than one injection mechanism is working within a certain range of doping. Much work is in progress on this subject, and much work has clearly to be done. The importance of the chemist's view to understand and predict electronic states has been well illustrated by Hoffmann and will not be repeated: it is sufficient here to show that even with a strongly simplified scheme of hole injection sequence is indeed possible to gain a rational understanding of the hole mobility of a 123 superconductor above T_c [14].

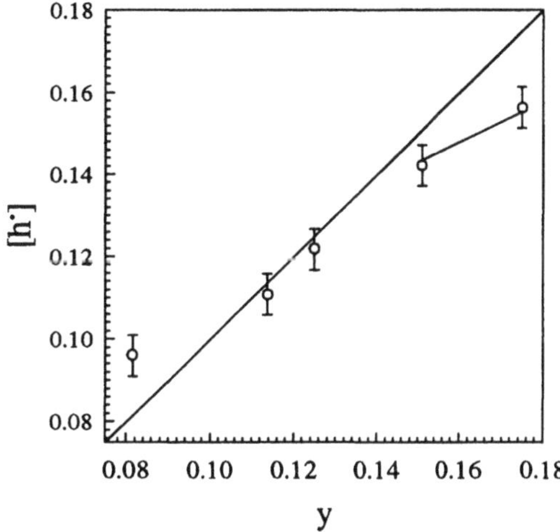

Fig. 10. Hole density *vs* oxygen excess in the BSCCO 2212 superconductor. The expected trend (diagonal line) corresponds to two holes per added oxygen, and is followed only in an intermediate doping range; injected electrons are present also in un-doped materials (y ~ 0), while fewer holes than expected are injected in heavily doped materials.

As a final illustration of the complexity of hole injection mechanisms in oxide superconductors and ability of the chemical point of view to provide a reliable understanding, we here quote some results for the so called BSCCO-2212 superconductor (a phase with reference $Bi_2Sr_2CaCu_2O_8$ stoichiometry). The investigation of this material has recently shown [15] that the amount of holes per copper atom does not vary as a function of extra added oxygen atoms (y) following the trend expected on the basis of defect equilibrium (eq. 3), which corresponds to two injected holes per added oxygen. The expected trend is followed only in an intermediate doping range; injected electrons are present also in undoped materials ($y \sim 0$), but less than two holes per added oxygen are injected in heavily doped materials (Fig. 10).

In these materials, a charge transfer process splits a Cu(II)-O band into two half bands. At $y = 0$, the upper half band is completely empty, but the lower half band is not completely filled because of the presence of a much narrower (more strongly localized) Bi-O band (left part of Fig. 11) which slightly overlaps and therefore is able to act as a source of more mobile holes. The process can be written in form of an internal redox reaction:

$$Bi^{3+} + Cu^{2+} \rightleftarrows Bi^{2+} + Cu^{3+} \quad \text{or} \quad Bi_{Bi}^{\times} + Cu_{Cu}^{\times} \rightleftarrows Bi_{Bi}^{'} + Cu_{Cu}^{\bullet}$$

with the additional specification that an electron in the Bi-O band is much more localized than a hole in the lower Cu-O half band. This amounts to an additional mechanism for the injection of itinerant holes and explains why hole doping is positive also at $y = 0$. In the intermediate range ($y \sim 0.12$), the amount of mobile holes in the lower Cu-O band is higher because of the increasing contribution of additional oxygen atoms: the atomic like states of these anions fall well below the Fermi level and therefore remove electrons from the Cu-O band (middle part of the figure). In the highest range of oxygen doping ($y > 0.14$), the interstitial oxygen atoms occupy sites with increasing energy (a feature which is reasonably explained on the basis of an increased Madelung potential) and the Fermi level is even more displaced to the bottom. Therefore, the more localized states related to the oxygen interstitials approach the Fermi level and each additional oxygen contributes on average with less than two holes to the overall density of mobile holes (right part of the figure.

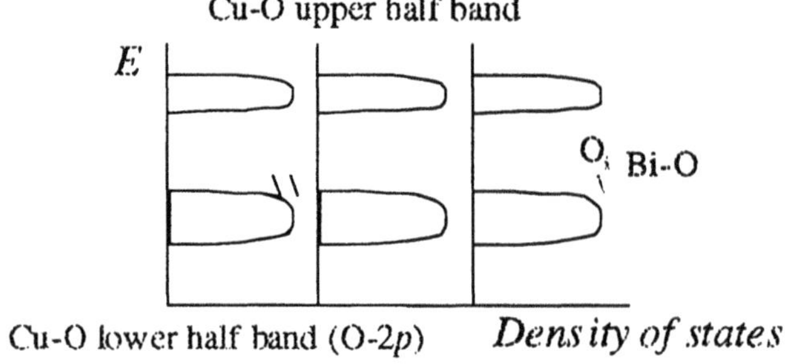

Fig. 11. Band picture of hole injection sequence in the BSCCO-2212 superconductor.

Synthesis

A solid state synthesis is typically performed by mixing into a pressed pellet a suitable heterogeneous mixture of reagent powders and submitting the pellet to several heating cycles under appropriate external conditions (T, oxygen partial pressure) with various intermediate grinding and re-pressing steps. A dramatic difference from typical chemical reactions in a fluid phase is the heterogeneous nature of these processes. The reagents are spatially separated from each other and the process is at the same time formation of a new chemical species (a chemical reaction) and a growth of a crystal structure (a phase transformation).

Notable features of this class of chemical reactions are that the reaction occurs at the boundary between *two* reagent grains, and that for each reacting couple at least one reagent *must* travel through the product to reach the other reagent (Fig. 12). After nucleation of the new phase, its growth is controlled by the movement of the two new interfaces (*i.e.* by the jump frequencies of the pertinent structural elements across these interfaces) or by energy (heat of reaction) transport or else by diffusion of the structural elements through the product phase. Typically, the last process becomes rate controlling when the product layer becomes thick 'enough'.

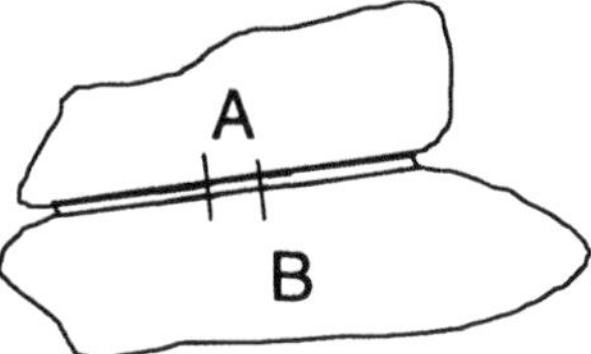

Fig. 12. Usually, a solid state chemical reaction is also a phase transformation. The product phase (black layer) is formed at the interface between the parent phases (A and B). At least one reagent must travel through the product phase to react.

In the theoretical and experimental approaches to solid state reactivity a key role is played by the geometrical setup usually referred to as chemical diffusion couples where *two* reagents in form of single crystals or sintered compacts are placed in contact with each other and allowed to react under isothermal and isobaric conditions. The reference theory of solid state kinetics [2, 3] describes the diffusion controlled growth of a product phase and shows how to investigate its mechanism using diffusion couple experiments. For instance, fig. 13 schematically describes the setup for a diffusion couple investigation of a reaction between two simple oxides (AO and B_2O_3) and the relations between experimental findings and mechanism. In such a solid state reaction, clarifying a mechanism means finding which ions move under the chemical potential gradients, which is the rate determining ion, and how its movement can be controlled using external conditions. For the cases of fig. 13, significant results of the treatment are that a) two stable planar interfaces are formed, b) the thickness (x) of the product layer follows a parabolic time law: $x^2 = k\ t$, c) the reaction rate is controlled by the second fastest ion (by the slowest of the moving ions), d) the practical rate constant (k) is related to the free energy of formation of the product from the reagent phases and to the diffusion coefficient of the rate determining ion.

Many extension of the basic treatment are available (for instance, extensions to cases where more than one product phase is stable, where the two reagents do not form a pseudobinary system, or a more complex reaction geometry is used), but there is no theoretical treatment which adequately predicts the practical reaction rates in a real solid state synthesis.

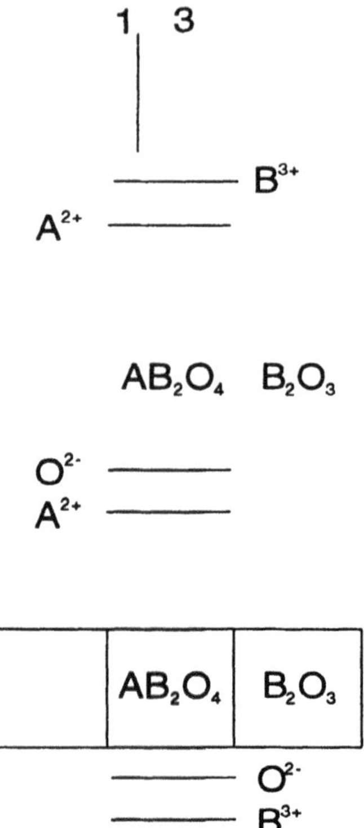

Fig. 13. Solid state reaction between an AO oxide and a B_2O_3 oxide under diffusion-controlled regime. The figures schematically show how the product layer growths under different reaction mechanisms (the arrows mark the position of the initial interface between parent reagents). In each of the three cases, two ions can be rate determining.

The problem is that a real solid state synthesis is not usually made under the simple conditions implied by the various theoretical treatments, but is made by reacting *powder* mixtures, because this process is typically faster by orders of magnitude. Even when reacting a powder mixture of only *two* reagents, the time evolutions of the product layers formed at the interfaces between different couples of reactant grains depend in badly predictable ways from *process* parameters such as initial shapes and sizes of the grains, size distributions, local values of the activity of gaseous components, and from the time evolution of these parameters.

Things are even worse when one considers the various reactions occurring in a powder mixture of *several* reactants: many different phases can then be formed by 'parallel' reactions at each interface between a couple of reagents, different couples should be taken into account at different times (reagent + reagent, intermediate product + reagent, intermediate product + intermediate product, plus whose possibly involving the final product), and a given phase can be produced by different reacting couples (typically with different rates). For all these reasons, a solid state synthesis is usually a very empirical subject and, in particular, the formation of the desired phase is frequently coupled to formation of spurious products.

In our opinion, however, it is possible to reach a rational approach to these important processes. The point to be stressed here is that the aim of a rational approach to solid state kinetics is not only to find an explicit expression for the time evolution of the amount of the product. From a practical point of view, what is important is to gain insights on the simple reactions occurring when processing a complex powder mixture and therefore to understand how these reactions affect formation of the desired final product or formation of spurious phases and how their rates can be controlled acting on the external conditions. In simple words, the important point is to understand how to plan a synthesis. For this seemingly restricted aim, the available theories do indeed give what is required for a rational approach.

To show the guidelines of the suggested approach, let us discuss the case of the formation of $YBa_2Cu_3O_x$. This materials is usually synthesized by reacting a 6:1:4 (by mole) mixture of CuO, Y_2O_3, and $BaCO_3$ or BaO_2 as suitable BaO precursors. Using chemical diffusion experiments on the various couples formed by the stable phases of the pseudoternary Cu, Ba, Y / O diagram, the following set of results have been obtained at $T = 900$ °C [16].

Table 1. Diffusion couple investigation of the reactivity in the Ba, Y, Cu / O system at 900 °C under air.

reaction	reagents	Mobile species	Product(s)	k/cm^{2}s^{-1}
A	$BaCO_3 + CuO$	Cu^{2+}, O^{2-}	$BaCuO_2$	2.6 10^{-11}
B	$Y_2O_3 + CuO$	Cu^{2+}, O^{2-}	$Y_2Cu_2O_5$	2.6 10^{-12}
C	$Y_2O_3 + BaCO_3$	-	-	-
D	$Y_2O_3 + BaCuO_2$	Ba^{2+}, Cu^{2+}, O^{2-}	211	1. 10^{-11}
			211+ 123 + 132	8. 10^{-11}
E	$Y_2Cu_2O_5 + BaCuO_2$	Ba^{2+}, Cu^{2+}, O^{2-}	123	6. 10^{-10}
F	$Y_2Cu_2O_5 + BaCO_3$	Ba^{2+}, (Cu^{2+}), O^{2-}	211	1. 10^{-10}

In the above table, 211, 123 and 132 denote Y_2BaCuO_5, $YBa_2Cu_3O_x$, and $YBa_3Cu_2O_x$, respectively. Reaction C does not form significant amounts of products under the reported external conditions, while for reaction D two product layers can be seen, and one of these actually is a three-phase mixture.

The above findings indicate that $YBa_2Cu_3O_x$ formation is due to the following three solid state reactions:

1. $BaCO_3 + CuO$ • $BaCuO_2$ (reaction A)
2. $Y_2O_3 + CuO$ • $Y_2Cu_2O_5$ (reaction B)
3. $Y_2Cu_2O_5 + 4\ BaCuO_2$ • $2\ YBa_2Cu_3O_{6.5}$ (reaction E)

Now, the fastest reaction between parent phases is reaction A. Therefore, its product ($BaCuO_2$) initially accumulates in the reacting mixture, because reaction B is an order of magnitude slower. In these conditions, the Y_2O_3 reagent is able to react not only with CuO (reaction B) but also with $BaCuO_2$ (reactions D, which is several times faster than reaction B). This results in a depletion of a reagent required for the final reaction of the sequence and formation of a well known spurious phase (211). Moreover, because of the close values of the rates of the competing reactions, conversion degree and effectiveness of the overall process are not controlled by basic variables (diffusion coefficients, free energies, ...), but by process variables (grain sizes, shapes, ...).

It is now clear that the drawback of the empirical synthesis is the *simultaneous* presence of $BaCuO_2$ and Y_2O_3 in the reacting mixture, and this drawback is easily removed if the three reactions of the 'ideal' sequence are *separately* performed as powder reactions in three successive processes. This has been done and the approach has been found successful [16].

The idea of performing an effective synthesis of a complex microscopic structure in several steps is a key idea in synthetic chemistry, and its application to solid state would not surprise. The surprising point, on the contrary, is its still scarce popularity. From the point of view of a modern synthetic organic chemist, most of the actual practice of solid state synthesis perhaps looks like trying to synthesize any complex organic molecule by mixing the corresponding stoichiometric amounts of the elements. Broadly speaking, the crude approach typically applied so far works because in many areas of solid state synthesis simple crystal structures and systems close to thermodynamic equilibrium are involved. Oxide superconductors are usually based on fairly more complex crystal structures, and a chemically more sensible approach is a real need.

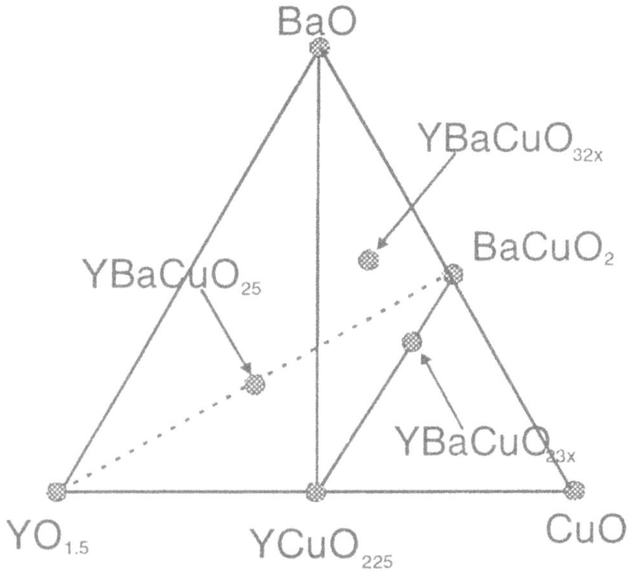

Fig. 14. Schematic phase diagram of the Y, Ba, Cu / O pseudo-ternary system. Full lines indicate the investigated reaction paths which correspond to pseudobinary cuts, the dashed lines is not a pseudobinary cut.

Coming back to the guidelines of the rational approach to synthesis of oxide superconductors, we note that all the above reactions for YBCO-123 formation are better discussed with reference to the equilibrium phase diagram of Fig. 14. In this kind of diagrams, a simple solid state reaction is implicitly described by a line connecting the two reagents, because the line describes a stoichiometric constraint.

Such a line is said to be a pseudobinary cut of the phase diagram when a phase lying along the line is thermodynamically stable with respect to a *mixture* (with the same overall composition) of phases lying on either sides of the line. For instance, the lines corresponding to reactions A, B, C, E, and F are pseudobinary cuts. The line between Y_2O_3 and $BaCuO_2$ is not a pseudobinary cut, as shown by formation of a mixture of phases (123+132) lying on either side of the line when reacting Y_2O_3 and $BaCuO_2$.

Another example of the effectiveness of this approach is given by the synthesis of the BSCCO-2212 superconductor (which is usually referred to as $Bi_2Sr_2CaCu_2O_8$, as said). A chemical diffusion study [17] of the Bi, Sr, Ca, Cu / O system under particular external conditions (T = 750 °C, 2:2:1:2 molecularity, carbonate precursors) shows that a seemingly promising sequence of reactions is:

1. $Bi_2O_3 + CaCO_3$ • $CaBi_2O_4$
2. $SrCO_3 + CuO$ • $SrCuO_2$
3. $CaBi_2O_4 + 2\ SrCuO_2$ • $Bi_2Sr_2CaCu_2O_8$

(using somewhat simplified stoichiometry for some phases).

However, the last reaction is shown ineffective by chemical diffusion experiments: a complex mixture of products is actually formed and no simple way for obtaining a single phase $Bi_2Sr_2CaCu_2O_8$ product has been found by changing the external conditions.

We now understand [18] that the most critical point in finding the desired reaction sequence towards a pure 2212 phase is the molecularity of the final product. Fig. 15 shows part of the composition tetrahedron of the system with the stability field of the phase and some lines investigated with diffusion couple experiments. The figure makes clear that the above reaction 3) is not effective because it does not cross the actual stability field of the final product, which does not enclose the ideal 2:2:1:2 composition point. Formation of a multi-phase mixture is then a consequence of thermodynamics. An effective manner of obtaining the desired superconductor is, for instance, the sequence ending with the reaction between a ternary (Sr, Ca, Bi / O) phase with monoclinic structure (TM in the figure) and the binary oxide $Sr_{14}Cu_{24}O_{41}$ (C in the figure).

To summarize the whole line of thinking, it is possible to gain a rational understanding of the whole set of chemical reactions occurring at each interface when processing of a complex powder mixture and to plan a rational solid state synthesis of a phase pure material.

Planning this synthesis corresponds to finding a particular sequence of solid state reactions. Each reaction of the sequence is performed separately from the others by reacting powder mixtures under the most appropriate external conditions (which concerns both basic variables: temperature, partial pressures,..., and processes variables: grain sizes, ...).

Each step *must* correspond to a pseudo-binary cut of the equilibrium phase diagram (under the specified conditions), so that only a single phase product is eventually formed: when the reaction does not correspond to a pseudo-binary cut, this cannot occur. The product of each step is either a candidate reactant of a further step or the desired final product.

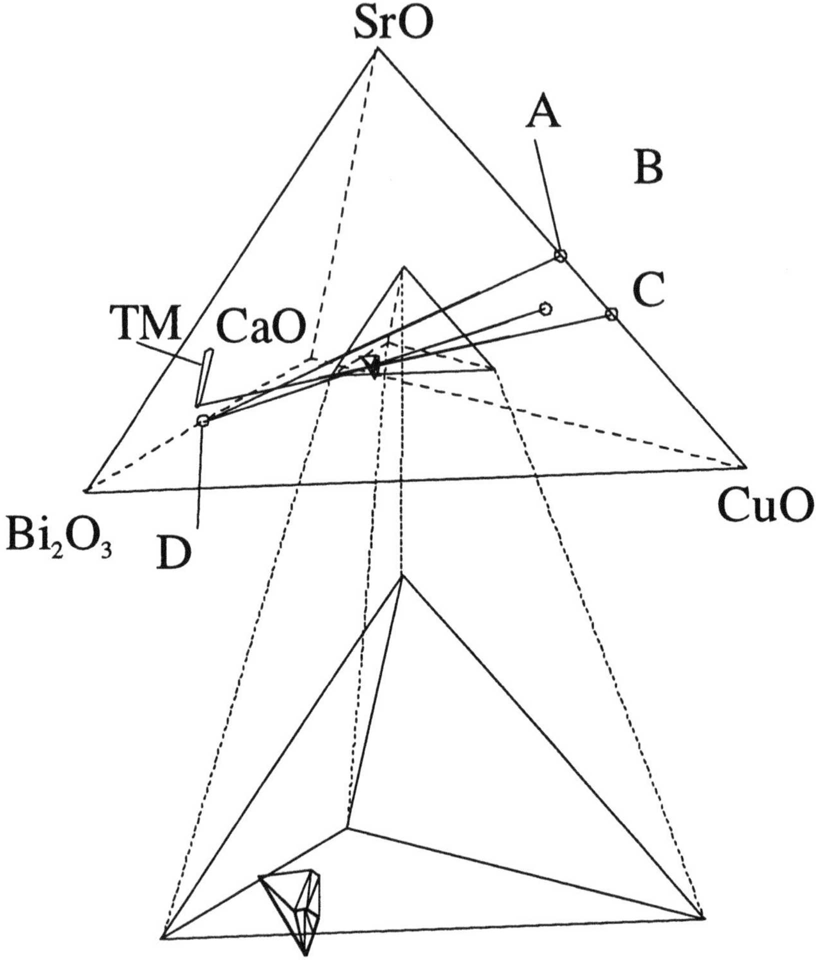

Fig. 15. Schematic phase diagram of the Bi,Sr,Ca,Cu/ O pseudo-quaternary system. The lower diagram shows the inner part of the whole diagram on an enlarged scale. TM is a ternary Bi,Sr,Ca/O phase with monoclinic structure, A is $SrCuO_2$, B is $Ca_{0.25}Sr_{0.75}CuO_2$, C is $Sr_{14}Cu_{24}O_{41}$, and D is $CaBi_2O_4$. The irregular polyhedron schematically indicates the actual stability field of the so-called "2212" phase under 1 atm oxygen pressure and 750 °C

Chemical diffusion experiments are a powerful tool for an overall understanding of the chemical reactivity in a multi-component system and for finding the desired path, and provide the required information about the feasibility (thermodynamics and kinetics) of each step, and therefore of the whole sequence.

References

[1] Roald Hoffmann, *Solids and Surfaces - A Chemist's View of Bonding in Extended Structures*, **1988**, VCH Publishers, Inc.
[2] F.A. Kroeger, *The chemistry of imperfect crystals*, **1964**, North-Holland.
[3] H. Schmalzried, *Solid State Reactions*, **1981**, Verlag Chemie.
[4] Per Kofstad, *Nonstoichiometry, Diffusion and Electrical Conductivity in Binary Metal Oxides*, **1972**, Wiley-Interscience.
[5] P. Ghigna, G. Spinolo, A. Filipponi, A.V.Chadwick and P.Hanmer, *Physica C*, **1995**, *246*, 345.
[6] M. Scavini, P. Ghigna, G. Spinolo, U. Anselmi Tamburini, and G. Flor, *Physica C*, **1995**, *251*, 89.
[7] H. Takagi, S.Uchida and Y. Tokura, *Phys. Rev. Lett.*, **1989**, *62*, 1197.
[8] P. Ghigna, G. Spinolo, M. Scavini, U. Anselmi Tamburini, and A.V. Chadwick, *Physica C*, **1995**, *253*, 147.
[9] P. Ghigna, G. Spinolo, E. Santacroce, S. Colonna, S. Mobilio, submitted.
[10] H. Zhang and H. Sato, *Phys. Rev. Lett.*, **1993**, *70*, 1697.
[11] D. Emin, *Phys. Rev. B.*, **1992**, *46*, 9419.
[12] M. Scavini, P. Ghigna, G. Spinolo, U. Anselmi Tamburini, G. Chiodelli, G. Flor, A. Lascialfari, and S. De Gennaro, *Phys. Rev. B*, **1998**, *58(4)*, 9385.
[13] H. Tolentino, F. Baudelet, A. Fontanine, T. Gourrieux, G. Krill, J. Y. Henry, and J. Roussat-Mignod, *Physica C*, **1992**, *192*, 115.
[14] G. Spinolo, P. Ghigna, G. Chiodelli, M. Ferretti, and G. Flor, *Z. Naturforsch.*, **1999**, *54a*, 95.
[15] P. Ghigna, G. Spinolo, G. Flor, and N. Morgante, *Phys. Rev. B*, **1998**, *57(21)*, 13426-13429.
[16] G. Flor, M. Scavini, U. Anselmi Tamburini, and G. Spinolo, *Solid State Ionics*, **1990**, *43*, 77.
[17] G. Spinolo, U. Anselmi Tamburini, P. Ghigna, and G. Flor, *Physica C*, **1993**, *217*, 347.
[18] G. Flor, P. Ghigna, U. Anselmi Tamburini, and G. Spinolo, *J. Solid State Chem.*, **1995**, *116*, 314.

Molecular Switches Based on the $[Ni^{II}(cyclam)]^{2+}$ Fragment

Valeria Amendola, Luigi Fabbrizzi, Maurizio Licchelli, Piersandro Pallavicini, Donata Sacchi

Dipartimento di Chimica Generale, Università di Pavia, viale Taramelli 12, I-27100 Pavia, Italy
e-mail: fabbrizz@unipv.it

Abstract. The nickel(II) complex of the 14-membered tetramine macrocycle 1,4,8,11-tetraaza-tetradecane (cyclam) is a versatile subunit, which can be employed to build up multicomponent molecular devices displaying different functions. In particular, the feasible Ni^{II}-to-Ni^{III} redox change has been used (i) to switch ON/OFF the emission of a fluorescent fragment covalently linked to cyclam, and (ii) to promote the translocation of the chloride ion between the Cu^{II} and Ni^{III} centres in a heterodimetallic complex. Moreover, taking profit from the temperature dependent high-to-low spin interconversion of the $[Ni^{II}(cyclam)]^{2+}$ complex, a fluorescent thermometric probe has been designed for measuring temperature in polar media, through a fast and reversible response.

1. The Design of a Molecular Switch

A property displayed by a molecular species in solution (e.g. colour, light emission, magnetic moment) can be made to change drastically on addition of an appropriate reagent. A well known example is provided by the acid-base indicator 2,4-dinitrophenol, **1**. When a base is added to the colourless acidic solution of **1**, a bright yellow colour develops. In fact, deprotonation of the -OH group (pK_A = 3.96) allows extended electron delocalisation over the entire molecular framework, thus seriously altering the energy of the MO levels. In particular, the energy difference between HOMO and LUMO is substantially reduced and, following deprotonation, light absorption moves from ultraviolet (colourless solution, pH < 3) to visible region (yellow, pH > 5). The process is reversible and addition of acid makes the yellow solution to decolorate instantaneously. It can be said that the colour is switched ON/OFF through sequential additions of base and acid. However, the envisaged system can hardly resemble a macroscopic switch of the everyday life (e.g. an electrical light switch).

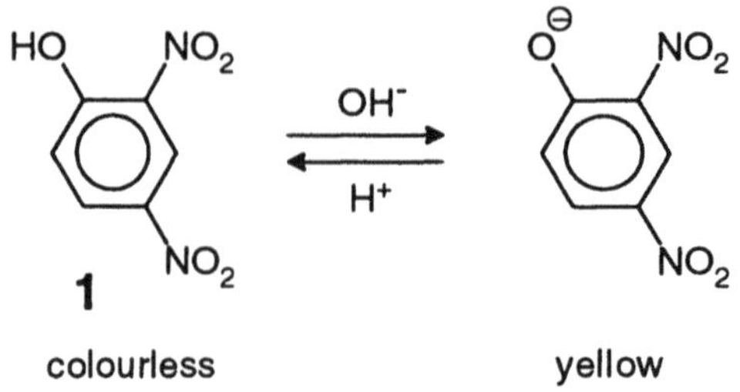

HO NO_2 $\xrightleftharpoons[H^+]{OH^-}$ $O^\ominus$ NO_2

NO_2 NO_2

1

colourless yellow

Fig. 1. The acid-base equilibrium of 2,4-dinitrophenol indicator (pK_A = 3.96). A fine variation of pH (from 3 to 5 and *vice versa*) switches ON/OFF the bright yellow colour.

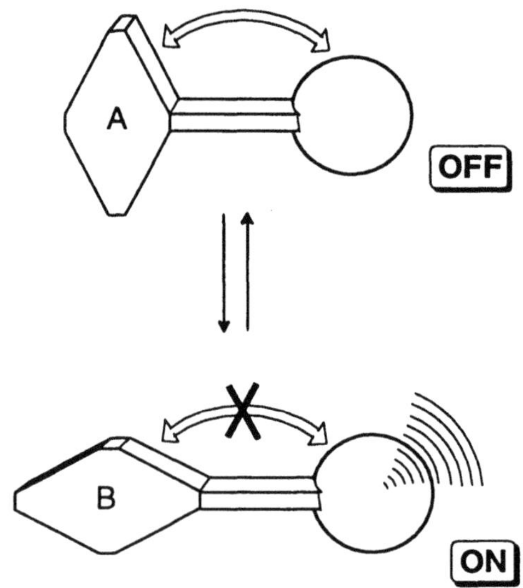

Fig. 2. The general behaviour of a two-component molecular switch. When in state A, the control unit interferes with the nearby active subunit, thus altering its properties. On the contrary, state B does not interfere and leaves the property of the active unit unchanged. It derives that a change from A to B, according to a fast equilibrium, switches ON/OFF the property of the active subunit.

More sophisticated molecular level switches, more similar to those of the macroscopic world, can be built up by assembling the appropriate molecular components in a covalently linked system [1]. In particular, the molecular fragment displaying the property whose magnitude has to be drastically varied (the so-called *active unit*) is covalently linked, through a given spacer, to another molecular fragment, which we will define as the *control unit*. The control unit must display the following properties: (i) it should exist in two states of comparable stability, A and B, which are interconnected by a fast and reversible equilibrium: the two states can be the protonated, LH^+, and neutral form, L, of an acid-base fragment, the A/B equilibrium being thus controlled by pH; or they can be the oxidised, Ox, and reduced form, Red, of a redox active fragment; in the latter case the A/B equilibrium is controlled by the redox potential, for instance the potential of the working electrode in an exhaustive electrolysis experiment; (ii) states A and B should interfere to a very different extent with the active unit through an inter-component process. For instance, it should happen that state A drastically modifies the property displayed by the nearby active subunit, whereas state B leaves it unchanged. In these circumstances, a change from A to B, promoted by an external input (pH change, variation of the redox potential) makes the chosen property to vary to a substantial extent, being switched ON/OFF at will by an external operator. The general behaviour of a two-component molecular switch is pictorially sketched in Figure 2.

The correspondence to the macroscopic world is especially close when the active subunit is a luminescent fragment and the chemical stimulus on the control subunit makes light emission to be turned ON/OFF [2]. A very simple yet efficient molecular switch of luminescence which is operated by an acid-base input is illustrated in Figure 3.

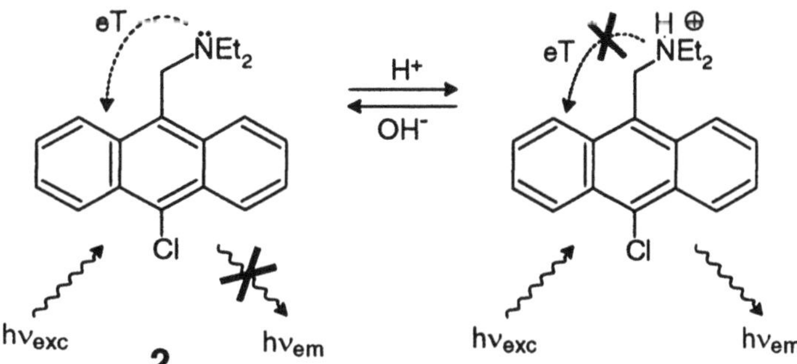

Fig. 3. A pH-controlled switch of luminescence. The protonated form, LH^+, of **2** displays the typical strong fluorescence of anthracene. In the neutral form, L, the emission is quenched due to the occurence of a photoinduced electron transfer process from the tertiary amine group to the anthracene fragment. Thus, fluorescence can be switched ON/OFF at will, through consecutive additions of acid and base.

The neutral form of anthrylamine **2** is not fluorescent. In fact, the typical emission of the anthracene fragment, expected at λ = 377 nm, is quenched due to the occurrence of an electron transfer (eT) process from the dialkylamine donor group $-NEt_2$ to the photoexcited fluorophore An*. When the amine group is protonated ($pH < pK_A = 7.7$), electrons are no longer available for the eT process, quenching of the fluorophore is prevented and the fluorescent emission is restored [3]. The behaviour of system **2** illustrates well the paradigm of a molecular switch: a fine variation of a bulk parameter (a pH change from 7.2 to 8.2, in the present case) induces a dramatic change of a given property (luminescence), according to a fast and reversible process which can be consecutively carried out at will, just as the repeated pressure of a finger on a macroscopic switch turns ON/OFF the light of a bulb.

A two-component molecular switch of luminescence which is operated by a redox input is illustrated in Figure 4. The luminescent unit is a $[Ru^{II}(bpy)_3]^{2+}$ fragment (bpy: 2,2'-bipyridine), which emits light at λ_{max} = 610 nm, through the decay of the ^{3}MLCT state $[Ru^{III}(bpy)^{-}(bpy)_2]^{2+}$. The redox active unit is a quinone, RO_2 / hydroquinone, $R(OH)_2$ system ($RO_2 + 2e^- + 2H^+ = R(OH)_2$). The RO_2 form displays pronounced electron transfer properties: it promotes the transfer of one electron from the nearby photoexcited luminophore, thus quenching its emission. On the other hand, no eT process takes place when the control unit is in its reduced form $R(OH)_2$. Thus, an electrochemical reduction performed on a non-luminescent solution of the covalently linked RO_2~$[Ru^{II}(bpy)_3]^{2+}$ system restores light-emission, due to reduction to hydroquinone form. Subsequent re-oxidation quenches luminescence again [4].

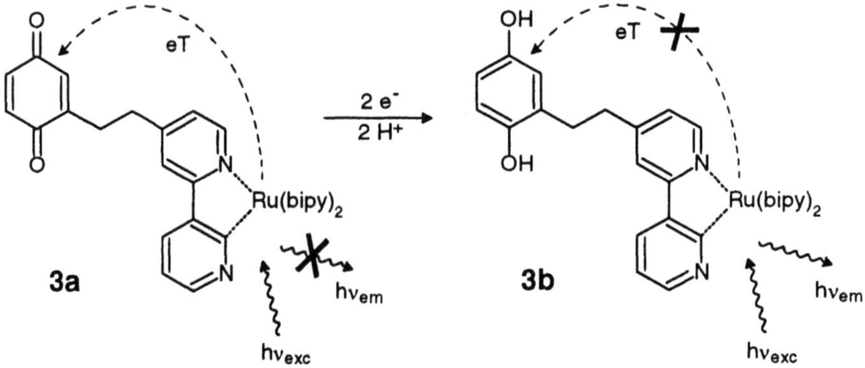

Fig. 4. A molecular switch of luminescence. The quinone fragment, RO_2, of system **3a** quenches the emission of the nearby luminescent fragment $Ru^{II}(bpy)_3]^{2+}$ due to a $Ru^{II}(bpy)_3]^{2+}$-to-RO_2 one electron transfer process. Such a photoinduced electron transfer process does not take place in the reduced form **3b**, $R(OH)_2$~$[Ru^{II}(bpy)_3]^{2+}$. Thus, the luminescence of the $[Ru^{II}(bpy)_3]^{2+}$ fragment can be switched ON/OFF by reducing/oxidising consecutively the quinone/hydroquinone unit, through the $RO_2 + 2e^- + 2H^+ = R(OH)_2$ half-reaction.

The two molecular switches discussed above are based on a control unit of organic nature: a dialkylamine group in **2** and a quinone/hydroquinone couple in **3**. We want now to demonstrate that transition metal ions, in a chosen coordinative environment, behave as efficient and versatile switches. This behaviour seems quite reasonable when switching derives from a redox input, due to the fact that transition metals are naturally prone to exchanging electrons through fast and reversible processes [5].

2. A versatile control unit: $[Ni^{II}(cyclam)]^{2+}$

Ni^{II}, as a first row transition metal ion, has a special affinity toward sp^3 nitrogen donor atoms, thus forming stable complexes with ammonia and amine containing multidentate ligands. Distinctive properties arise when Ni^{II} is incorporated into the 14-membered tetramine macrocycle **4**, cyclam. In particular, the $[Ni^{II}(cyclam)]^{2+}$ complex displays an extreme kinetic inertness towards metal extrusion. In fact, it persists in a strongly acidic solution almost indefinitely, even if demetallation and simultaneous protonation of the amine groups would be very favoured from a thermodynamic point of view. For instance, in 1 M $HClO_4$, Ni^{II} cyclam has a lifetime of 30 years [6]. The very high kinetic stability reflects the mechanical resistance of the cyclam amine groups to detach themselves from the metal centre and to be exposed to the H^+ ions of the solution. Inertness toward demetallation is a useful feature in a switching context, as addition of acid, to pH=2 and lower (a typical stimulus for a pH-sensitive switch) would leave the metal-tetramine system intact.

Another unique feature of the $[Ni^{II}(cyclam)]^{2+}$ complex is its magnetic bistability in polar media [7]. In particular, $[Ni^{II}(cyclam)]^{2+}$ in water (or MeCN, or DMSO) exists in two forms of comparable stability, which differ for their electronic properties and stereochemistry: (i) an octahedrally elongated form, $[Ni^{II}(cyclam)S_2]^{2+}$, in which the four nitrogen atoms of cyclam span the equatorial plane and two solvent molecules S occupy the axial sites: the complex has two unpaired electrons (high-spin form) and exhibits a blue-violet colour; (ii) a square species, $[Ni^{II}(cyclam)]^{2+}$, with only the macrocycle coordinated to the metal: the two highest energy electrons are now paired (low-spin form) and the colour is yellow, due to a relatively intense d-d band centred at 450 nm, molar absorbance ε = 65 M^{-1} cm^{-1}, in 0.1 M $NaClO_4$, at 25°C. The two forms undergo a fast and reversible interconversion equilibrium, which is strongly temperature dependent:

$$\underset{\text{blue, high-spin}}{[Ni^{II}(cyclam)S_2]^{2+}} \rightleftharpoons \underset{\text{yellow, low-spin}}{[Ni^{II}(cyclam)]^{2+} + 2S}$$

In particular, a temperature increase favours the formation of the yellow, low-spin species; a temperature decrease favours the formation of the blue, high-spin form. The endothermicity of the high-to-low spin conversion essentially reflects the loss of the two axial Ni^{II}-S bond, when moving from the octahedral to the the square stereochemistry. At 25°C, in aqueous solution, there exists 25% of the low-spin species, and 75% of the high-spin species: however, the solution appears yellow, owing to the higher absorbance of the low-spin complex. Distribution varies with solvent's nature and is also affected by structural modifications on the cyclam backbone (e.g. presence of alkyl substituents on nitrogen and carbon atoms).

Moreover, coordination of Ni^{II} by cyclam favours the access to the otherwise elusive Ni^{III} state [8]: in fact, the Ni^{II}-to-Ni^{III} oxidation process takes place at a moderately positive potential: 0.71 V vs. NHE in 1 M HCl. The trivalent complex which forms has a d^7 low-spin electronic configuration and exhibits a *trans*-

octahedral stereochemistry, e.g. *trans*-$[Ni^{III}(cyclam)Cl_2]^+$, in presence of Cl^-. The easy attainment of the Ni^{III} state is associated to the strong coordinative interactions exerted by the macrocycle: they raise the metal centred antibonding level of Ni^{II}, thus making the abstraction of one electron from it especially easy.

Inertness toward demetallation, temperature dependent high-low-spin inter-conversion, and feasible Ni^{II}/Ni^{III} oxidation process are all convenient features one can profit from for designing molecular switches which can be operated by a variety of stimuli.

3. Redox switches of fluorescence

Linking a $[Ni^{II}(cyclam)]^{2+}$ moiety to a light-emitting fragment can give rise to a redox switch of luminescence, if one oxidation state quenches the nearby photo-excited luminophore and the other does not. A system of this type can be easily obtained through the single step template synthesis illustrated in Figure 5.

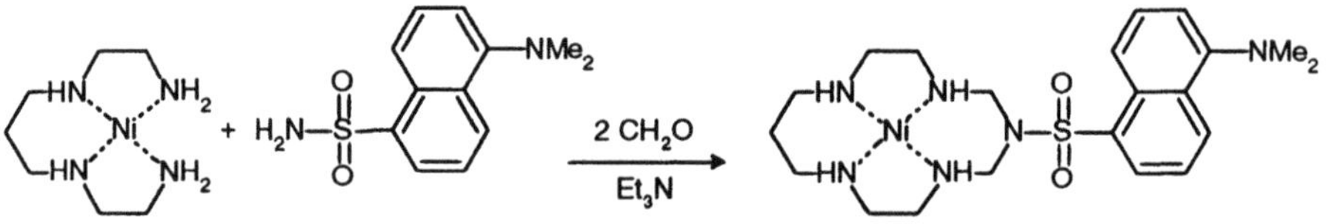

Fig. 5. The template synthesis of a two-component system suitable for fluorescence redox switching. Dansylamide acts as a locking fragment closing the tetramine preoriented by the coordination to the metal centre. The amide nitrogen atom does not interact with the metal and plays only an architectural role.

In particular, the open-chain tetramine 2.3.2-tet, preoriented through the square coordination to the Ni^{II} centre, is cyclised by the light-emitting dansylamide molecule, which acts as a locking fragment (in presence of two formaldehyde molecules, which preliminary give Schiff-base condensation at the primary amine groups of the chelating agent). The 14-membered macrocycle that forms is not exactly the same as cyclam, as the middle $-CH_2-$ group of one of the two trimethylenic chains of cyclam has been replaced by a sulphonamide nitrogen atom. However, the complex displays the typical macrocyclic properties described in the previous Section [9].

In an MeCN solution, at 25°C the metal centre of the two-component system $[Ni^{II}(5)]^{2+}$ exists mostly as yellow (square-planar, low-spin) form, while the adjacent subunit displays the typical emission in the visible region expected for a dansyl fragment, Dns (λ_{max} = 505, see Figure 6). The emission is due to the radiative decay of a charge transfer excited state, Dns*, involving the naphthalene fragment and the sulphonamide substituent.

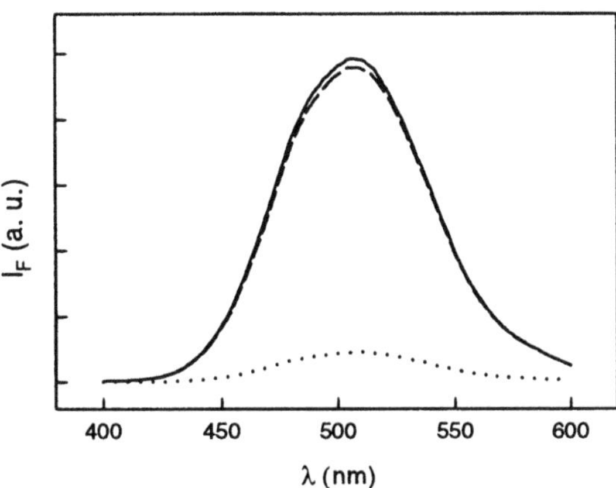

Fig. 6. Redox switching of the fluorescent emission of the dansyl subunit in the two-component system $[Ni^{II}(\mathbf{5})]^{2+}$. Solid line: emission spectrum of an MeCN solution of $[Ni^{II}(\mathbf{5})]^{2+}$; dotted line: emission spectrum of the same solution after exhaustive electrolysis (working electrode potential set at 0.23 V vs. Fc^{+}/Fc): the $[Ni^{III}(\mathbf{5})]^{3+}$ form is present in solution; dashed line: emission spectrum of the same solution, which has been electrolysed again at a potential E = -0.07 V: the $[Ni^{II}(\mathbf{5})]^{2+}$ species is formed on reduction, and fluorescent emission is fully restored [10].

The Ni^{II} centre, in the $[Ni^{II}(\mathbf{5})]^{2+}$ two-component system, undergoes a one electron oxidation in an MeCN solution, made 0.1 M in Bu_3BzNCl, at $E_{1/2}$ = 0.08 vs. Fc^{+}/Fc, as shown by voltammetric investigations. A green solution of the Ni^{III} derivative can be obtained in a controlled potential electrolysis experiment, by setting the potential of the working macro-electrode (a platinum gauze) at 0.23 V vs. Fc^{+}/Fc. The electrolysed solution is no longer fluorescent, the dansyl emission band being almost completely quenched (see Figure 6). However, if the potential of the working electrode is set at -0.07 V, the yellow solution of the Ni^{II} derivative forms again and fluorescence is fully restored (see the dashed line in Figure 6). Thus, light emission of the $[Ni^{II}(\mathbf{5})]^{2+}$ system can be switched OFF/ON at will, through the Ni^{II}/Ni^{III} couple, by using a variation of the electrode potential as an input [10].

Fluorescence quenching in the oxidised form $[Ni^{III}(\mathbf{5})]^{3+}$ has to be ascribed to the occurrence of of a Dns*-to-Ni^{III} eT process. The eT nature of the quenching process has been demonstrated by carrying out spectrofluorimetric investigations on a butyronitrile (BuCN) solution of the Ni^{III} derivative, frozen at the liquid nitrogen temperature. On freezing, the the solvent molecules are immobilised and cannot move to follow the rearrangement of the electrical charges associated to the eT process (from Ni^{III}~Dns* to Ni^{II}~$Dns^{\cdot+}$). As a consequence, the eT process is no longer allowed, and the photoexcited fragment can dissipate the absorbed photonic energy in the typical radiative mode: fluorescence is revived. Indeed, the glassified BuCN solution of the $[Ni^{III}(\mathbf{5})]^{3+}$ derivative, at 77K, shows full dansyl emission, thus demonstrating the eT nature of the fluorescence quenching process at room temperature.

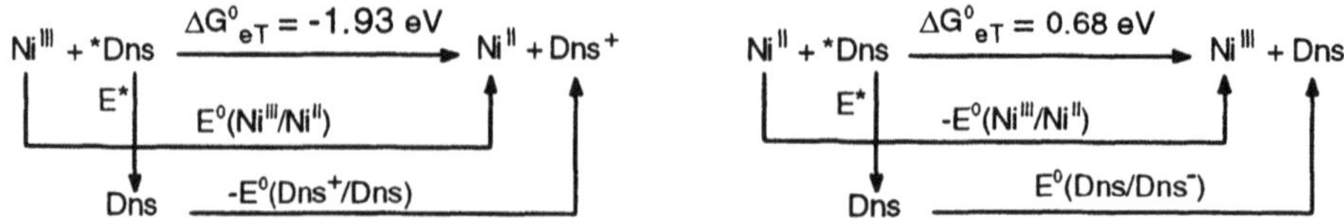

Fig. 7. Thermodynamic cycles for the calculation of the ΔG°_{eT} values associated to eT processes for $[Ni^{III}(\mathbf{5})]^{3+}$ (which occurs, left, fluorescence OFF) and for $[Ni^{II}(\mathbf{5})]^{2+}$ (which does not occur, right, fluorescence ON).

The occurrence of the Dns*-to-Ni^{III} eT process can be accounted for on a thermodynamic basis. In particular, the free energy change associated to the eT process, ΔG°_{eT}, can be calculated through the thermodynamic cycle reported in Figure 7, by combining the pertinent photophysical, E^{0-0}(Dns*), given by the energy of the emission band, and electrochemical quantities, $E_{1/2}(Ni^{III}/Ni^{II})$ and $E_{1/2}(Dns^{+}/Dns)$, obtained from the cyclic voltammetry profiles. In the present case, ΔG°_{eT} is distinctly negative, -1.93 eV, thus making the eT process responsible for fluorescence quenching strongly favoured (state: OFF). On the contrary, the occurrence of a Ni^{II}-to-Dns* eT process is thermodynamically prevented (ΔG°_{eT} = 0.68 eV), thus accounting for the unperturbed light emission of the reduced form $[Ni^{II}(\mathbf{5})]^{2+}$ (state: ON).

Similarly to the macroscopic world, molecular light-bulbs can be replaced, if desired. For instance, the dansyl fragment, emitting in the visible, can be replaced by a plain naphthalene fragment, which gives a genuine π-π* excited state and emits in the UV region (λ_{max} = 360 nm), by using naphthalenesulphonamide as a locking fragment in a template reaction of the type illustrated in Figure 5. Also in the present case, fluorescence can be switched ON (Ni^{II}) / OFF (Ni^{III}) through the metal centred redox couple (the real switch), both electrochemically (in a controlled potential electrolysis experiment on an MeCN solution) and chemically (in an MeOH solution, by consecutive addition of $Na_2S_2O_8$ for Ni^{II}-to-Ni^{III} oxidation, and of $NaNO_2$ for Ni^{III}-to-Ni^{II} reduction).

Further switching systems based on different metal centred redox couples can be designed by following the same two-component approach. A convenient redox change to be used for switching purposes is Cu^{II}/Cu^{I}. However, the tetramine coordinative environment of cyclam and cyclam-like macrocycles cannot be employed as a host in the present case, since amine coordination strongly stabilises Cu^{II} with respect to Cu^{I} and the Cu^{II}-to-Cu^{I} reduction process takes place at a too negative potential. In order to provide comparable stability to the two oxidation states, amine nitrogen atoms should be replaced by thioethereal sulphur atoms. A cyclic structure of the ligand should be maintained to impart kinetic stability and to prevent escaping of the metal ion into the solution. In particular, the 14-membered macrocycle 1,4,8,11-tetrathia-cyclotetradecane, thiacyclam, forms stable complexes with both Cu^{II} and Cu^{I} and the Cu^{II}/Cu^{I} redox change takes place at 0.25 V vs. Fc^{+}/Fc [11].

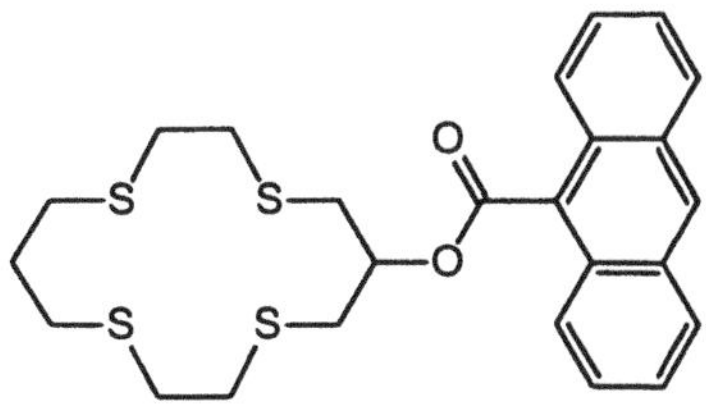

6

In the two-component assembly **6**, a thiacyclam subunit has been covalently linked to the light-emitting anthracene fragment through an ester group acting as a spacer [12]. The fluorescent emission of the 9-anthracenoic ester subunit (λ_{max} = 470 nm, charge transfer excited state An(CT)*) can be switched ON/OFF through the Cu^{II}/Cu^{I} redox couple, in an MeCN solution. In particular, the Cu^{II} centre quenches the emission of the nearby fluorophore, due to a thermodynamically favoured An(CT)*-to-Cu^{II} eT process. On the other hand, the Cu^{I}-to-An(CT)* eT process is poorly favoured and does not take place. Thus, the $[Cu^{I/II}(\mathbf{6})]^{+/2+}$ provides a further example of an ON (reduced metal) / OFF (oxidised metal) molecular switch of luminescence, operating under an electron transfer regime.

4. A fluorescent molecular thermometer

The two-component approach described in the previous Section has been used to obtain, at the molecular level, another common object of the everyday life: a thermometer. In order to assemble a fluorescent probe of temperature, the light emitting unit has to be linked to a molecular fragment sensitive to temperature. In particular, the temperature sensitive unit, TSU, should exist in two different states of comparable stability A and B (again, a *bistable* system). A and B should be in equilibrium and the equilibrium should be affected by temperature, being distinctly eso- or endo-thermic. As a second requirement, A and B should have a different perturbing effect on the proximate fluorophore: state A quenches fluorescence and state B does not (or *vice versa*). In these circumstances, one could modulate the intensity of the fluorescent emission by changing the relative concentration of A and B by a temperature variation. As a consequence, the temperature of the solution could be measured from the fluorescence intensity, I_F, on the basis of a I_F vs. temperature calibration curve.

Fortunately, a TSU building block is already in our hands: the $[Ni^{II}(cyclam)]^{2+}$ complex. It has been mentioned in Section 2 that in a polar medium $[Ni^{II}(cyclam)]^{2+}$ can exist in two forms of different spin state: a high-spin and a low-spin form which are related by a fast and reversible interconversion equilibrium. The high-to-low-spin conversion is distinctly endothermic (in an aqueous solution, $\Delta H°$ = 5.4 kcal mol^{-1}).

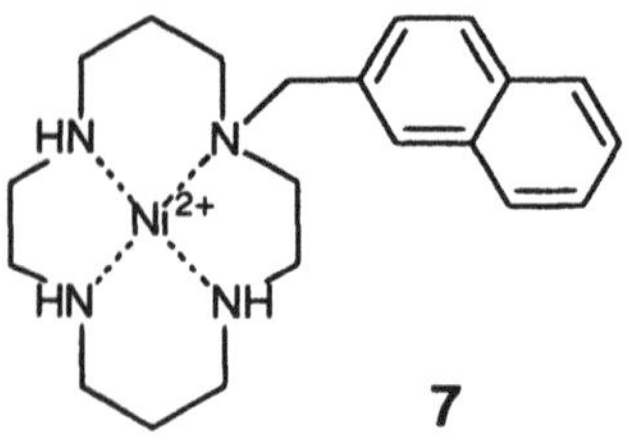

7

On these premises, the $[Ni^{II}(cyclam)]^{2+}$ fragment was linked through a $-CH_2-$ group to the classic naphthalene fluorophore, to give the two-component assembly **7** [13]. It has now to be verified whether the two metal spin states have a different perturbing effect on the nearby photoexcited naphthalene fragment. This can be assessed by studying the emission behaviour of complex salts of formula $Ni^{II}(7)X_2$. In a $Ni^{II}(7)X_2$ salt, in the solid state, the Ni^{II} centre exhibits an octahedrally elongated coordination geometry, in which the two X^- anions occupy the two apical positions. The spin-state of the metal centre is determined by the coordinating tendencies of X^-: (i) strongly coordinating anions (e.g. Cl^-) stabilise the high-spin state (Ni^{II} 'feels' an *octahedral* stereochemistry); (ii) weakly coordinating anions stabilise the low-spin state (Ni^{II} 'feels' a *square* stereochemistry). The complex salts $Ni^{II}(7)X_2$ were dissolved in $CHCl_3$, where, owing to solvent's non-coordinating tendencies, they keep intact their donor set and maintain their spin state. Thus, measuring the emission spectra of $CHCl_3$ solutions of different $Ni^{II}(7)X_2$ salts allows one to evaluate the effect of the Ni^{II} spin state on the emission properties of the nearby naphthalene fragment. Quite interestingly, this effect is substantially different for high- and low-spin complexes.

Table 1. Quantum yields, Φ, values determined for different salts of $[Ni^{II}(7)]^{2+}$ complex in a $CHCl_3$ solution [13]. Quantum yield for plain naphthalene, in an EtOH solution, is 0.21.

anion	Cl^-	NO_3^-	NCS^-	ClO_4^-	$CF_3SO_3^-$
Φ	0.007	0.009	0.005	0.019	0.020

Table 1 reports the quantum yields of a series of $Ni^{II}(7)X_2$ complex salts of different spin multiplicity. High-spin complexes ($X^- = Cl^-$, NO_3^-, NCS^-) show a quantum yield 3-4 times lower than low-spin complexes ($X^- = ClO_4^-$, $CF_3SO_3^-$). It should be noticed that the quantum yield is in any case distinctly lower than observed with plain naphthalene, thus indicating that the metal centre has a more or less pronounced quenching effect on the nearby excited fluorophore. It has been observed that the quenching mechanism has an energy transfer nature (fluorescence intensity being not revived when the solution is frozen to 77K). In particular, it has been hypothesised that the energy transfer takes place through a double electron exchange (Dexter mechanism). Quantum yield values in $CHCl_3$ solution show that such a process is remarkably more efficient for the high-spin metal centred than for the spin paired complex, to which a more compact electronic structure has to be ascribed [13].

On these bases, one should expect that the spin-state effect on fluorescence should be observed also in a polar and coordinating medium, where the axial position are not longer occupied by the X^- anions, but by solvent molecules and where the high- and low-spin species coexist according to a temperature dependent equilibrium. Indeed, the emission intensity of a solution of a $Ni^{II}(7)(ClO_4)_2$ complex in an MeCN solution varies with temperature and, in particular, it increases with increasing temperature. Figure 10 displays the family of emission spectra recorded over the 27-65°C range.

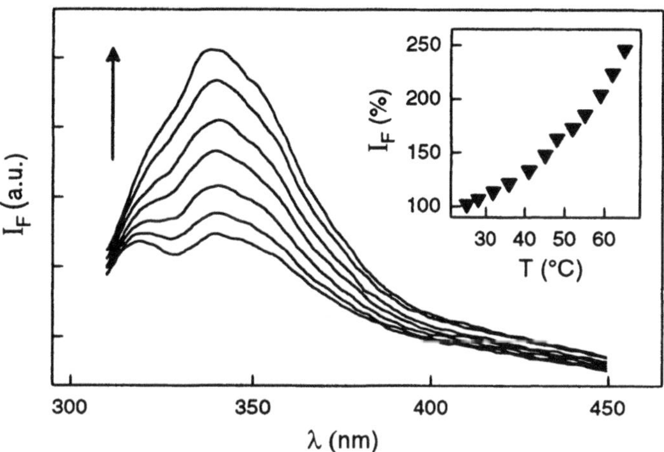

Fig. 8. Emission spectra of an MeCN solution 10^{-5} M in the $Ni^{II}(7)(ClO_4)_2$ complex salt recorded over the 27-65°C range (λ_{exc}= 290 nm). On increasing temperature, fluorescence increases, owing to the conversion of the high-spin to the low-spin state of the Ni^{II} centre. The inset shows the variation of the fluorescence intensity at 335 nm, I_F, with temperature and represents the calibration curve of the molecular fluorescent thermometer $Ni^{II}(7)(ClO_4)_2$.

The spin conversion equilibrium taking place in the MeCN solution is illustrated by the following equation:

$$[Ni^{II}(7)(MeCN)_2]^{2+} \rightleftharpoons [Ni^{II}(7]^{2+} + 2MeCN$$

At lower temperatures the $[Ni^{II}(cyclam)(MeCN)_2]^{2+}$ complex, high-spin, dominates, showing the more pronounced quenching effect on the nearby naphthalene fluorophore. Due to the endothermic nature of the high-to-low-spin conversion, on raising the temperature the concentration of the low-spin form, $[Ni^{II}(cyclam)]^{2+}$, increases, causing a concomitant enhancement of the fluorescent emission. The inset of Figure 8 shows the variation of the maximum of the emission intensity with temperature. Such a diagram represents a sort of calibration curve of the $[Ni^{II}(7)]^{2+}$ system, which behaves as a thermometric probe. The response of the thermometer is fully reversible and extremely fast, as the fluorescent emission change is related to the release-uptake of solvent molecules from-to the axial positions of the metal and to the consequent electron rearrangement, a process taking place well below the ms time scale.

Systems of this type can be profitably used for measuring the temperature of biological fluids, including intracellular ones. In particular, fluorescence microscopy measurements would allow one to determine temperature inside a cell in which the fluorescent probe had been injected, with both spatial and temporal resolution. In this context, it should be noted that most biochemical reactions are accompanied by a more or less pronounced heat effect: thus, local temperature monitoring could provide a means to detect *where* and *when*, inside the cell, the envisaged chemical process takes place. The advantage of systems like $[Ni^{II}(7)]^{2+}$ is that, due to the high sensitivity of the fluorescent signal, they can be used at very low, non-invasive concentration levels. In particular, temperature changes of a solution of $[Ni^{II}(7)]^{2+}$ can be distinctly perceived at a 10^{-7} M concentration. However, the rather lipophilic system $[Ni^{II}(7)]^{2+}$ itself cannot be used as a thermometric probe for biological fluids due to its insolubility in water. The obstacle can be circumvented by appending to the cyclam framework a more hydrophilic fluorogenic fragment than naphthalene.

5. Translocating anions by means of the $[Ni^{II,III}(cyclam)]^{2+,3+}$ switch

Bistability can have also a topological nature. Topological bistability is observed, for instance, when a molecular or supramolecular system contains a mobile portion which can be moved with respect to to the rest of the assembly, which remains motionless [14]. Motion can be induced by an external stimulus (pH change, variation of the redox potential, illumination) and gives rise to two (or more) states, say A and B, of different topology and comparable stability. The externally driven movement of the mobile part produces *mechanical work* and the whole system is therefore defined a molecular *machine* [15, 16]. First examples of machines operating at the molecular level refer to supramolecular assemblies: rotaxanes [17] and catenanes [18].

In the Stoddart's rotaxane, a π-accepting wheel is made to oscillate between two unequivalent π-donating stations present on the axle by varying, through either a redox process or an acid-base reaction, the donor properties of one station [17]. In Sauvage's catenate, two intertwined rings are coordinated to a copper centre: one ring contains a 2,2'-bipyridine (bpy) bidentate moiety, the other ring contains both a bpy and a 2,2':6',2"-terpyridine (terpy) tridentate moiety [18]. When the metal centre is in the Cu^{I} oxidation state (preferred coordination number, CN = 4), the two bpy subunits are coordinated to the metal. On oxidation to Cu^{II} (preferred CN = 5), the ring containing both bpy and terpy moieties rotates in order to offer to the metal its tridentate subunit. On subsequent reduction to Cu^{I}, the ring makes a further half-turn to bring the bpy subunit again on the Cu^{I} centre.

Many other example of externally controlled oriented motions have been described since the above mentioned pioneering investigations. Very intererestingly, a monolayer of Stoddart rotaxane molecules containing two equivalent π-donating stations in its axle, sandwiched between two metal electrodes, has been shown to operate much more efficiently than conventionally wired logic gates [19].

There exists another way to produce oriented motions at the molecular level: translocating a particle (e.g. an ion) from a given site to another of a fixed molecular framework, following a defined pathway [20]. We will consider in the following the translocation of an anion between two metal centres for which it displays selective coordinating tendencies, a process driven by a change of the redox potential. In the envisaged system, two different metal centres M_1 and M_2 are hosted by appropriate receptors, and the two receptors are covalently linked through a spacer. Both M_1 and M_2 are coordinatively unsaturated, i.e. they have room available for the coordination of the anion to be translocated: X^-. Moreover, one of the metal centres, let us say M_2, is redox active and may exist in two consecutive oxidation states M_2^{n+} and $M_2^{(n+1)+}$, which have a comparable stability. The essential prerequisite for translocation of X^- between M_1 and M_2 to occur is that the binding affinity of the anion decreases along the series: $M_2^{(n+1)+} > M_1 > M_2^{n+}$. In these circumstances, the redox equilibrium (1) is established:

$$\overset{X}{\overset{|}{M_1}}\!\sim\!\sim M_2^{n+} \rightleftharpoons M_1\!\sim\!\sim \overset{X}{\overset{|}{M_2}}{}^{(n+1)+} + e^- \qquad (1)$$

In particular, when M_2 is in its reduced form, M_2^{n+}, X^- prefers to stay on M_1; but when the M_2 centre is oxidised, X^- moves to $M_2^{(n+1)+}$, the cation for which it displays the highest affinity. Thus, it will be possible to move the X^- particle from M_1 to M_2 and *vice versa*, at will, by operating the $M_2^{(n+1)+}/M_2^{n+}$ switch, through a variation of the redox potential. If the M_1–X and M_2–X coordinative bonds are labile (a rather common feature for 3d metal ions), the redox change on the M_2 centre will induce the simultaneous displacement of X^- in either direction.

We considered that the $[Ni^{II/III}(cyclam)]^{2+/3+}$ system could be a convenient redox switch to drive the intramolecular translocation of an anion. In fact, $[Ni^{II}(cyclam]^{2+}$, in its low-spin state is square and does not want to bind any further ligand, including a coordinating anion; when in its high-spin state, the Ni^{II} complex of cyclam displays octahedral stereochemistry, but even the most coordinating anion, if added according to a 2:1 stoichiometry, cannot compete for axial binding with the solvent molecules, which are present in overwhelming amount. Thus, it can be concluded that the Ni^{II}-cyclam complex, whether in high- or low-spin state, exhibits a poor affinity (if not nil) towards any X^- anion. On the other hand, the oxidised complex (d^7, low-spin) has a pronounced tendency to bind X^- anions, to give a *trans*-octahedral species: $[Ni^{III}(cyclam)X_2]^+$. We need now a metal complex M_1 coordinatively unsaturated and displaying a moderate affinity towards the envisaged anion. A good candidate is the Cu^{II} complex with the tripodal tetramine tris(2-aminoethyl)amine (tren). Tren imposes to the bound metal centre a trigonal bipyramidal stereochemistry, leaving one of the axial sites available for X^- coordination. In particular, in an MeCN solution, the stability of the 1:1 adduct with the Cl^- anion decreases along the series: $[Ni^{III}(cyclam)]^{3+}$ >> $[Cu^{II}(tren)]^{2+}$ >>

$[Ni^{II}(cyclam)]^{2+}$, which matches the required affinity sequence for the redox driven translocation between the Cu^{II} and Ni^{II}/Ni^{III} centres.

On these premises, the ditopic ligand **8** was synthesised, in which a cyclam and a tren subunit are covalently linked by the 1,4-xylyl spacer [21]. Ni^{II} and Cu^{II} ions can be addressed to their reserved place by taking profit from the macrocyclic effects, either *thermodynamic* or *kinetic*. In fact, on reaction of 1 equiv. of $Ni^{II}(ClO_4)_2$, the metal centre chooses the cyclam cavity, rather than the tren subunit, in order to profit from the *thermodynamic* macrocyclic effect (number and type of donor atoms being the same, metal complexes of cyclic ligands are orders of magnitude more stable than the corresponding ones with open-chain counterparts [22]). Then, 1 equiv. of $Cu^{II}(ClO_4)_2$ is added to the same solution. Cu^{II} forms more stable complexes than Ni^{II}, in particular with amine ligands, and is therefore entitled to displace Ni^{II} from the privileged cyclam site. However, this event cannot occur, due to the well known resistence of any transition metal complex of cyclam towards demetallation (the *kinetic* macrocyclic effect [23]). Therefore, Cu^{II} is forced to make do with the tren subunit and the $\{[Cu^{II}(tren)]\sim[Ni^{II}(cyclam)]^{4+}$ heterodimetallic complex is formed, which can be isolated and characterised.

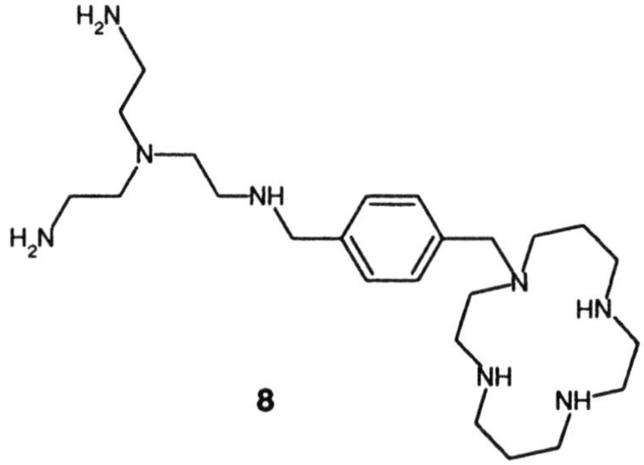

In an MeCN solution of $\{[Cu^{II}(tren)]\sim[Ni^{II}(cyclam)](CF_3SO_3)_4\}$, in view of the lack of competition of the counteranion, the axial site of the $[Cu^{II}(tren)]^{2+}$ trigonal bipyramidal moiety is occupied by a solvent molecule. On addition of Cl^-, the pale blue solution takes an intense blue-green colour, due to the formation of the $[Cu^{II}(tren)Cl]^+$ chromophore (ligand-to-metal charge transfer band, centred at 470 nm). The log*K* value for the 1:1 adduct formation is 5.7. This means that, in an MeCN solution 10^{-3} M both in $\{[Cu^{II}(tren)]\sim[N^{II}(cyclam)](CF_3SO_3)_4$ and in Cl^-, 95% of the added chloride ion is bound to the Cu^{II} centre, whereas 5% is dispersed in the solution (notice that the $[Ni^{II}(cyclam)]^{2+}$ subunit, either in its square form or in its *trans*-octahedral form, $[Ni^{II}(cyclam)(MeCN)_2]^{2+}$, does not compete at all for the anion). If a controlled potential electrolysis experiment is carried out on the solution (with the potential of the working macroelectrode set at 0.4 V vs. Fc^+/Fc), 1 mol of electrons per mol of heterodimetallic complex is consumed and an intense green-yellow colour develops. The band at 315 nm is due to the $[N^{iIII}(cyclam)Cl]^{2+}$ chromophore. This indicates that, on Ni^{II}-to-Ni^{III} oxidation, the Cl^- anion moves from copper to nickel centre. 100% of chloride ions are now bound to Ni^{III} centres, due to the very high constant for the 1:1 adduct formation: log$K > 8$. This value has been evaluated from the titration with a standard Cl- solution of a solution of the $\{[Cu^{II}(tren)]\sim[Ni^{III}(cyclam)]\}^{5+}$ complex, which had been obtained through exhaustive electrolysis. In particular, it was observed that even on addition of an excess of chloride (up to 5 equiv.), only one anion coordinates the metal centre. This is in contrast with that observed in the case of the plain cyclam complex, which gives a stable $[Ni^{III}(cyclam)Cl_2]^+$ species. It is possible that in the $\{[Cu^{II}(tren)]\sim[Ni^{III}(cyclam)]\}^{5+}$ conjugate the xylyl substituent on one of the nitrogen atoms of the cyclam ring sterically prevents the binding of a second Cl^- to the Ni^{III} centre.

If the same solution is then subjected to the controlled potential electrolysis experiment (with the working electrode set at the potential of 0.0 V vs. Fc^+/Fc), 1 mol of electrons per mol of heterodimetallic complex is consumed and the blue-green colour of the reduced complex forms again, thus indicating that on Ni^{III}-to-Ni^{II} reduction the chloride anions moves back to the copper centre. It is therefore possible to translocate back and forth the Cl^- anion between the two metal centres by operating the Ni^{II}/Ni^{III} redox couple, which again behaves as a switch. The reversible translocation process is illustrated by the half reaction reported below and its occurrence is signalled to the outside by a sharp colour change of the solution: from blue-green , $[Cu^{II}(Cl)\sim Ni^{II}]$, to bright green-yellow, $[Cu^{II}\sim Ni^{III}(Cl)]$.

$$\underset{\displaystyle Cu^{II}\sim\!\sim Ni^{II}}{\overset{Cl}{|}} \rightleftharpoons Cu^{II}\sim\!\sim \underset{\displaystyle Ni^{III}}{\overset{Cl}{|}} + e^- \qquad (2)$$

The process can be repeated indefinitely, by setting the potential of the platinum gauze used as a working electrode alternatively at 0.4 and 0.0 V vs. Fc^+/Fc. The intimate mechanism of the redox driven translocation process is illustrated by the square scheme in Figure 9. Starting from a solution of the $Cu^{II}(Cl)\sim Ni^{II}$ complex (beginning from the upper left corner of the square and moving clockwise, see Figure 9), we have first a pure oxidation step, $\mathbf{E}_1$, in which Ni^{II} is oxidised to Ni^{III} (and a solvent molecule probably goes to to fill the vacant coordinative position on the Ni^{III} centre). Then, step $\mathbf{T}_1$, the Cl^- ion 'moves' from Cu^{II} to Ni^{III}. However, process $\mathbf{T}_1$ is too fast to be distinctly detected, for instance by electrochemical methods, and $\mathbf{E}_1$ and $\mathbf{T}_1$ steps appear as simultaneous (this is due to the very high substitutional lability of both Cu^{II} and Ni^{III} centres). A question may arise whether the Cl^- ion moving on Ni^{III} comes from the proximate Cu^{II} centre (intramolecular process) or from a Cu^{II} centre belonging to to a different heterodimetallic complex (intermolecular process). From a mechanistic point of view, the intramolecular process could involve folding of the two halves of the $Cu^{II}(Cl)\sim Ni^{II}$ conjugate on the 1,4-xylyl bridge, followed by anion exchange at the axial positions. On the other hand, the intermolecular process should result from the collision of the $Cu^{II}(Cl)$ half of a conjugate with Ni^{III} half of a different conjugate, according to a bimolecular process. The probability of the intramolecular process to occur is extremely higher than that of the intermolecular one. Consider for instance that the

Cu^{II}-Cl fragment moves in a sphere whose centre is Ni^{II} and whose radius is 7.5 Å (i.e. the Cu^{II}-Ni^{II} distance, as calculated through molecular modelling). Notice that a Cu^{II}-Cl fragment in a volume of 1766 $Å^3$ (the volume of the sphere) has a concentration 0.94 M, i.e. much higher that the $5x10^{-4}$ M solution used in the spectro-electrochemical experiments. This suggests that the intramolecular anion exchange is extremely much probable that the intermolecular collision. From another point of view, one could observe that in a $5x10^{-4}$ M solution the average distance between two distinct molecules is 1380 Å, to be compared to the 7.5 Å which separe Cu^{II} and Ni^{II} in the heterodimetallic complex. The much lower distance points towards the much more probable occurrence of the intramolecular anion translocation, compared to the intermolecular anion transfer.

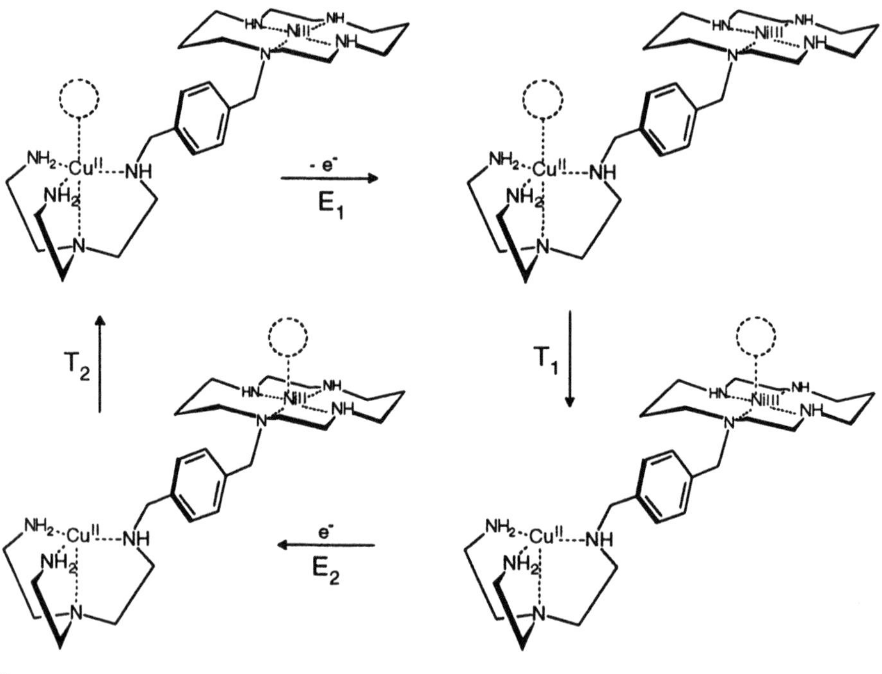

Fig. 9. The square scheme illustrating the redox driven translocation of the chloride ion between copper and nickel centres in the $\{[Cu^{II}(tren)]\sim[Ni^{II}(cyclam)]\}^{4+}$ heterodimetallic complex. The two steps $\mathbf{E_1}$ (Electron transfer) and $\mathbf{T_1}$ (anion Translocation) can be only ideally separated and are detected as simultaneous. The same is observed for the back-translocation process, $\mathbf{E_2} + \mathbf{T_2}$.

The neat predominance of the intra- over the inter-molecular mechanism is also supported by thermodynamic arguments. The $\Delta G°$ value associated to the translocation process, $\mathbf{E_1} + \mathbf{T_1}$, can be obtained from the difference of the Ni^{III}/Ni^{II} half-wave potentials measured in absence and in presence of Cl^-, and is remarkably negative: -11.5 kcal mol^{-1}. This value should be compared to the $\Delta G°$ value associated to the Cl^- transfer from an individual $[Cu^{II}(tren)Cl]^+$ species to an individual $[Ni^{III}(cyclam)]^{3+}$ species, a process which is necessarily intermolecular. In order to make a correct comparison, the *N*-substituted derivatives of tren, **9**, and of cyclam, **10**, were considered.

Thus, the following anion transfer equilibrium was investigated:

$$[Cu^{II}(\mathbf{9})(Cl)]^{+} + [Ni^{III}(\mathbf{10})]^{3+} = [Cu^{II}(\mathbf{9})]^{2+} + [Ni^{III}(\mathbf{10})Cl]^{2+} \quad (3)$$

whose $\Delta G°$ value (-4.0 kcal mol^{-1}) is much less negative than that observed for the translocation equilibrium taking place with the heterodimetallic complex. It derives that the intermolecular process (3) is much less favoured than process (2). By measuring electrode potentials at varying temperature, it was possible to evaluate the $\Delta H°$ and $T\Delta S°$ contribution to $\Delta G°$ for both reaction (2) and (3) (see pertinent values in Table 2).

$\Delta H°$ values are comparable for the two equilibria. This seems reasonable because the $\Delta H°$ quantity mainly includes bonding terms and in both processes (2) and (3) a Cu^{II}-Cl bond is broken and a Ni^{III}-Cl bond is formed, in a similar coordinative environment. On the other hand, the $T\Delta S°$ term strongly disfavours the intermolecular anion transfer equilibrium (3) with respect to (2). It has been pointed out before that the intramolecular mechanism should be extremely favoured with respect to the intermolecular mechanism, owing to the greater probability to occur. This probability effect seems to be reflected in the large difference of the entropy terms of process (2), which is hypothesised intramolecular, and process (3), which is surely intermolecular.

Table 2. Thermodynamic parameters determined for equilibria (2) and (3).

	$\Delta G°$ (kcal mol^{-1})	$\Delta H°$ (kcal mol^{-1})	$\Delta S°$ (cal mol^{-1} K^{-1})
Eq. (2)	-12	-12	0
Eq. (3)	-4	-12	-28

The square scheme in Figure 9 is completed by the reduction step $\mathbf{E_2}$ and by the anion back translocation step $\mathbf{T_2}$. Again, there is no way to investigate the two steps separately.

Other conjugate systems can be designed for anion translocation between different metal centres. Each pair of metal ions requires its own appropriate receptor and has its own topology. For instance, no other pair of transition metal ions than Cu^{II} and Ni^{II} allows redox driven anion translocation within the tren~cyclam conjugate. Even simple swapping of the two metals prevents the process. In fact, Cu^{II}, when encircled by cyclam, can have access to the Cu^{III} state, even if at a rather positive potential. However, the Cu^{III} centre, for having a d^8 electronic configuration, likes being square and does not have any affinity for anion binding, thus altering the required sequence of binding tendencies: $M_2^{(n+1)+} > M_1 > M_2^{n+}$.

Moreover, no other anion than Cl^- can be translocated within the $\{[Cu^{II}(tren)]\sim[Ni^{II}(cyclam)]\}^{4+}$ system. In particular, anions possessing distinctive donor tendencies (NCS^-, Br^-) have also more or less pronounced reducing tendencies and undergo oxidation before the Ni^{II}-to-Ni^{III} process takes place. Anions resistant to the oxidation (ClO_4^-, NO_3^-) do not exhibit satisfactorily high binding tendencies towards the $[Cu^{II}(tren)]^{2+}$ subunit. Cl^- is unique in that it is a rather poor reducing agent, but it is also a good ligand for Cu^{II} and Ni^{III} tetramine complexes.

6. Conclusions

Cyclam, the prototype of poly-aza macrocycles, was obtained as a by-product of the synthesis of linear tetramines in 1936 [24], i.e. decades before the development of macrocyclic chemistry. Perhaps the most versatile of cyclam metal complexes, $[Ni^{II}(cyclam)]^{2+}$, was synthesised in 1965 through reaction of a nickel(II) salt with the free ligand [25], and was structurally characterised, as the high-spin $[Ni^{II}(cyclam)Cl_2]$ complex salt, in the same year [26], in the decade in which macrocyclic chemistry tumultuously developed. Most of $[Ni^{II}(cyclam)]^{2+}$ distinctive properties (inertness [27], easy oxidation to the tervalent state [28], thermodynamic stability [29], spin crossover in solution [30]) were established and interpreted during the 70's. Meanwhile, $[Ni^{II}(cyclam)](ClO_4)_2$ could be obtained in multigram amount through the one-pot metal template synthesis settled by Barefield in 1972 [31], a procedure which later made cyclam commercially available [32]. In the following decade, further interesting properties of $[Ni^{II}(cyclam)]^{2+}$ and of its derivatives were discovered, which include its role as electron mediator in the electrochemical [33], and photochemical reduction of CO_2 to CO [34], and as a carrier for the counter-transport of electrons and anions across a liquid membrane [35]. More recently, the $[Ni^{II}(cyclam)]^{2+}$ subunit has been used to build up multicomponent covalently linked systems aimed to perform functions at the molecular level. This Chapter has illustrated a few examples in this area.

The design of elaborated multicomponent devices displaying sophisticated activity is a preliminary step to the development of a technology based on single molecules, and will probably represent one of the more exciting topics of the chemistry of the first decade of the third millennium. The very simple $[Ni^{II}(cyclam)]^{2+}$ fragment, due to the unique combination of electronic and stereochemical features, remains a convenient piece one can profit from in order to design molecular level devices whose activity involves redox changes, electron transfer, and signal generation.

Acknowledgements

This work was supported by the Ministry of University and Research (MURST, Progetto *Dispositivi Supramolecolari*)

References

[1] L. Fabbrizzi and A. Poggi, *Chem. Soc. Rev.*, **1995**, *24*, 197.
[2] L. Fabbrizzi, M. Licchelli, P. Pallavicini, *Acc. Chem. Res.*, **1999**,
[3] R. A. Bissell, A. P. de Silva, H. Q. N. Gunaratne, P. L. M. Lynch, G. E. M. Maguire, K. R. A. S. Sandanayake, *Chem. Soc. Rev.* 1992, 187.
[4] V. Goulle, A. Harriman, J.-M. Lehn, *J. Chem. Soc., Chem. Comm.*, 1993, 1034.
[5] R. Bergonzi, L. Fabbrizzi, M. Licchelli, C. Mangano, *Coord. Chem. Rev.*, **1998**, *170*, 31.
[6] E. J. Billo, Inorg. Chem. **1984**, *23*, 236.
[7] L. Sabatini and L. Fabbrizzi, *Inorg. Chem.*, **1979**, *18*, 438.
[8] F. V. LoVecchio, E. S. Gore, D. H. Busch, *J. Am. Chem. Soc.*, 1974, *96*, 3109.
[9] F. Abbà, G. De Santis, L. Fabbrizzi, M. Licchelli, A. M. Manotti Lanfredi, P. Pallavicini, A. Poggi, and F. Ugozzoli, *Inorg. Chem.*, **1994**, *33*, 1366.
[10] G. De Santis, L. Fabbrizzi, M. Licchelli, N. Sardone and A. H. Velders, *Chem.- Eur. J.*,**1996**, *2*, 1243-1250.
[11] D. B. Rorabacher, M. M. Bernardo, A. M. Q. Van de Linde, G. H. Leggett, B. C. Westerby, M. J. Martin, L. A. Ochrymowicz, *Pure Appl. Chem.* **1988**, *60*, 501.
[12] G. De Santis, L. Fabbrizzi, M. Licchelli, C. Mangano, D. Sacchi, *Inorg. Chem.*, **1995**, *34*, 3581.
[13] M. Engeser, L. Fabbrizzi, M. Licchelli, and D. Sacchi, *Chem. Commun.*, **1999**, 1191.
[14] J.-M. Lehn, *Supramolecular Chemistry, Concepts and Perspectives*, VCH: Weinheim, 1995.
[15] V. Balzani, M. Gómez-López, J. F. Stoddart, *Acc. Chem. Res.* **1998**, *31*, 405.
[16] J.-P. Sauvage, *Acc. Chem. Res.*, **1998**, *31*, 611.
[17] R. A. Bissell, E. Córdova, A. E. Kaifer, J. F. Stoddart, *Nature*, **1994**, *369*, 33.
[18] A. Livoreil, C. O. Dietrich-Buchecker, J. P. Sauvage, *J. Am. Chem. Soc.* **1994**, *116*, 9399.
[19] C. P. Collier, E. W. Wong, M. Belohradsky, F. M. Raymo, J. F. Stoddart, P. J. Kuekes, R. S. Williams, J. R. Heath, *Science*, **1999**, *285*, 391.
[20] V. Amendola, L. Fabbrizzi, M. Licchelli, C. Mangano, P. Pallavicini, L. Parodi, A. Poggi, *Coord. Chem. Rev.*, **1999**, *190-192*, 649.
[21] L. Fabbrizzi, F. Gatti, P. Pallavicini, E. Zambarbieri, *Chem.-Eur. J.*, **1999**, *5*, 682.
[22] D. K. Cabbiness, D. W. Margerum, *J. Am. Chem. Soc.***1969**, *91*, 6540.
[23] D. K. Cabbiness, D. W. Margerum, *J. Am. Chem. Soc.* **1970**, *92*, 2151.
[24] J. Van Alphen, *Rec. Trav. Chim. Pys-Bas*, **1936**, 55, 835.

[25] B. Bosnich, C. K. Poon, M. L. Tobe, *Inorg. Chem.*, **1965**, *4*, 1102.
[26] B. Bosnich, R. Mason, P. J. Pauling, G. B. Robertson, M. L. Tobe, *Chem. Commun.*, **1965**, 97.
[27] D. H. Busch, *Acc. Chem. Res.*, 1978, 11, 392.
[28] D. P. Rillema, J. F. Endicott, E. Papacostantinou, *Inorg. Chem.*, **1971**, *10*, 1739.
[29] F. P. Hinz, D. W. Margerum, *Inorg. Chem.*, **1974**, *13*, 2941.
[30] A. Anichini, L. Fabbrizzi, P. Paoletti, and R. M. Clay, *Inorg. Chim. Acta*, **1977**, *24*, L21.
[31] E. K. Barefield, *Inorg. Chem.*, **1972**, *11*, 2273
[32] E. K. Barefield, E. Wagner, A. W. Herlinger, A. R. Dahl, *Inorg. Synth.*, **1976**, *16*, 220.
[33] M. Beley, J.-P. Collin, R. Ruppert, J.-P. Sauvage, *J. Am. Chem. Soc.*, **1986**, *108*, 7461.
[34] J. L. Grant, K. Goswami, L. O. Spreer, J. W. Otwos, M. Calvin, *J. Chem. Soc., Dalton Trans.*, **1987**, 2105.
[35] G. De Santis, M. Di Casa, M. Mariani, B. Seghi, L. Fabbrizzi, *J. Am. Chem. Soc.*, **1989**, *111*, 2422.

Molecular Conformations in Organic Monolayers Affect Their Ability to Resist Protein Adsorption

M. Grunze and A. Pertsin

Angewandte Physikalische Chemie, Universität Heidelberg, Im Neuenheimer Feld 253, D-69120 Heidelberg, Germany
e-mail: Michael.Grunze@urz.uni-heidelberg.de

Abstract. In this article we review and discuss experimental and theoretical work which demonstrates that the surface properties of oligo(ethylene glycol) (OEG) terminated self-assembled monolayers (SAMs) are determined by the molecular conformation of the OEG moieties. This conclusion was drawn from comparison of OEG derivatized alkanethiolate SAMs on gold and silver substrates. The lateral packing density on Au allows the OEG moieties to assume a helical or "amorphous" conformation, whereas on Ag the higher packing density forces the OEG tails into a planar "all-trans" conformation. Atomistic force field calculations provide a deeper insight in this density driven transition. Using a variety of proteins in solution as a probe, it was shown that the helical and amorphous conformers are inert towards protein adsorption, whereas the planar conformer is not. Measurements of the force/distance relationship with appropriately derivatized Atomic Force Microscopy (AFM) cantilevers confirm the dependence of the sign of the force on the molecular conformation. The inertness of the helical conformer can be rationalized in terms of *ab initio* Hartree-Fock calculations on the interaction of water with the helical and all-trans conformers.

Introduction

The preparation of surfaces inert to protein adsorption is of paramount importance in basic research on the interaction of cells with artificial surfaces and in the manufacturing of implantable medical devices. Empirically, the ability of an organic surface to resist adsorption of proteins and cells is related to its surface energy, conveniently measured by its wettability [1]. In general, the more hydrophilic a surface, the more likely it will resist the non-specific adhesion of macromolecules. However, on most hydrophilic surfaces inertness is not quantitative, requiring more robust coatings for biomedical applications.

A common method for the preparation of an inert surface is by coating it with end-grafted hydrophilic polymer brushes. The resistance of surfaces coated with poly(ethylene glycol) (PEG), $(\text{-O-CH}_2\text{CH}_2\text{-})_n$ towards non-specific protein and cell adsorption is well explained by the "steric repulsion" theory [2,3] which associates the inertness of the polymer brushes with their high conformational freedom and with the unfavorable free energy change of the system when the polymer brushes are compressed and dehydrated. This theoretical concept has been expanded and refined in the last years [4,5] and has been successfully applied to interpret Surface Force Apparatus (SFA) measurements on the interaction between polymer brushes and proteins [6]. In the "steric repulsion" theory the polymer is treated as a free chain of hard balls coupled by entropic springs, void of any chemical properties characteristic of the ethylene oxide moieties comprising the PEG chain. Consequently, the driving forces for solvation and dehydration in these theories have to be entropic, manifesting the prevailing view that in general the inertness of organic surfaces is related to a steric repulsion effect. The behavior of the polymer brushes also depends critically on the grafting density and molecular weight of the polymer, and the inertness is normally lost at physiological temperatures [6].

This perception is, however, challenged by the ability of oligo(ethylene glycol) (OEG) terminated alkanethiol self-assembled monolayers (SAMs) to resist protein adsorption, as shown in a series of systematic experiments by Prime and Whitesides [7]. The surfaces are not very hydrophilic and comprise densely packed molecules with constrained conformational freedom, as compared to the end-grafted polymer chains. This makes the "steric repulsion" mechanism to account for the observations unlikely. In some of these films the OEG chain consists of only two ethylene glycol units, suggesting that another mechanism than the steric repulsion effect in end-grafted polymer films may cause the observed inertness.

In this article we discuss our work to derive a physico-chemical model to explain the dependence of the protein and cells resistance of OEG derivatized SAMs on the molecular conformations in the OEG chains. First we will summarize the experiments, followed by a section on our theoretical work, and finally discuss the findings emphasizing the need for further work to establish a general model for inert surfaces.

Experimental work

In our studies to understand the ability of (OEG) derivatized undecanethiol SAMs to resist protein adsorption [8] we investigated three different OEG termini: a tri(ethylene glycol) with an $-OCH_3$ or –OH end group (EG3-OMe and EG3-OH, respectively), a OH-terminated hexa(ethylene glycol) (EG6-OH) and a tri(ethylene glycol) with a $-CH_2-O-CH_3$ side chain (EG[3,1]-OMe) [8].

The experiments involved both SAMs formed on polycrystalline gold and silver surfaces. For EG3-OMe and EG3-OH, they revealed an unexpected and in retrospect fortunate correlation between the packing density and the molecular conformation of the OEG units. The FTIRAS spectra of both EG3-OMe and EG3-OH on the two metal surfaces recorded in dry air revealed small but significant differences, most importantly a consistent shift of the CH_2 wagging mode from 1350 cm^{-1} on the Au surface to 1320 cm^{-1} on the Ag surface, respectively (Fig 1).

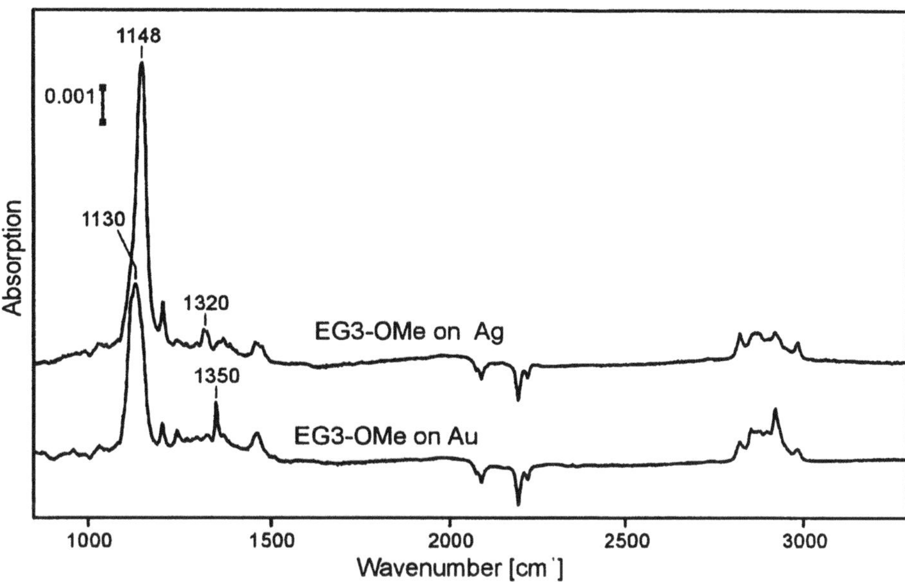

Fig. 1. FTIRAS spectra of EG3-OMe terminated SAMs on Ag and Au

As discussed in detail in Ref. 8, this shift is indicative of the presence of helical and amorphous[1] conformers in the Au-supported monolayers, and of planar "all-trans" conformations in the film on the Ag substrate, respectively. From the experiments, it was concluded that the transition from the helical to planar conformation on going from Au to Ag is caused by the higher packing density of the Ag supported monolayer due to the formation of an incommensurate solid phase with nearly upright chains and a smaller lattice spacing (4.6-4.7 Å [9] versus 5.01 Å on Au [10], which corresponds to a density difference of 8-10 %). The model derived from the experiments is schematically shown in Fig.2. It also depicts the tilt angles of the alkane and ethylene oxide chains with respect to the surface normal as determined from the spectroscopic experiments.

[1] The helical structure of PEG is characteristic of the crystalline state of the polymer. In the helical structure of PEG the sinuousness is 2/7 and the bonds of the backbone are arranged in a trans-gauche-trans (tgt) order, where the gauche angle is rotated uniformly with respect to the -C-C-O plane over the length of the helix, either clock-or counterclockwise (+ or -). Introduction of gauche rotations other than those characteristic for the uniform helix leads to a gauche defect and ultimately to an amorphous conformation in which the sense of rotation is arbitrary. We use the word amorphous here to indicate the presence of non uniform gauche rotations along the molecular axis, but we do not imply a truly random orientation which would require a larger space per molecule than available in the densely packed SAMs.

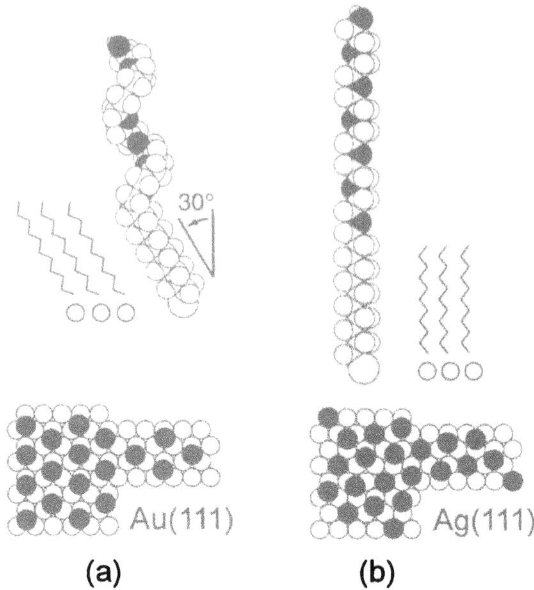

Fig. 2. The conformation and orientation of chains and also the arrangement of the sulfur headgroups in the EG6-OH terminated alkanethiol SAMs on Au and Ag

Obviously, the model refers to the orientation and conformation in an idealized homogenous SAM, ignoring the increased area available for molecules near defects in the closely packed layers or at domain boundaries. That indeed the actual density in the films is less than expected for a homogeneous SAM is confirmed by XPS measurements [8]. Using a calibration of the attenuation of the XPS intensities of the respective substrate metals covered by alkanethiolate SAMs HS $(CH_2)_nCH_3$ (n=5-21), the relative coverage of the OEG derivatized monolayers was established. It was found that the films on Au and Ag had an effective thickness of about 79 % and 85 % of the theoretical thickness, respectively. This confirms a higher defect density and/or domain boundary concentration in the OEG SAMs as compared to the alkanethiolate SAMs and implies that the monolayers can not only consist of helical (on Au) or planar all-trans (on Ag) conformers but that due to the less stringent confinement other conformers are present.

That the conformational composition is, to a variable degree, non homogeneous depends on the quality of the polycrystalline substrates and on the preparation variables (e.g. immersion time and purity of the solvents, in particular the possibility of water contamination in the ethanolic thiol solutions which may partly solvate the OEG moieties). As an example, we show in Fig. 3 a series of infrared spectra of EG3-OMe films on Au surfaces recorded after variable immersion times of the substrate in the thiol solution. These spectra reveal an variable degree of crystallinity in the films as shown by the sharpness and shape of the bands, in particular the C-O-C stretching vibration near 1139 cm^{-1}. Also, the CH_2 wagging mode, indicative for the helical or all-trans conformers increases in intensity with increasing homogeneity.

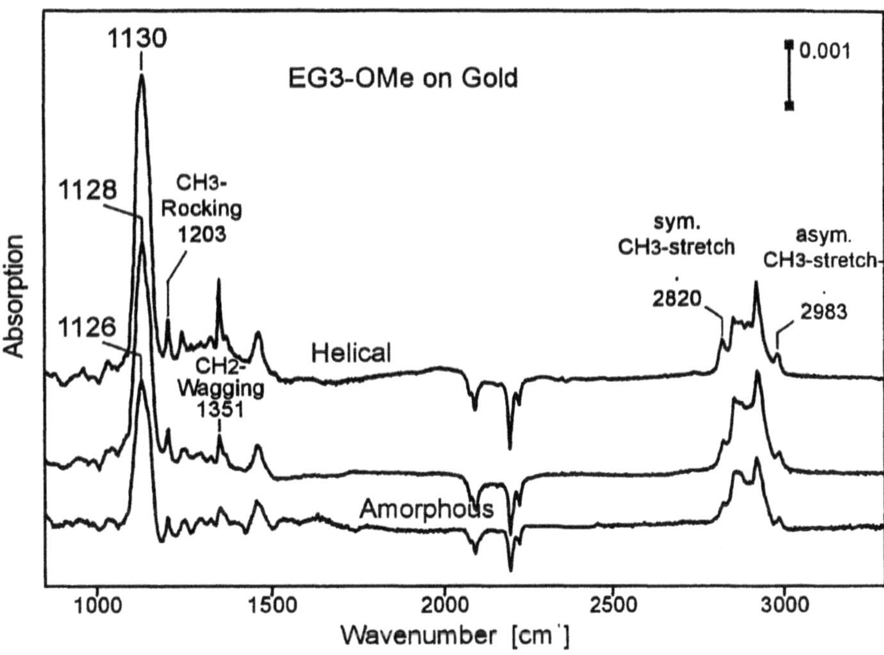

Fig. 3. FTIR spectra of EG3-OMe terminated alkanethiol SAMs on gold

Depending on coverage on Au the monolayer film will consist of amorphous and helical conformers, whereas on silver the densely packed crystalline planar "all-trans" population is diluted with helical/amorphous species. How the variability in composition affects protein adsorption will be discussed in the next section.

Protein Adsorption Studies

The work by Prime and Whitesides [7] involved a systematic study of human plasma protein adsorption on the OEG terminated undecanthiol thiol monolayers as a function of dilution with a hydrophobic undecanthiolate SAM. They measured, by ellipsometry, the thickness of the adsorbed protein layer using the pure hydrophobic undecanethiolate film as a reference for an adsorbing surface. Films with two or more ethylene glycol units on an undecylthiol spacer are completely resistant towards adsorption of the human plasma proteins tested (fibrinogen, pyruvate kinase, lysozyme or ribonuclease A). Protein resistance was exhibited not only by the pure EGn-OH with n = 1, 2, 4, 6, and EG6–OMe surfaces, but also by films which were diluted with undecanthiol up to a certain mole fraction. For these mixed OEG-SAMs Prime and Whitesides reported a chain-length dependence for protein repulsion. In general, SAMs with longer OEG chains can be more diluted with alkane thiols until their protein resistance is lost. An OEG-terminated SAM with 17 polydisperse ethylene glycol units can be diluted by up to 70 - 80 % undecanethiol until the layer starts to adsorb proteins, whereas for the shorter EG3-

OMe SAMs the resistance to fibrinogen adsorption is lost at about 35% dilution. Hence, the inertness is not sensitive to defects as far as an ethylene glycol chain length dependent average distance between the OEG molecules is not exceeded.

Assuming that as in pure alkanethiolate SAMs the molecules in the mixed layers are homogeneously distributed and arranged in a $\sqrt{3}\times\sqrt{3}$ R30^{o} structure, the nearest neighbor and next nearest neighbor distance is about 5 Å and 7.5 Å, respectively. These distances correspond in the hexagonal structure to an area per molecule of 21.4 Å^2 and 32.5 Å^2, respectively, i.e. the later is large enough for the OEG moieties to assume more relaxed structures in the mixed layers. There are no spectroscopic data reported in the study by Prime and Whitesides [7], but we speculate that the molecular conformations in these films were ranging between mostly helical in the pure EG monolayers to amorphous in the films diluted with undecanethiol. As we showed in a recent publication [11], conformers containing non-uniform gauche rotations are stabilized in water as compared to the helical form, which is found to be the lowest energy configuration in vacuum and air.

We repeated the protein adsorption studies on the OEG terminated films on Au and Ag surfaces with fibrinogen, which is a typical high molecular weight human plasma protein, and Mfep-1, the adhesion protein of the common black mussel *Mytilus edulis* [2]. We found that the helical and amorphous conformers of EG3-OMe, EG3-OH, EG6-OH and (EG[3,1]-OMe), assembled on the gold surfaces, are inert towards adsorption of both proteins, whereas the planar all-trans conformers of EG3-OMe and EG3-OH present on the silver substrates are not [8,12]. Due to the subtle balance between the lattice energy and the intramolecular stabilization of the helical conformation in the non aqueous solution from which the films are prepared, the longer chain EG6-OH molecules cannot be compressed into a planar all-trans structure on silver (see the theoretical section) and therefore are also present in a helical/amorphous form on silver rendering the surface inert towards adsorption. In the case of EG[3,1]-OMe, intramolecular steric constrains prevents ordering into the planar phase on the silver substrates, also rendering the Ag surface resistant to protein adsorption.

In the case of EG3-OMe (and EG3-OH) on Ag we found a high variability in the relative concentration of the amorphous versus the planar conformers allowing to make a correlation between protein resistance and surface composition. As a convenient measure for the degree of non uniform gauche rotations we used the vibrational frequency and intensity of the C-O-C stretching vibration [12]. The frequency and intensity increases with increasing concentration of the planar conformer in the film, as shown for eight different samples in Fig 4. Immersion into a buffered fibrinogen solution for 15 min leads to a variable degree of

---

[2] The foot protein (Mefp-1) of *M. edulis* is a prominent protein in the byssus -a fibrous holdfast structure- and is used by marine mussels for underwater adhesion to hard surfaces. Purified Mefp-1 is a flexible, rod-shaped molecule in solution with a mass of 110 kDa . It is the main component in formulations for cell and tissue attachment (Cell-Tak, BioGlue) because it has excellent adhesive strength, is nontoxic, and does not interfere with cell division and spreading. Mefp-1 consists of more than 50% hydroxy-containing amino acids in its primary structure. Eighty percent of the primary sequence consists of decapeptide repeats that have conserved lysines at positions 2 and 10, hydroxylated prolines at positions 3, 6 and 7, and 3, 4-dihydroxyphenylalanines (DOPA) at positions 5 and 9 .

adsorption as reflected in the appearance of the Amide I and Amide II bands, ranging from 1.8 % of a monolayer for the film with the highest concentration of amorphous conformers, to 25% of a monolayer for the most crystalline film.

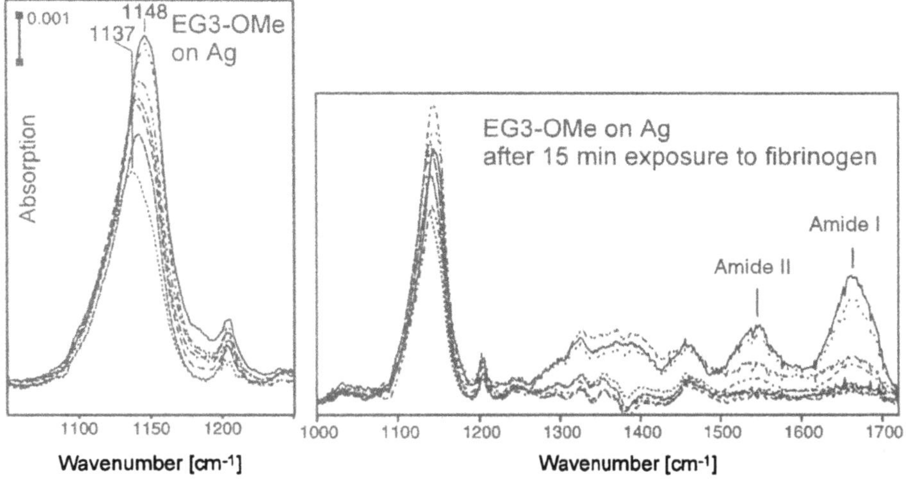

Fig. 4. FTIRAS spectra of EG3-OMe terminated alkanethiol SAM on Ag.

Table 1 summarizes these results, and also shows that immersion of the films into the buffered protein solution leads to a decrease in intensity and small but noticeable shifts of the C-O-C stretching vibration to lower wavenumbers,

sample	Fibrinogen adsorption [% of monolayer]	Frequency and intensity of C-O-C stretching vibration before (after) fibrinogen adsorption	
		Frequency [cm^{-1}]	Intensity [arb. units]
1	1.8	1137.4 (1137.0)	4330 (3950)
2	2.8	1141.9 (1141.3)	5020 (4650)
3	2.2	1140.7 (1139.6)	5990 (5220)
4	2.0	1141.5 (1140.3)	6160 (5310)
5	7.3	1146.4 (1143.7)	6980 (6130)
6	9.0	1143.4 (1141.9)	6960 (4030)
7	21.0	1148.0 (1145.6)	7450 (n.a.)
8	25.7	1146.6 (1144.6)	7420 (n.a.)

Table 1. Frequency and intensity of the C-O-C stretch before and after fibrinogen adsorption on EG3-OMe/Ag SAM. Fibrinogen coverage was derived from the intensities of the Amide I and Amide II bands as described in Ref. 8.

indicative of relaxation's in the film from the "all-trans" conformers into conformations containing gauche bonds. This observation is consistent with the fact that for unconstrained OEG chains the amorphous conformers are stabilized in aqueous solution with respect to the helical and all-trans conformers [11]. Since there is a correlation between the relative concentration of the planar species with the amount of fibrinogen adsorbed, we can also conclude that irreversible adsorption on the EG3-OMe surfaces depends on the *local* composition, i.e. it occurs only in the film areas containing "all- trans" conformers.

The inertness of the helical and amorphous surfaces is independent of the nature of the protein, and of the terminal functionality of the OEG moieties. Replacing the methoxy group by a hydroxyl functionality did not change the correlation between conformation and resistance, i.e. the effect of inertness seems to be independent of the contact angles measured with the tensile drop method (63° on the $-OCH_3$ and 35° on the -OH terminated surfaces, respectively).

The difference of the conformers with respect to their fibrinogen adsorption characteristics was later verified by force/distance measurements between differently derivatized AFM tips and the OEG terminated SAM surfaces on Au and Ag [13]. The first experiments involved the measurement of the force/distance curves for a silicon nitride tip coated with a fibrinogen layer. The OEG derivatized SAMs on Au exhibited a long-range repulsion (a repulsive force was sensed at about 60 nm above the surface). A shorter range attractive force was measured over the silver surface, again in agreement with the adsorption experiments from solution. However, the thickness of the fibrinogen layer on the cantilever is ill defined, introducing large uncertainties in the measured ranges of interaction. Therefore, the tip surfaces were coated with a gold film and derivatized with a hydrophobic hexadecanethiol SAM as a model for the hydrophobic domains of proteins. We found that the repulsive interaction between helical and amorphous conformers on gold and a hydrophobic tip is also of extremely long-range (60 nm) in deionized water, and is reduced to a hard-wall repulsion of about 4-6 nm range with increasing electrolyte concentration, consistent with the decrease in Debye length (see Fig 5). In these AFM model experiments both surfaces are electrically connected and neutral, *i.e.* no electrostatic surface charges are present which could explain the long-range interaction. The long-range repulsion over tens of nanometers which is not caused by a Coulomb type electrostatic repulsion of point charges can not be explained by present theories. Note also, that repulsion in the force/distance experiments is not observed in other solvents than water [13].

In the same experiment the interaction of a hydrophobic tip with an end-grafted PEG brush was measured [13] in order to confirm that the long-range repulsion measured on the SAM surface is unique to these systems and distinguishes them from the end-grafted PEG surfaces (Fig 5). Quite unexpected, the denser packed films exhibit a longer range repulsion than the polymeric brushes. The range of the repulsive forces observed for the thiol-coupled PEG graft of molecular weight 2000 ($n \approx 45$) is about 6 nm (approximately the brush thickness) and is not scaled to the Debye length, emphasizing that indeed a different mechanism is responsible for resistance of protein adsorption to the oligomeric quasicrystalline layers and the less densely packed polymeric brush surfaces, respectively.

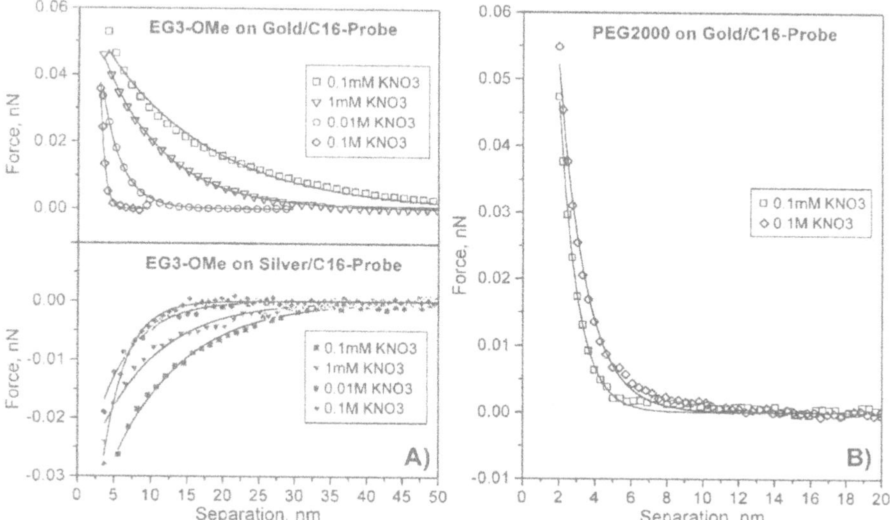

Fig. 5. Force/distance curves for the interaction of a hydrophobic AFM probe with EG3-OMe terminated SAM on Au and Ag (A) and with an end-grafted PEG brush (B)[13].

To summarize the experimental studies, the results obtained by immersing the helical/ amorphous or planar all-trans conformers in a protein solution and the force /distance measurements using derivatized AFM tips suggest, that there is a unique correlation between molecular conformation and inertness, and that the resistance to adsorption of these OEG derivatized SAMs does not depend on the nature of the protein nor the water contact angle. We emphasize, however, that the correlation refers to the conformations measured in dry air. In particular in the less densely packed films containing defects, other conformers than those identified in the dry state may be energetically and entropically more stable in water.

Computational studies

As we discussed in the previous section, our FTIRAS experiments [8] with OEG terminated SAMs detected a distinct change in the OEG conformation on going from the Au to the Ag substrate. The interpretation of this change in terms of specific conformers was, however, to a certain extent speculative because of the variable reproducibility of the spectra. To elucidate the detailed SAM structure on Au and Ag and to put our inferences from FTIRAS to an independent test, we resorted to molecular mechanics simulations [14]. The particular SAMs studied were the ones terminated by EG3-OMe moieties. To find out low-energy SAM configurations, we thoroughly explored the configurational space of the SAMs by combining the methods of stochastic global search and local minimization. The intra- and intermolecular interactions in the SAMs were calculated using a classical atomistic force field fitted to *ab initio* MP2 level results for a low-molecular OEG homologue [15]. The SAM lattice on Au and Ag was assumed to have the same symmetry but different period (to reflect the difference in areal density between the SAMs).

For the Au-supported SAM, the global search found a lot of low-energy configurations differing both in the conformation of the OEG tail and in the arrangement of the molecules on the metal surface. In the lowest energy configuration, the OEG tail assumed a $(tgt)_3$ conformation like a helix (though the deviations from the ideal 7/2 helix were substantial). Compared to the lowest-energy all-trans $(ttt)_3$ configuration, the best $(tgt)_3$ structure was 0.5 kcal/mole lower in energy. To appreciate the origin of this difference, it is instructive to refer to Table 1, which shows the energetics of various conformers containing tgt EG units, relative to the corresponding all-trans conformers for isolated OEG chains and SAMs.

tgt[a]	tgt	ttt tgt	$(ttt)_2$tgt	$(tgt)_3$	$(tgt)_3$ in SAM
0.15	0.14	-0.02	-0.02	-0.3	-0.5

Table 2. Energies of low energy OEG conformers containing tgt units, relative to the corresponding all-trans forms, kcal/mole. [a]*Ab initio* results for dimethoxyethane [15]

For a single EG unit, the force field calculation gives nearly the same result as the *ab initio* calculation for dimethoxyethane: The tgt conformer is 0.14 kcal/mole higher in energy than the all-trans one. When one or two all-trans EG units are attached to the tgt one, the resulting chain becomes energetically favored. An analysis of the individual components of the conformational energy shows that the tgt conformer is mostly stabilized due to the electrostatic interaction with the neighboring unit, which favors the tgt conformer over ttt because of a higher dipole moment of the former. The stabilization of the tgt conformer is particularly pronounced when all the EG units assume the tgt conformation. Further stabilization of the tgt conformer occurs in the SAM because of the intermolecular interaction between different chains. Thus the calculation results suggest that the cause of the so called "gauche effect" is not in the conformational properties of the EG unit itself, but in the "self-stabilizing" effect of the neighboring units in the same and neighboring chains.

To simulate the transition from the Au to Ag substrate, the minimum energy SAM structures were calculated as a function of the lattice period c starting from $c = 5$ Å, the value for the Au-supported SAM, and down to values characteristic of the denser Ag-supported SAM [10]. When c reached ~4.6 Å, both the helical and all-trans configurations experienced a structural transformation, so that the molecules initially tilted by ~30° from the surface normal became nearly upright, as observed experimentally on Ag. As the lattice spacing was decreased, the difference in lattice energy between the helical and all-trans configurations reduced until the energy curves intersected and the all-trans structure became more favorable (Fig. 6). Compared to the experimentally observed lattice period of the SAM on Ag (4.6-4.7 Å [9]), the calculated transition lattice period is 0.2-0.3 Å too small. This result is not unexpected since static lattice energy calculations always underestimate lattice periods because of the neglect of thermal expansion. As a consequence, the whole phase diagram of the system shifts to higher densities.

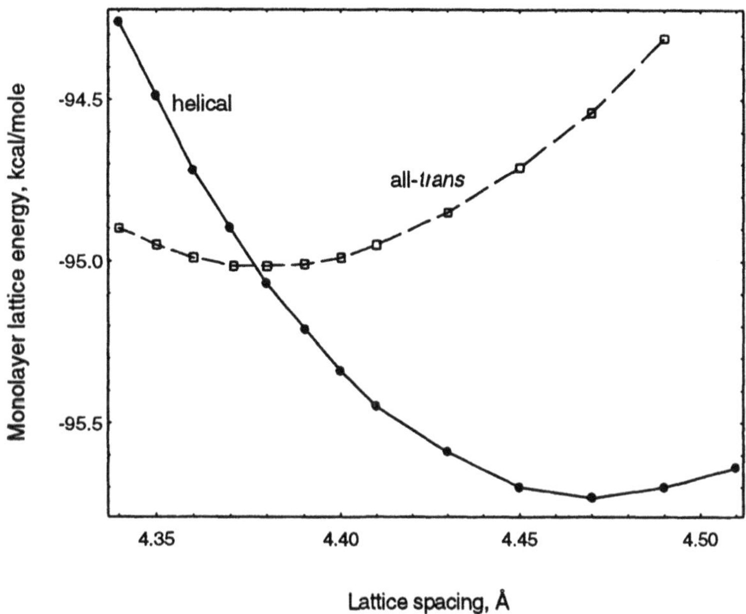

Fig. 6. Lattice energy of the helical and all-trans SAM configurations as a function of the lattice spacing

The reasonableness of the calculated SAM configurations was further tested through calculation of the vibrational spectra of the OEG tail. The results are summarized in Table 3. Only modes whose transition dipole moments are preferentially parallel to the surface normal are listed. The data refer to the region of 1000-1400 cm^{-1}, where spectral changes are observed in going from the Au to Ag substrate [12]. The corresponding experimental results are given in parentheses in the first column. The major arguments put forward in our experimental work [8] in favor of the transition from the helical to all-trans conformation were: (1) the disappearance (at least, a drastic decrease in intensity) of the CH_2-wagging mode at 1350 cm^{-1}, (2) the appearance of a new CH_2-wagging band at 1325 cm^{-1}, and (3) a shift of the strongest CO-stretching vibration from 1130 to 1145 cm^{-1}. The corresponding calculated modes (at 1376, 1327, 1122 and 1145 cm^{-1}) well reproduce this behavior.

[a]ν, cm^{-1}	$(d\mu/dQ)_z$	potential energy distribution, %	[b]assign-ment
helical configuration			
[c]1382 (1370 w)	1.6	88 CCH	W CH_2
1376 (1350 s)	1.9	88 CCH	W CH_2
1368	-0.2	86 CCH	W CH_2
1287 (1297 w)	-3.2	93 CCH	T CH_2
1239 (1244 m)	-0.1	98 CCH	W CH_2
1224	-0.5	86 CCH	W CH_2
1210 (1204 s)	-0.3	16 C-O, 60 CCH	T CH_2
1157 (1145 sh)	-0.4	46 C-O, 17 C-C, 20 CCH	νC-O
1122 (1130 s)	-1.4	61 C-O, 16 C-C	νC-O
1110 (1114 sh)	0.9	52 C-O, 17 C-C	νC-O
1094	1.3	46 C-O, 36 CCH	νC-O
1049	-1.1	18 C-O, 31 C-C, 30 CCH	νC-C
1018	-1.0	10 C-O, 30 C-C, 30 CCH	νC-C
all-trans configuration			
1380 (1370 w)	-0.8	100 CCH	W CH_2
1358 (1356 w)	1.0	80 CCH	W CH_2
1327 (1325 m)	3.5	88 CCH	W CH_2
1209 (1204 s)	-0.4	24 C-O, 50 CCH	W CH_3
1126	0.1	72 C-O, 25 C-C	νC-O
1145 (1145 vs)	-1.3	84 C-O, 9 C-C	νC-O
1114 (1114 sh)	-0.8	50 C-O, 30 C-C	νC-O
1090	0.6	59 C-O, 36 C-C	νC-O
1035	-0.5	27 C-O, 17 C-C, 15 CCO	νC-O

Table 2. Vibrational frequencies, projections of transition dipole moments onto the surface normal, potential energy distributions, and mode assignments in the region of 1000-1400 cm^{-1} for the helical and all-trans SAM configurations. [a]Given in parentheses are the frequencies of experimental FTIRAS bands [8] (s, strong; m, medium; w, weak; vs, very strong; sh, shoulder). [b]W and T denote wagging and twisting modes, respectively. [c]The most characteristic frequencies of both configurations are shown underlined

Models to explain the adsorption behavior of proteins

From the experiments described by Prime and Whitesides [7], and our spectroscopic investigation of the Au and Ag supported monolayers [8,12], it follows that the ability to resist protein adsorption depends on the lateral density of the OEG chains in the SAMs, which in turn determines the molecular conformation. Prime and Whitesides [7] noted that protein resistance in the undecanethiol diluted OEG derivatized SAMs can be empirically correlated to the OEG concentration in the layers by $\chi_{res} = k\ n^{-0.4}$ where n is the number of EG moieties and χ_{res} is the minimum mole fraction of OEG terminated alkanethiols necessary to render the surface inert. This relationship is similar to the equation given by de Gennes [16] $\phi \approx \rho\ n^{-2/5}$ for the average number ϕ of monomer groups per unit area found at a certain distance from the surface in relation to of the density ρ of grafted polymer chains. Although some assumptions of the cited theory are certainly not fulfilled in simple OEG SAMs, such as missing interactions between the grafted chains and a minimum number n of at least 100 monomer units, the agreement of the exponents of the two equations suggests that an OEG monolayer needs a certain *minimum* density of EG segments per volume in order to become resistant.

Our experiments show, that there is also a *maximum* limit to the density in inert films, which coincides with the transition from the helical to the planar "all-trans" conformation. Hence, we have to distinguish between three different regimes, i.e. the protein adsorbing planar "all-trans" structure, the resistant helical/amorphous SAM, and the diluted OEG SAM which again adsorbs proteins. Here we compare the planar "all-trans" with a helical SAM, not discussing the transition from the amorphous to the diluted adsorbing monolayer, where the loss of inertness has been explained by the too low concentration of the OEG chains [4].

That the undiluted monolayers adsorbed on gold are unique follows from the observation that the repulsive force measured with the hydrophobic AFM tips exceeds by far the range of repulsive interactions consistent with a steric repulsion mechanism as discussed for end-grafted PEG layers, and that there is no obvious correlation with the hydrophilicity of the SAMs in the protein adsorption measurements, nor the chemical nature or size of the protein used as a probe. The only common ingredient in the experiments with the Au supported monolayers is the solvent water.

Considering that water is essential we concluded [17] that the observed difference in protein resistance is due to a difference in the way the helical and planar "all-trans" conformers interact with water. To test the validity of this suggestion, we explored the microscopic structure of the SAM/water interphase region using *ab initio* 6-31G* SCF calculations [17].[3] The systems considered represented clusters containing up to 12 EG3-OMe terminated alkanethiol strands and up to 20 water molecules. We found that because of the three-dimensional structure of the helical OEG strand, a water molecule can interact with two successive O atoms of the strand to form a very strong bridge-mode hydrogen bond. Because of the zigzag structure of the planar OEG, there is no such bridge-mode hydrogen bond of water on it. For a hexagonal lattice of planar OEG strands with a lattice constant of 5 Å there is not enough room to allow water molecules to move down to form a strong hydrogen bond with the top O atoms of the OEG strands.

---

[3] Some selected MP2 calculations were also done to check the accuracy of our results

By contrast, for a similar lattice of helical OEG strands, its open structure and its surface easily accommodates water molecules to its topmost O atoms. With a larger period (beyond 5.5 Å) for the hexagonal lattice of helical OEG strands, water molecules can move further down between the OEG strands to form very strong single and double bridge-mode hydrogen bonds. This later situation is most likely realized at defects and domain boundaries in the Au supported monolayers.

A remarkable property of the helical SAM phase found in these static calculations was not only its ability to strongly bind water, but also to allow the water molecules in contact with the SAM surface to assume a density (expressed by the pair correlation function of the water oxygen atoms) similar to bulk water. Shown in Fig. 7 is the electrostatic potential of a cluster of four helical OEG strands, which act on a single water molecule approaching the SAM surface. For comparison, Fig. 8 shows the electrostatic potential created by the same four OEG strands together with six water molecules adsorbed in their lowest energy positions. One can clearly see a close similarity between the kidney-shaped potentials in Figs. 7 and 8. It is this similarity of the local field at the bare OEG surface to that inside the water adlayers and ultimately to bulk water that in our hypothesis facilitates a interface without a weak boundary layer between the two materials

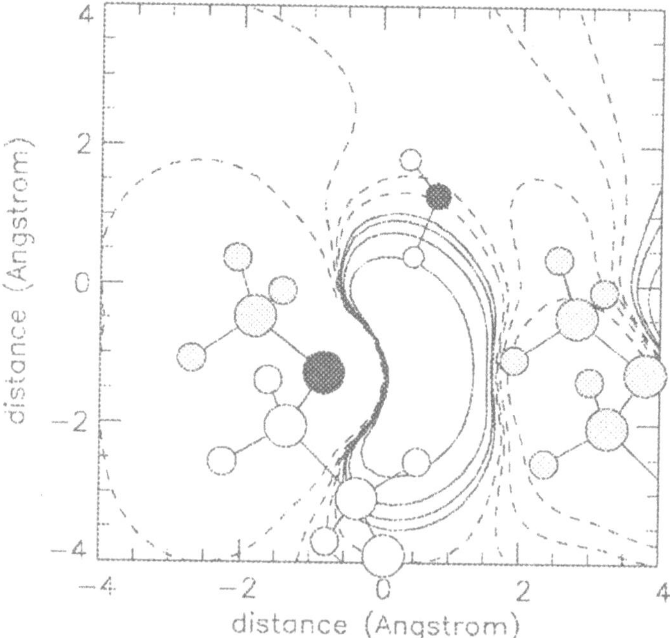

Fig. 7. Electrostatic potential of a cluster of four helical OEGs separated by 5 Å acting on a water molecule, in a plane perpendicular to the surface. Solid lines are for -1.5, -1.0, and -0.5 eV, and dashed lines for 0.5, 1.0, and 1.5 eV. Atoms depicted as full, shaded, and open circles are in, below, and above the plane, respectively.

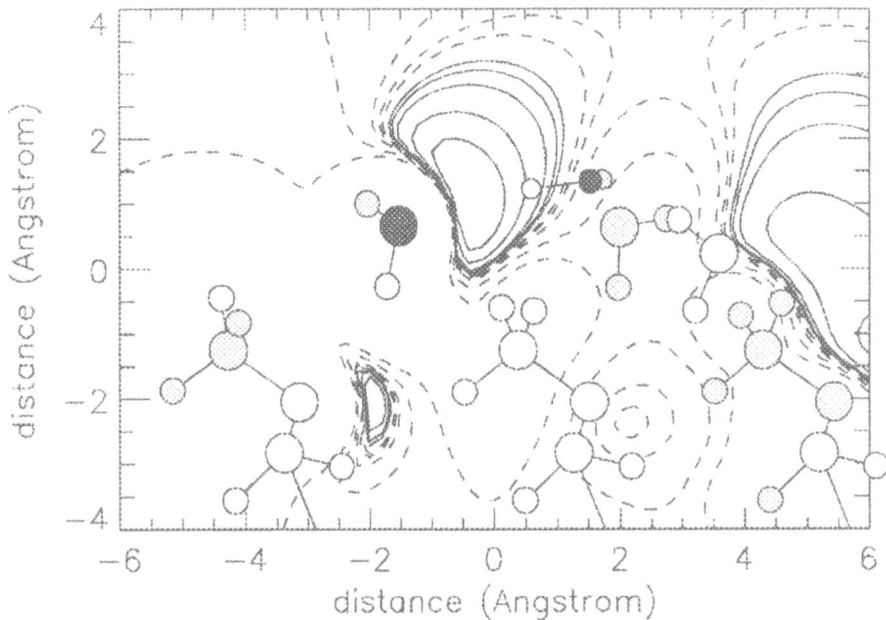

Fig. 8. Electrostatic potential produced by a cluster of four helical OEGs (as in Fig. 7) together with three adsorbed water molecules, acting on a fourth water molecule. Otherwise as Fig. 7.

In our model, the ability of the helical SAM to form strong hydrogen bonds with water molecules, without constraining the ability of the water molecules to interact also with their next neighbors as in the bulk liquid, generates an tightly bound interphase water layer at the SAM surface preventing the surface from direct contact with protein molecules. By contrast, no interphase water layer capable of hindering protein adsorption forms near the surface of the planar SAM phase, and so the latter is not protein resistant.

The above described theoretical models provides an explanation for the protein resistance of a crystalline lattice of helical OEG conformers. However, as discussed above, the SAMs contains other conformers due to defects and domain boundaries, and thus the calculations on the crystalline OEG surfaces can only be considered as a first approximation to explain the observed inertness. Also, this model can hardly explain the long-range repulsive forces measured by AFM. In order to verify or falsify the concept of a "template" for liquid water, experimental evidence and numerical simulations for the existence of a protective water layer on the SAM surfaces has to be obtained. The relevant NMR and neutron reflectivity measurements, as well as grand canonical Monte Carlo simulations of the SAM/water interphase region, are currently under way.

Closely related to the above-discussed computational work are our calculations of the response of OEG molecules to mechanical stretching and electrostatic fields [11,18,19]. Studies of the stretching behavior of OEG were undertaken in connection with recently reported force-extension measurements on individual

PEG chains grafted to a substrate and elongated using an AFM tip in different solvents [20]. The measurements showed that in the range of intermediate forces the force-extension relationship is strongly dependent on the solvent surrounding the molecule: In PBS buffer, the PEG chains are substantially stiffer than in the non-polar hexadecane solvent. Our theoretical work on stretching of OEG chains in water and vacuum agrees almost quantitatively with the experimental observations. We find, that the solvated OEG units are more stable than those in vacuum, and that amorphous hydrated OEG units are energetically stabilized with respect to the hydrated helical conformers. This implies for our SAM studies that water is likely to penetrate the OEG terminated SAM and restructures the organic surface.

Effects of external electrical fields on the structure of the SAM

Our computational studies of the effect of electrostatic fields on OEG molecules were stimulated by experimental results which suggest that local electrostatic fields such as those generated by polar solvent molecules, ions, or a charged AFM tip can induce substantial conformational changes in thin organic films, in particular, in OEG terminated SAMs. In order to understand if and to what extent electrostatic fields can influence the conformation and structure in such films, we undertook *ab initio* and force field calculations of the field effects. We considered the simplest kind of the field, namely an external homogeneous electrostatic field $\boldsymbol{F}$.

We began with *ab initio* HF calculations of the effect of an external homogeneous field on the helical and planar conformers of a single OEG molecule. At low field strengths (less than 0.5 V/Å), the electrostatic field applied along the molecular axes affects mainly the distribution of the electron density along the molecule, though the helical conformer also shows perceptible conformational changes at $F > 0.3$ V/Å. At higher fields, up to 1.5 V/Å, the helical conformer experiences substantial conformational changes, whereas the planar all-trans conformer remains practically intact. As the field strength is further increased, both of the conformers undergo field dissociation.

The conformational changes experienced by the helical conformer in the range 0.3 – 1.5 V/Å are very much akin to electrostriction: With increasing field strength, the dihedral angles change so that the overall length of the molecule reduces. For EG3, for instance, the distance between the terminal carbon atoms decreases form 10 Å at $F = 0$ to 8 Å at $F = 1$ V/Å. The electrostriction leads to formation of an additional intramolecular hydrogen bond, which further stabilizes the helical conformer. Substantial changes are also observed in the orientation of the terminal methoxy groups, which are either stretched or bent depending on the field polarity. Clearly, the response of a single OEG molecule to electrostatic fields may well differ from that of the same molecule in a SAM because of the interaction of this molecule with its neighbors. To clear up this point, we performed force field calculations of the equilibrium structure of an EG3-OMe terminated alkanethiol SAM as a function of an electrostatic field, F_z, applied perpendicular to the SAM surface. The effect of field was included in the potential energy simply by adding the terms $q_i F_z z_i$, where q_i is the charge on atom i and z_i is the projection of the position vector of this atom on axis z. To make allowance for the electronic polarization, the values of q_i were taken as obtained in our *ab initio* calculations at the corresponding field. The equilibrium structure of the SAM was determined in two ways: (1) by minimization of the static lattice energy (to give the equilibrium structure at absolute zero) and (2) through Monte Carlo (MC) simulations of the SAM at room temperature.

The field-induced changes in the equilibrium configuration of the SAM at $T = 0$ can be appreciated from Fig. 9, which shows the conformation of the EG3-OMe

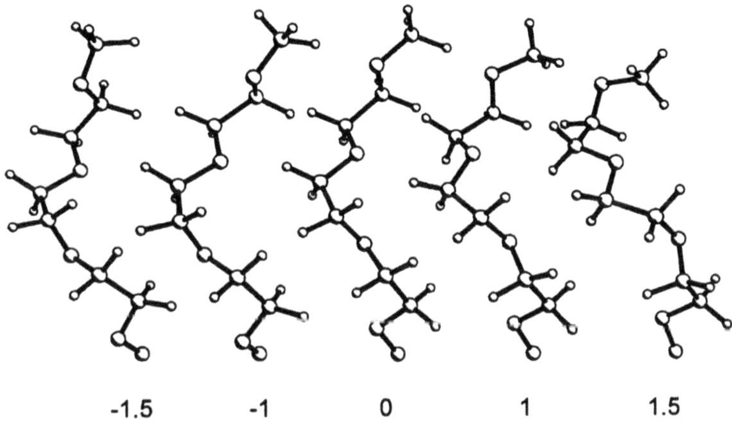

Fig. 9. Conformation of the OEG tail in the SAM at $T = 0$ as a function of the field strength, V/Å.

tails in the SAM as a function of the applied field. It is seen that the tail gradually shrinks with increasing electrostatic field, just as it is observed with single OEG molecules. One can also see that positive fields force the terminal methoxy group to tilt closer to the surface plane thus exposing its oxygen atom. By contrast, at negative fields the orientation of the methoxy group becomes closer to the surface normal, so that its oxygen atom buries deeper in the SAM.

Our MC simulations at various fields showed, in general, similar trends in the conformation of the OEG tails and in the orientation of the terminal methoxy groups. What was new is a substantial disordering of the SAM structure produced by negative fields (inward the surface). This disordering can be appreciated from a snapshot of the SAM in Fig. 10. One can see that the applied field affects not only molecular conformation but also the orientation of the alkanethiol chains around their axes.

The observed ability of external electrostatic fields of the order of 1 V/Å to make the topmost oxygen atoms of the SAM buried or exposed depending on the field polarity implies a potential possibility of developing SAM coated surfaces with electrically controlled (switchable) properties. Local fields of similar strength can also be created by solvent molecules, so that similar solvent effects on the SAM structure are quite likely.

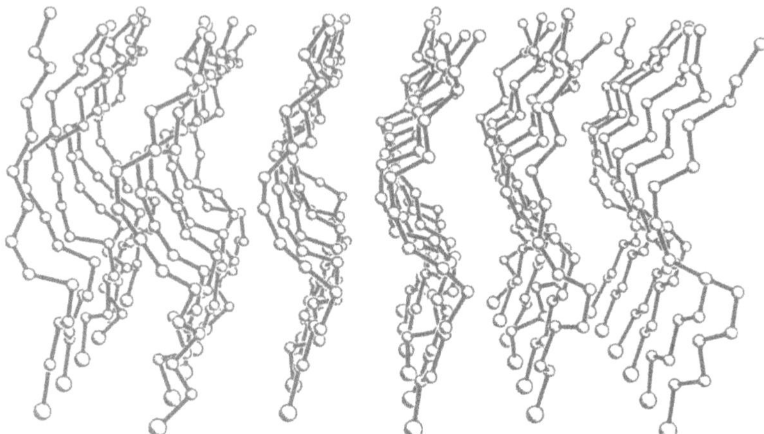

Fig. 10. A snapshot of a 36 molecules fragment of the SAM at $T = 298$ K and $F = -1$ V/Å. (The hydrogen atoms are not shown for clarity)

Conclusions

The dependence of inertness of the OEG terminated SAMs on density, and thus molecular conformation, has been well established by the experiments and can in general be related to the way the helical and amorphous or the planar all-trans moieties interact with water. *Ab initio* Hartree Fock (HF) calculations on the interaction of water with the helical and amorphous or planar all-trans conformer [17] emphasize the basic differences between the two surfaces: The helical surface provides the hydrogen bond acceptor sites for water to form strong hydrogen-bridge bonds, and the resulting three dimensional electrostatic fields generated by the polar moments in the hydrated OEG surface provide a "template" for a protective liquid water film, effectively preventing the macromolecules to attach to the substrate via van der Waals forces. In contrast, the molecular conformation of the planar "all-trans" form does not provide efficient enough hydrogen bridge bonding sites, and hence no protective water film is formed. This model can help to rationalize the observed conformation dependent macroscopic properties of the organic film, although it certainly can not explain the long-range repulsive forces measured with AFM. Direct experimental evidence for the protective water layer on the SAM surfaces has to be obtained to support the model, and Neutron Reflectivity measurements are currently on the way to probe the density profile of OEG tails and water in the interface of the SAM with the bulk liquid.

Diluting the OEG SAMs with alkanethiols maintains protein resistance up to a critical OEG concentration, above which inertness is lost. In this coverage regime the conformational composition of the films is less defined, and it might well be that the mechanism of protein resistance is related to entropic forces as might be indicated by the dependence of protein resistance on OEG density. This possibility could be investigated in more detail by force/distance measurements to see if the range of repulsive forces is consistent with a steric repulsion effect.

The results described here are important for several reasons. First, they demonstrate that the macroscopic behavior of a surface is controlled and ultimately can be manipulated by the conformation of the molecular constituents. Second, the range of repulsion exerted by the uncharged densely packed short chain molecules exceeds the range of steric repulsion characteristic of grafted PEG surfaces, suggesting that there is another and so far unexplained mechanism rendering surfaces inert. Third, our theoretical models emphasize the contributions quantum mechanics can make (but also shows the present limits) to understand complex systems relevant to biological interfaces. Finally, the results discussed here provide new design principles for biocompatible surfaces.

Acknowledgements

This work would not have been possible without the experimental contributions by P. Harder, M. Buck, G. Whitesides and P. Laibinis, and the experiment and theoretical collaboration with K. Feldman, G. Hähner, N. Spencer and H. J. Kreuzer respectively. Financial support was obtained from the Deutsche Forschungsgemeinschaft, the Office of Naval Research, and the Fond der Chemischen Industrie.

References

1 E.A. Vogler, *Adv. Coll. Interf. Sci.* **1998,** *74*, 69.
2 J.M. Harris, *Poly(Ethylene Glycol) Chemistry*: PLenum: New York, **1992**.
3 S.I. Jeon, J.H. Lee, J.D. Andrade, P.G. de Gennes, *J. Colloid Interf. Sci.* **1991**, *142*, 149. S.I. Jeon, J.D. Andrade, J.D. *Ibid.* **1991**, *142*, 159.
4 I. Sleifzer, *Curr. Opin. Solid State Mat. Sci.* **1996**, *2*, 337.
5 A. Halperin, *Langmuir* **1999**, *15*, 2525.
6 D.E. Leckband, S. Sheth, A. Halperin, *J. Biomaterials Sci., Polymer Ed.* (in press).
7 K.L. Prime, G.M. Whitesides, *J. Am. Chem. Soc.* **1993**, *115*, 10714.
8 P. Harder, M. Grunze, R. Dahint, G.M. Whitesides, P.E. Laibinis, *J. Phys. Chem. B* **1998**, *102*, 426.
9 P. Fenter, P. Eisenberger, J. Li, N. Camilone, III, S. Bernasek, G. Scoles, T.A. Ramanarayanan, K.S. Liang, *Langmuir* **1991**, *7*, 2013. M.G. Samant, C.A. Brown, J.G. Gordon, II, *Langmuir* **1993**, *9*, 1082.
10 N. Camilone, C.E.D. Chidsey, G.-Y. Liu, G. Scoles, *J. Chem. Phys.* **1993**, *98*, 3503.
11 H. J. Kreuzer, R. L. C. Wang, M. Grunze, *New J. Phys.* **1999**, 1,1-1.16
12 P. Harder, Dissertation, Universität Heidelberg, **1999**.

13 K. Feldman, G. Hähner, N. D. Spencer, P. Harder, M. Grunze, *J. Am. Chem. Soc.***1999**,*121,*10134.
14 A. J. Pertsin, M. Grunze, L. A. Garbuzova, *J. Phys. Chem. B* **1998**, *102*, 4918.
15 R.L. Jaffe, G.D. Smith, D.Y. Yoon, *J. Phys. Chem.* **1993**, *97*, 12745. G.D. Smith, R.L. Jaffe, D.Y. Yoon, *J. Phys. Chem.* **1993**, *97*, 12752.
16 P.G. de Gennes, *Macromolecules* **1980**, *13*, 1069
17 R. L. C. Wang, H. J. Kreuzer, M. Grunze, *J. Phys. Chem. B* **1997**, *101*, 9767
18 R. L. C. Wang, H. J. Kreuzer, M. Grunze, A. J. Pertsin, submitted to PCCP, 1999.
19 A. J. Pertsin, M. Grunze, H. J. Kreuzer, R. L. C. Wang, submitted to PCCP, 1999.
20 F. Oesterhelt, M. Rief, H. E. Gaub, *New J. Phys.* **1999**, *1*, 6.1

Electrochemical and Structural Aspects of Metallofullerenes

Piero Zanello

Dipartimento di Chimica dell'Università di Siena, Via Aldo Moro, 53100 Siena, Italy

e-mail: zanello@unisi.it

Abstract. The presence of a metal inside or a metal fragment outside the cage of fullerenes should increase the attention towards these spectacular molecules as far as either their molecular structures or their ability to exchange electrons are concerned. The present review paper gives an updated survey of the redox aptitude of either exohedral or endohedral metallofullerenes. As far as possible structural details are illustrated for those molecules the electrochemistry of which has been studied. Literature data are also reported for those complexes, which have been characterised only from the crystallographic point of view.

Fullerenes are molecules that not only possess quite aesthetically appealing molecular geometries, Figure 1 [1], but they also exhibit highly spectacular redox aptitude, in that they display a rich series of one-electron transfers. For instance, as illustrated in Figure 2, C_{60} and C_{70} undergo up to six, reversible, one-electron reductions [2a]. Substantially similar responses are given by the higher fullerenes C_{76}, C_{78} [2b], C_{82} [2c], and C_{84} [2d].

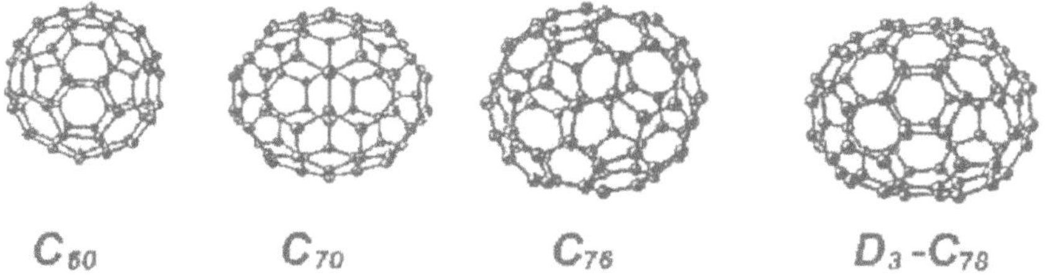

Fig. 1. The molecular structure of a few fullerenes. (Adapted from Ref. 1)

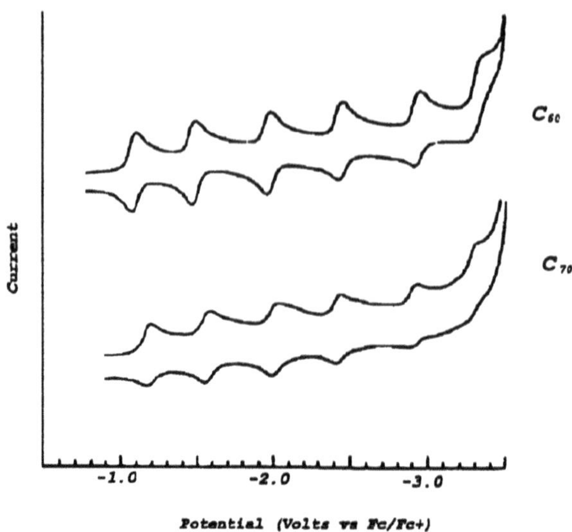

Fig. 2. Cyclic voltammetric responses recorded at a glassy carbon electrode on MeCN-toluene solutions of: (a) C_{60}; (b) C_{70}. (Adapted from Ref. 2a)

Indeed, such remarkable electron-transfer ability was not unexpected in that highly conjugated arene systems commonly exhibit high redox propensity. This is for instance the case of decacyclene, Figure 3 [3].

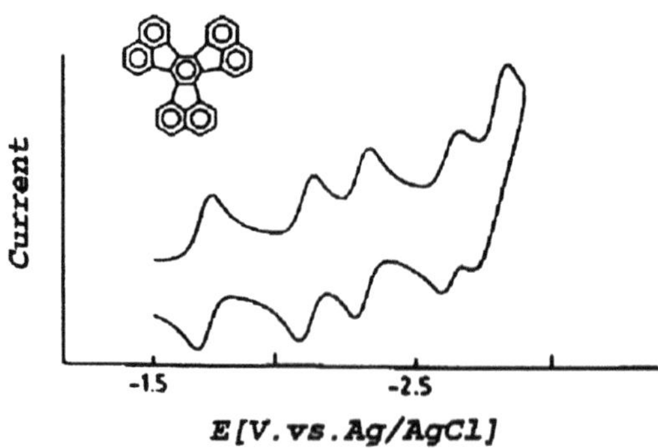

Fig. 3. Cyclic voltammogramm of decacyclene in THF solution. (Adapted from Ref. 3)

Theoretical calculations have well accounted for the ability of fullerenes to accept electrons [4]. For instance, C_{60} possesses two triply-degenerate nonbonding LUMOs, which make it potentially able to accept up to 12 electrons, Figure 4.

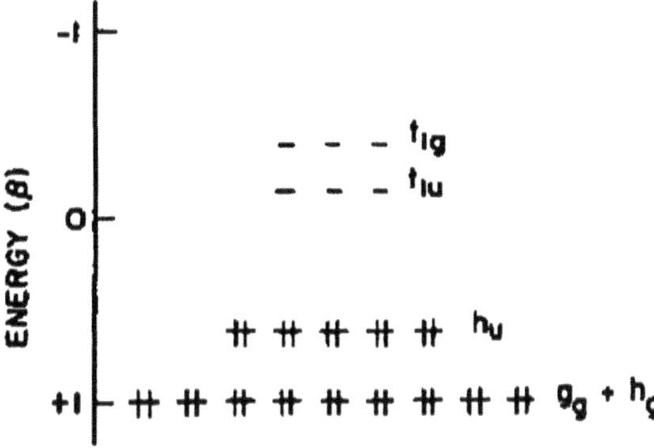

Fig. 4. MO energy level diagram for the frontier orbitals of C_{60}.

On this basis one can ask what happens from the structural viewpoint to fullerenes when they accept electrons. It is conceivable that at least in C_{60}, the progressive population of the degenerate t_{1u} orbitals should cause Jahn-Teller distortion (in order to remove the degeneracy). As a matter of fact, accurate X-ray analysis concerned with the mono- and dianions $[C_{60}]^-$ and $[C_{60}]^{2-}$ lended support to such an hypothesis. Relative to the perfectly spherical geometry of the neutral C_{60} [5a], the monoanion seems to display a slightly axially compressed geometry [5b], whereas the dianion seems to exhibit a slight axial elongation [5c,d], if any [5e]. As a consequence, as compiled in Table 1, the typical C/C bond lengths, even if at the margins of statistical significance, display variations which agree with the common assumption that the t_{1u} orbitals are antibonding with respect to the 6:6 carbon/carbon double bonds and bonding with respect to the 5:6 carbon/carbon single bonds, Scheme 1.

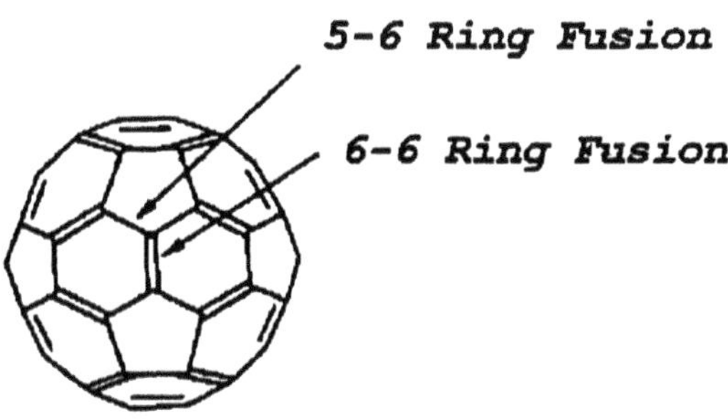

Scheme 1.

Table 1. Variations of the C/C bond lengths (in Å) on passing from C_{60} to its anions

Species	Carbon/carbon bond type	
	6:6	6:5
C_{60}	1.383	1.453
$[C_{60}]^{-}$	1.389	1.449
$[C_{60}]^{2-}$	1.399	1.446

That being stated, let us discuss the metal complexes of fullerenes. A few review papers have recently dealt with this topic [6], but none of them have given a systematic survey of the redox ability of metallofullerenes.

1. Exohedral metallofullerenes

Fullerene derivatives in which a metal fragment(s) is placed outside the spheroidal carbon cage are defined as *exohedral* metallofullerenes. Starting in 1991, a number of exohedral metallofullerenes have been characterised, but few of these have been studied from the electrochemical viewpoint. The first exohedral metallofullerene which was crystallographically resolved was the osmium adduct $C_{60}[OsO_4(4\text{-}tert\text{-}butylpyridine)_2]$, which is illustrated in Figure 5 [7].

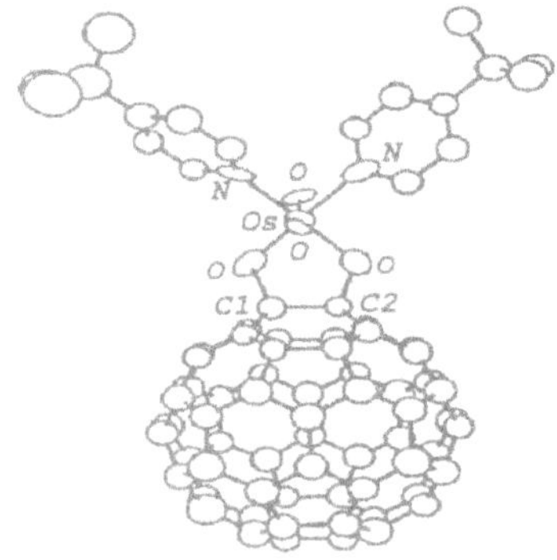

Fig. 5. X-Ray structure of $C_{60}[OsO_4(4\text{-}tert\text{-}butylpyridine)_2]$. C1-C2 = 1.62 Å. (Adapted from Ref. 7)

It is useful to note that the coordination of the metal fragment to the fullerene ligand occurs at the 6:6 carbon/carbon double bond, but (i) there is no direct bond between the metal and C_{60}; (ii) with respect to the uncoordinated 6:6

carbon/carbon bond distances, which are around 1.39 Å, the coordinated one (hereafter indicated as: C1-C2) increases to 1.62 Å.

Following this study, a rich series of Vaska-type complexes have been structurally characterised by Balch's group, namely the mono-adducts ($\eta^2 C_{60}$) [$Ir(CO)Cl(PPh_3)_2$] [8a], (η^2-C_{60})[$Ir(CO)Cl(bobPPh_2)_2$] (bob = 4-benzyloxybenzyl) [8b], (η^2-$C_{60}O$)[$Ir(CO)Cl(PPh_3)_2$] [8c], (η^2-$C_{60}O$)[$Ir(CO)Cl(AsPh_3)_2$] [8d], (η^2-$C_{60}O_2$)[$Ir(CO)Cl(PPh_3)_2$] [8e], as well as the bis-adducts C_{60}[Ir(CO) $Cl(PMe_2Ph)_2]_2$ [9a], $C_{60}[Ir(CO)Cl(PMe_3)_2]_2$ and $C_{60}[Ir(CO)Cl(PEt_3)_2]_2$ [9b]. Related complexes are: (η^2-C_{60})[$Rh(CO)H(PPh_3)_2$] [10a], (η^2-C_{60})[Rh(acac)(3,5-$Me_2py)_2$] (acac = 2,4-pentanedionate) [10b], $C_{60}[Ir_2(OMe)(OPh)(1,5\text{-}COD)_2]$ [10c] and $C_{60}[Ir_2Cl_2(1,5\text{-}COD)_2]_2$ [10d]. At variance with (C_{60})[OsO_4(4-*tert*-butylpyridine)$_2$], in these complexes the metal atoms are directly bound to the fullerene moiety through η^2-6:6 coordination. The same holds for the complexes: (η^2-C_{60})[$Ru(NO)Cl(PPh_3)_2$] [11a], (C_{60})[$Ru_2(\mu\text{-}Cl)(\mu\text{-}X)(C_5Me_5)_2$] (X = H, Cl) [11b]. Unusual bonding modes of the metal fragment to C_{60} are present in (C_{60})[(C_4H_4)$Fe(CO)_3$] [12a], Tl(η^5-$C_{60}Ph_5$) [12b], and (C_{60})[$Si(SiMe_3)_2$] [12c].

The first electrochemical report on crystallographically resolved derivatives was devoted to the series (η^2-C_{60})[$M(PR_3)_2]_n$ (M = Ni, Pd, Pt; R = Ph, Et; n = 0-4) [13a]. The molecular structures of (η^2-C_{60})[$Pt(PPh_3)_2$] and (η^2-C_{60})[$Pd(PPh_3)_2$] are shown in Figure 6 [13b,c].

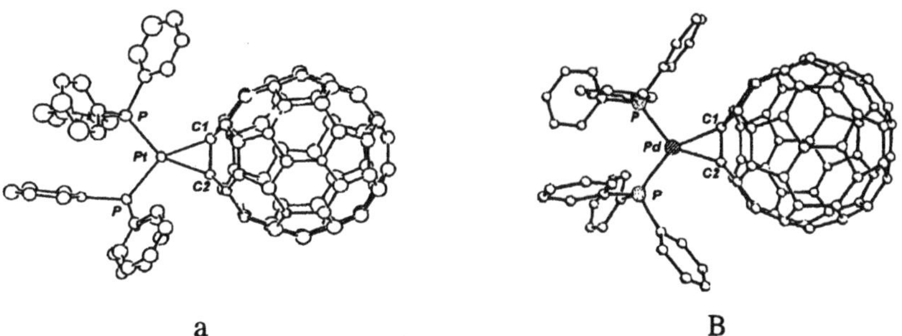

Fig. 6. X-Ray structures of: (a) (η^2-C_{60})[$Pt(PPh_3)_2$]. C1-C2 = 1.50 Å, Pt-C1 = 2.14 Å, Pt-C2 = 2.12 Å; (b) (η^2-C_{60})[$Pd(PPh_3)_2$]. C1-C2 = 1.45 Å, Pd-C1 = 2.12 Å, Pd-C2 = 2.09 Å. (Adapted from Refs. 13b, c)

The molecular structure of the hexa-adduct (η^2-C_{60}) [$Pt(PEt_3)_2]_6$ is also known [13d].

Figures 7 and 8 show the cyclic voltammetric profiles of (η^2-C_{60})[$Pt(PR_3)_2$] (R = Ph, Et) and (η^2-C_{60})[$M(PEt_3)_2$] (M = Ni, Pd, Pt), respectively [13a]. The starred peaks just reveal the presence of free fullerene or fulleride anions, which, arising from impurities or partial dissociation of the adducts, are often present in the electrochemical profiles of these complexes.

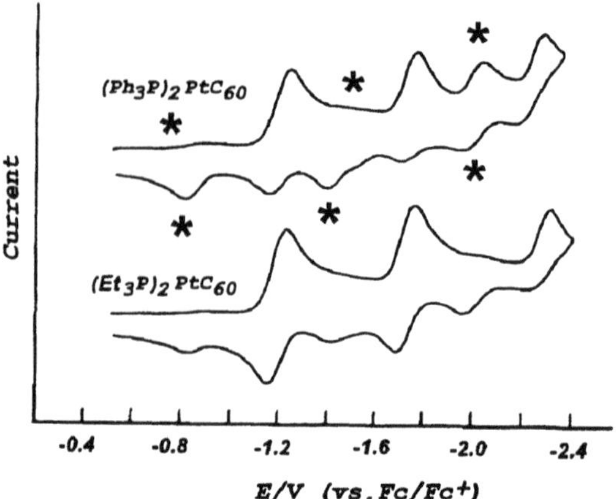

Fig. 7. Comparison between the cyclic voltammograms of $(\eta^2\text{-}C_{60})[Pt(PPh_3)_2]$ and $(\eta^2\text{-}C_{60})[Pt(PEt_3)_2]$ in THF solution. Pt working electrode. Scan rate 0.2 Vs^{-1}. (Adapted from Ref. 13a)

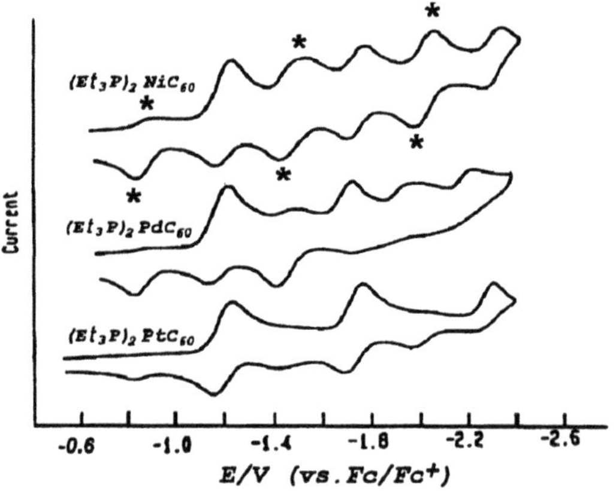

Fig. 8. Comparison between the cyclic voltammograms of $(\eta^2\text{-}C_{60})[Ni(PEt_3)_2]$, $(\eta^2\text{-}C_{60})[Pd(PEt_3)_2]$, and $(\eta^2\text{-}C_{60})[Pt(PEt_3)_2]$ in THF solution. Pt working electrode. Scan rate 0.2 Vs^{-1}. (Adapted from Ref. 13a)

As shown, the presence of the metal fragment makes (i) the sequential reductions to shift towards more negative potential values with respect to free fullerene (the extent of the shifts is compiled in Table 2) [13a]; (ii) the second and third reductions tend to cause decomplexation of fullerene.

Table 2. The reduction potentials (V, *vs.* Fc/Fc+) of the fullerene moiety in the series (η^2-C_{60})[M(PR_3)$_2$] in THF solution.

Complex	$E^{o'}_{0/-}$	$E^{o'}_{-/2-}$	$E^{o'}_{2-/3-}$
(η^2-C_{60})[Pt(PPh_3)$_2$]	-1.21	-1.75	-2.23
(η^2-C_{60})[Pt(PEt_3)$_2$]	-1.20	-1.73	-2.27
(η^2-C_{60})[Pd(PEt_3)$_2$]	-1.18	-1.69	-2.23[b]
(η^2-C_{60})[Ni(PEt_3)$_2$]	-1.20	-1.74	-2.32
C_{60}	-0.86	-1.44	-2.00

[b] Irreversible process

It is noteworthy that the mean negative shift is around 0.35 V, independently of the nature of either the phosphine or the metal, supporting the C_{60}-centered character of the reduction processes.

The same authors also prepared and studied the electrochemical behaviour of the multiadducts (C_{60})[Pt(PEt_3)$_2$]$_n$ (n = 1 - 4). Table 3 shows the negative shifts observed in these complexes with respect to free C_{60}.

Table 3. The reduction potentials (V, *vs.* Fc/Fc+) of the fullerene moiety in (C_{60})[Pt(PEt_3)$_2$]$_n$ in THF solution.

Complex	$E^{o'}_{0/-}$	$E^{o'}_{-/2-}$	$E^{o'}_{2-/3-}$
(C_{60})[Pt(PEt_3)$_2$]	-1.20	-1.73	-2.27
(C_{60})[Pt(PEt_3)$_2$]$_2$	-1.51	-	-
(C_{60})[Pt(PEt_3)$_2$]$_3$	-1.93	-	-
(C_{60})[Pt(PEt_3)$_2$]$_4$	-2.31[b]	-	-
C_{60}	-0.86	-1.44	-2.00

[b] Irreversible process

As Figure 9 shows, the cathodic shift linearly increases with the number of metal fragments [13a].

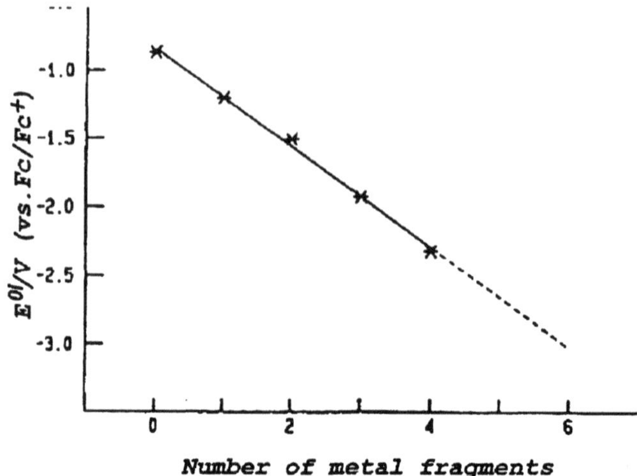

Fig. 9. Variation of the reduction potential of the first reduction step within the family $(C_{60})[Pt(PEt_3)_2]_n$. (Adapted from Ref. 13a)

Concomitantly to this report, an electrochemical study dealt with the indenyliridium(I) complex $(C_{60})[(\eta^5\text{-}C_9H_7)Ir(CO)]$, the X-ray structure of which is now in progress [14]. As Figure 10 shows, the complex displays two sequential reversible reductions, which are centered on the fullerene moiety, followed by an irreversible reduction, as well as an irreversible oxidation, likely involving the iridium(I) fragment. With respect to free C_{60}, the two fullerene-centered reductions occur at potentials more negative by 0.08 V, *i.e.*, the cathodic shift is notably lower than that observed in the preceding platinum complex.

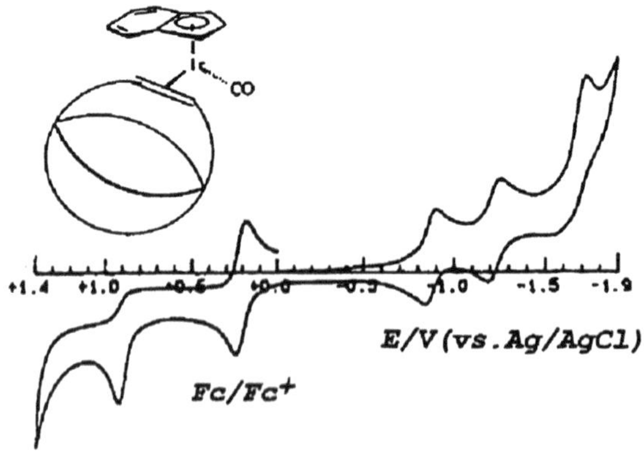

Fig. 10. Cyclic voltammogram of $(C_{60})[(\eta^5\text{-}C_9H_7)Ir(CO)]$ in CH_2Cl_2 solution. Pt working electrode. Scan rate 0.1 Vs^{-1}. (Adapted from Ref. 14)

Following these initial reports, the number of electrochemical studies on these new spectacular types of molecules started to grow.

(η^2-C_{60})[Pd(PPh_3)$_2$], the crystal structure of which is illustrated in Figure 6b, displays a redox propensity in toluene-acetonitrile solution (9:1 v/v) substantially similar to that of (η^2-C_{60})[Pt(PPh_3)$_2$], affording fullerene-centred reductions cathodically shifted relative to free C_{60}; the shift of the first reduction is 0.23 V [15].

Figure 11 shows the electrochemical response given by the tungsten complex (η^2-C_{60})[W(CO)$_3$(dppe)] (dppe = 1,2-diphenylphosphinoethane) together with its X-ray structure [16]. In this case, three reversible, fullerene-centered, one-electron reductions are detectable, which are shifted towards more negative potential values by 0.17 V relative to free C_{60}.

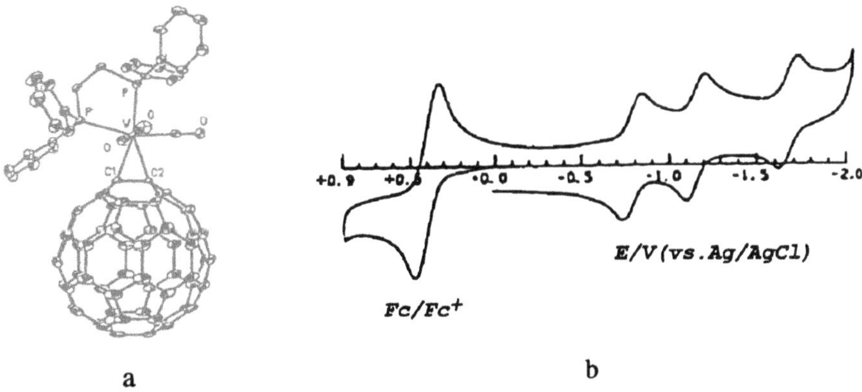

Fig. 11. X-Ray structure of (η^2-C_{60})[W(CO)$_3$(dppe)] (C1-C2 = 1.50 Å, W-C1 ≈ W-C2 = 2.29 Å) (a) and its cyclic voltammetric response in MeCN/toluene solution. Glassy carbon electrode. Scan rate 0.2 Vs^{-1} (b). (Adapted from Ref. 16)

As Figure 12 shows, the same electrochemical behaviour is exhibited by the (X-ray characterised) [16b] analogue (η^2-C_{60})[Mo(CO)$_3$(dppe)], as far as the cathodic processes are concerned (even if in this case the presence of free C_{60} is rather abundant), displaying a negative shift by 0.19 V relative to free fullerene [17a].

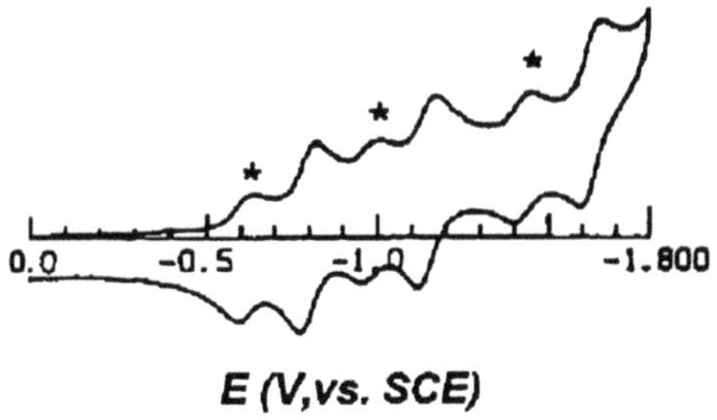

Fig. 12. Cyclic voltammetric response of (C_{60})[Mo(CO)$_3$(dppe)] in CH_2Cl_2 solution. Pt working electrode. Scan rate 0.2 Vs^{-1}. T = -10 °C. (Adapted from Ref. 17a)

The related tungsten complex (η^2-C_{60})[W(CO)$_3$(dppb)] shown in Scheme 2 (dppb = 1,2-diphenylphosphinobenzene) [17a] also exhibits an electrochemical profile in which the sequential reductions of the fullerene moiety are shifted towards more negative potential values by 0.17 V relative to free C_{60}, Figure 13 [17b].

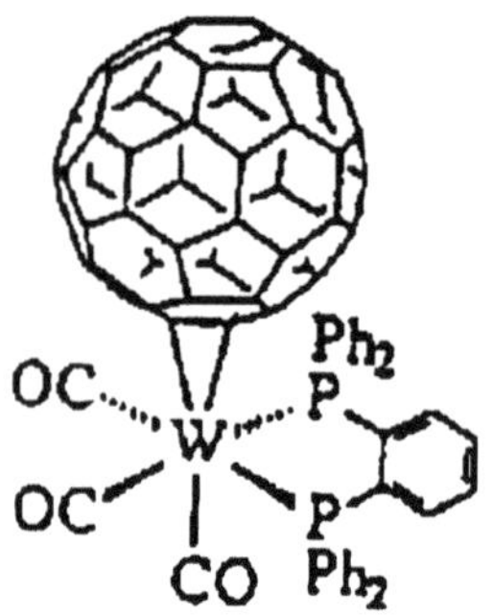

Scheme 2.

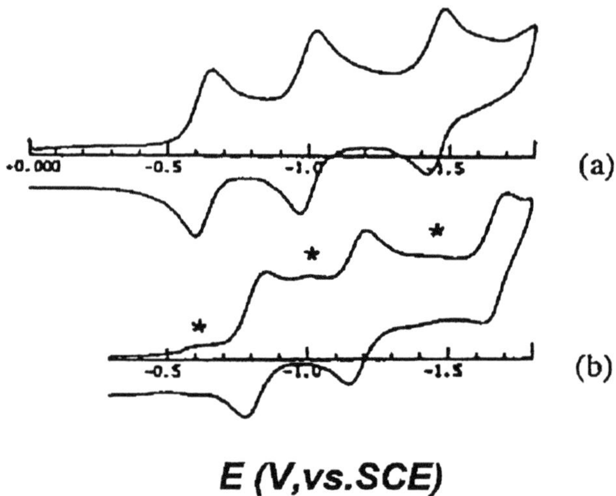

Fig. 13. Comparison between the cyclic voltammetric response of (C_{60})[W(CO)$_3$(dppb)] (b) and that of C_{60} (a) in CH_2Cl_2 solution. Pt working electrode. T = -10 °C. Scan rates 0.2 Vs^{-1}. (Adapted from Ref. 17b)

Figure 14 shows the molecular structure of (η^2-C_{60})[W(CO)$_2$(phen)(dbm)] (phen = 1,10-phenanthroline; dbm = dibutylmaleate) [18a], whose cyclic voltammetric response is shown in Figure 15.

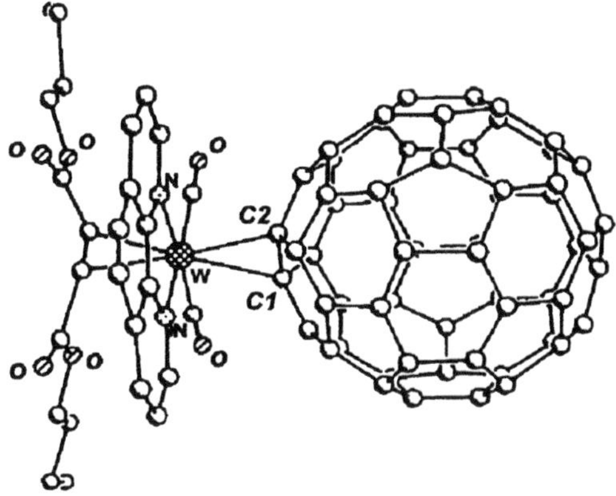

Fig. 14. X-Ray structure of (η^2-C_{60})[W(CO)$_2$(phen)(dbm)]. C1-C2 = 1.43 Å, W-C1 ≈ W-C2 = 2.30 Å. (Adapted from Ref. 18a)

The first three fullerene-centred one-electron reductions of the complex possess features of chemical reversibility and are shifted towards more negative potential values by 0.15 V relative to free C_{60} [18b].

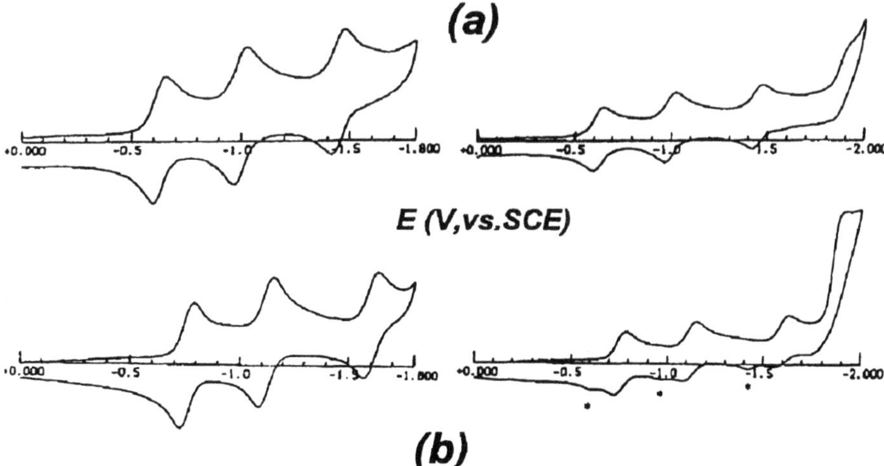

Fig. 15. Comparisons between the cyclic voltammetric responses of C_{60} (a) and (C_{60})[W(CO)$_2$(phen)(dbm)] (b) in CH_2Cl_2 solution at different potential windows. Pt working electrode. T = -10 °C. Scan rates 0.2 Vs^{-1}. (Adapted from Ref. 18b)

In contrast to the fourth one-electron reduction step exhibited by C_{60}, the tungsten complex exhibits a fourth irreversible metal-centred two-electron reduction, which, arising from the following ECE mechanism:

$$[M^0(\eta^2\text{-}C_{60})]^{3-} \xrightarrow{+e-} [M^{-1}(\eta^2\text{-}C_{60})]^{4-} \xrightarrow{k} [M]^- + [C_{60}]^{3-} \xrightarrow{+e^-} [C_{60}]^{4-}$$

M = $W(CO)_2$(phen)(dbm)

accounts for the appearance of free fulleride reoxidations in the backscan (starred peaks).

The same behaviour is exhibited by the analog $(C_{60})[Mo(CO)_2(phen)(dbm)]$ [17b]. Figure 16 compares the responses given by the three complexes $(C_{60})[Mo(CO)_2(phen)(dbm)]_n$ (n = 1-3) [18c]. As previously pointed out for the series $(C_{60})[Pt(PEt_3)_2]_n$ [13a], the di- and tri-molybdenum derivatives suffer the presence

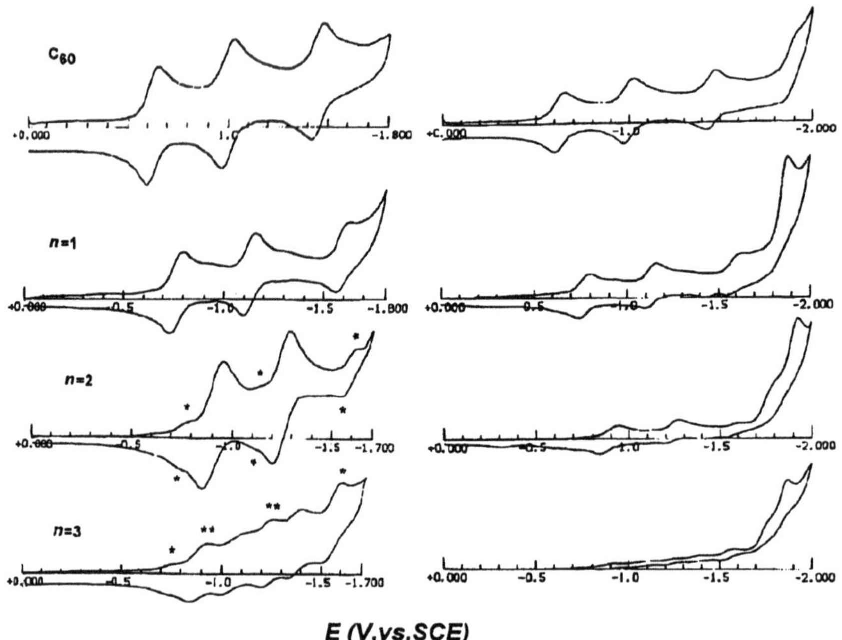

Fig. 16. Cyclic voltammetric responses of $(C_{60})[Mo(CO)_2(phen)(dbm)]_n$ (n = 1-3) in CH_2Cl_2 solution at different potential windows. Pt working electrode. T = -10 °C. Scan rate 0.2 Vs^{-1}. (Adapted from Ref. 18c)

of the corresponding mono- (monostarred peaks) and di-molybdenum (bistarred peaks) derivatives, which likely arise from either difficult purification or equilibria taking place in solution. The relevant electrode potentials are compiled in Table 4.

Table 4. Formal electrode potentials (V, *vs.* SCE) for the redox changes exhibited by the metallofullerenes $(C_{60})[MoL]_n$ (L = $(CO)_2$(phen)(dbm)) in dichloromethane solution, T = -10 °C.

Complex	$E^{o\prime}_{0/-}$	$E^{o\prime}_{-/2-}$	$E^{o\prime}_{2-/3-}$	$E^{o\prime}_{3-/4-}$
C_{60}	-0.63	-1.00	-1.45	-1.88
$(C_{60})[MoL]$	-0.77	-1.13	-1.60	-1.86[a]
$(C_{60})[MoL]_2$	-0.91	-1.24	-1.78	-1.92[a]
$(C_{60})[MoL]_3$	-1.04	-1.37	-1.86[b]	-

[a]Peak potential for metal-centred ECE mechanism; measured at 0.2 Vs^{-1}; [b]peak potential for irreversible processes, measured at 0.2 Vs^{-1}.

It is evident that the progressive insertion of $Mo(CO)_2$(phen)(dbm) fragments into the fullerene moiety makes the electrode potential of the C_{60}–centered reductions shift towards more negative potential values by 0.15 V *per* metal fragment.

A similar cathodic behaviour is exhibited by complex $(\eta^2\text{-}C_{60})$[(triphos)RhH] (triphos = $CH_3C(CH_2PPh_2)_3$), the molecular structure of which is shown in Figure 17, together with its (partial) cyclic voltammetric behaviour. In this case the fullerene-centered reductions are negatively shifted by 0.29 V with respect to free C_{60} [19].

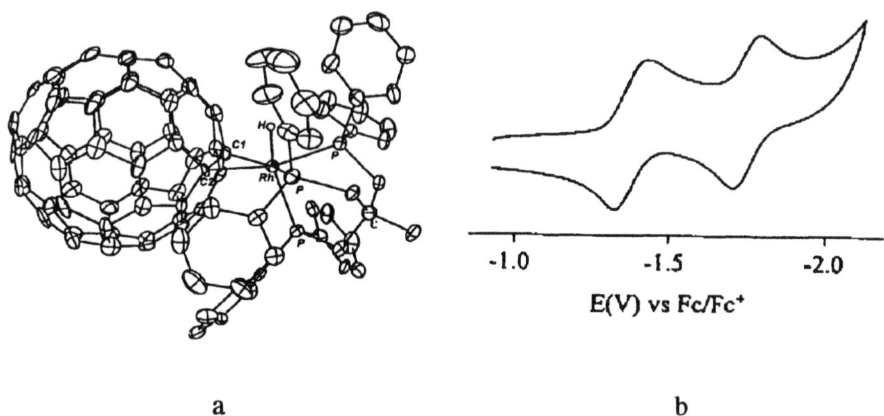

Fig. 17. X-Ray structure of $(\eta^2\text{-}C_{60})$[(triphos)RhH] (C1-C2 = 1.46 Å, Rh-C1 ≈ Rh-C2 = 2.16 Å) (a) and its cyclic voltammetric response in 1,2-dichlorobenzene solution. Platinum electrode. Scan rate 0.1 Vs^{-1} (b). (Adapted from Ref. 19)

Another series of complexes, which were also studied by electrochemical techniques, is constituted by $(C_{60})[(S_2)Fe_2(CO)_6]_n$ (n = 1-3) [20]. Figure 18 shows the X-ray structure of the monoadduct $(\eta^2\text{-}C_{60})[(S_2)Fe_2(CO)_6]$ [20b]. Table 5 summarizes the reduction potentials of the metal fragments [20b]. At variance with the preceding derivatives, these are connected to the fullerene ligand *via* a disulfur bridge, exerting a slight electron-withdrawing effect, which causes a slight shift of the fullerene-centered reductions towards less negative potential values relative to free C_{60}.

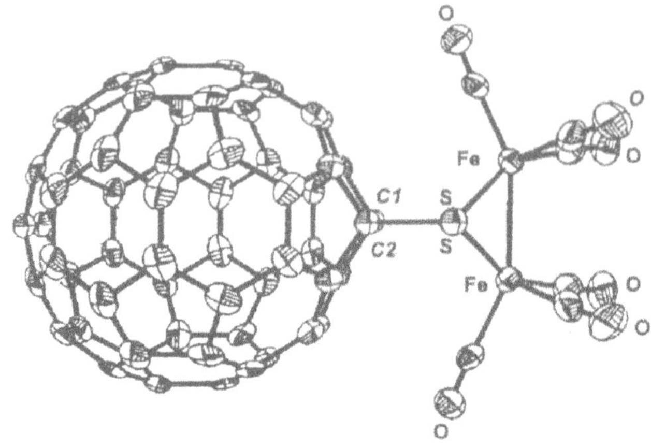

Fig. 18. X-ray structure of $(\eta^2\text{-}C_{60})[(S_2)Fe_2(CO)_6]$. C1-C2 = 1.54 Å. (Adapted from Ref. 20b)

Table 5. The reduction potentials (V, *vs.* Ag/AgCl) of the fullerene moiety of $(C_{60})[(S_2)Fe_2(CO)_6]_n$ in 1,2-dichlorobenzene solution.

Complex	$E^{o'}_{0/-}$	$E^{o'}_{-/2-}$	$E^{o'}_{2-/3-}$	$E^{o'}_{3-/4-}$
$C_{60}[(S_2)Fe_2(CO)_6]$	-0.45	-0.83	-1.28	-1.76
$C_{60}[(S_2)Fe_2(CO)_6]_2$	-0.45	-0.80	-1.30	-1.79
$C_{60}[(S_2)Fe_2(CO)_6]_3$	-0.41	-0.83	-1.29	-1.75
C_{60}	-0.47	-0.85	-1.31	-1.77

Anodic shifts of fullerene reductions have also been found in most bipyridine- or terpyridine-ruthenium(II) complexes such as those shown in Scheme 3, in which the metal is once again not directly coordinated to the fullerene [21].

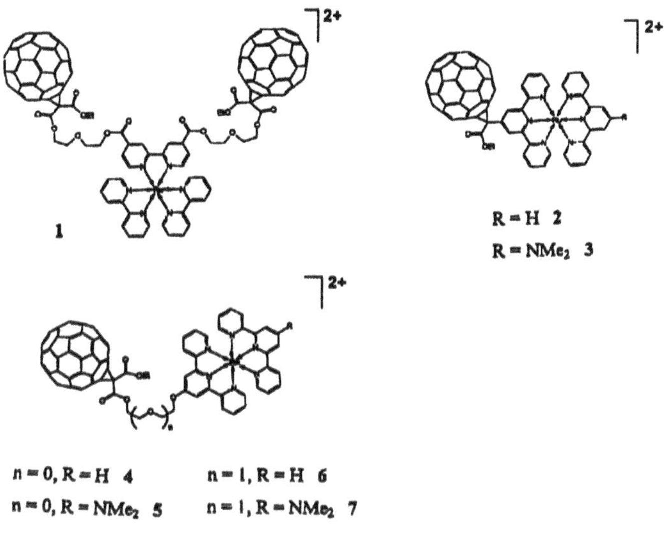

Scheme 3.

Also in these cases the positive shifts of the fullerene reductions (Figure 19, Table 6) are rather small with respect to the corresponding steps in their free ligands. In these cases however, one cannot neglect the contribution of the dipositive charge on the metal complexes to the electrostatic favouring of the reduction processes relative to the neutral ligands.

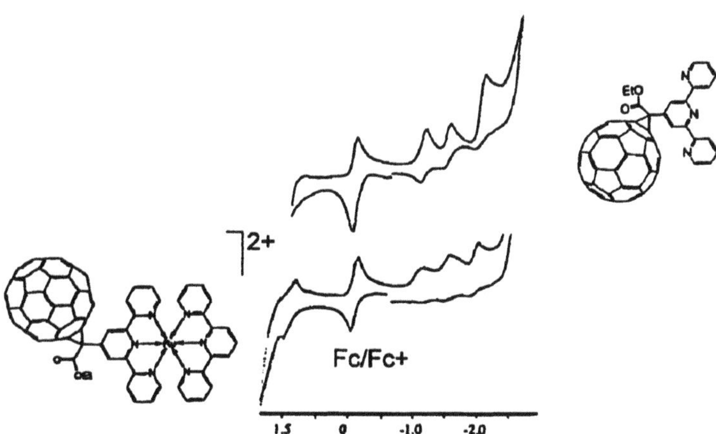

Fig. 19. Comparison between the cyclic voltammetric response of complex **2** and that of its fullerene ligand in CH_2Cl_2 solution. Glassy carbon electrode. Scan rate 0.1 Vs^{-1}. (Adapted from Ref. 21)

Table 6. The reduction potentials (V, *vs.* Fc/Fc^+) of the fullerene moiety of the ruthenium complexes illustrated in Scheme 3. Dichloromethane solution.

Complex	E°' 1st	E°' 2nd	E°' 3rd
1	-1.10	-1.48	-1.92
Free ligand	-1.04	-1.42	-1.93
2	-1.00	-1.40	-1.83
3	-1.02	-1.42	-1.80
Free ligand	-1.05	-1.41	-1.92
4	-1.03	-1.42	-1.80
5	-0.99	-1.43	-1.82
Free ligand	-1.11	-1.50	-1.90
6	-1.02	-1.43	-1.91
7	-1.01	-1.40	-1.90
Free ligand	-1.08	-1.46	-1.91

On the other hand, in the related phenanthroline-silver(I) monocation shown in Scheme 4, the fullerene centered reductions are essentially localised at the same potential value as those of the free ligand, Table 7 [22].

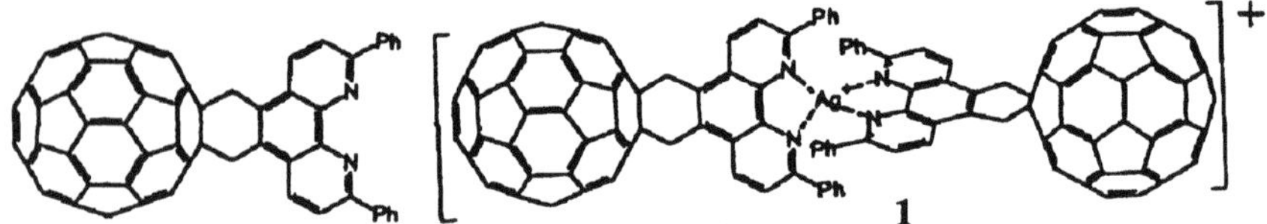

Scheme 4.

Table 7. The reduction potentials (V, *vs.* Ag/AgCl) of the fullerene moiety of the silver(I) complex illustrated in Scheme 4. Dimethylformamide solution.

Complex	E°' 1st	E°' 2nd	E°' 3rd
1	-0.34	-0.80	-1.44
Free ligand	-0.34	-0.79	-1.42

Based on these examples we can conclude that the shift of the reduction potentials of the fullerene-centered reductions with respect to free fullerene is governed by two not easily separable effects: (i) the extent of saturation of the η^2-coordinated carbon/carbon double-bond, which makes the reductions more difficult; (ii) the inductive effects exerted by the metal fragment(s), which can favour or disfavour the reduction processes. The two effects can add or compensate each other. In this way, the measurement of this shift in redox potentials through electrochemical techniques offers a simple, but powerful tool to evaluate the overall electronic factors governing the interaction of the metal fragments with the fullerene ligands.

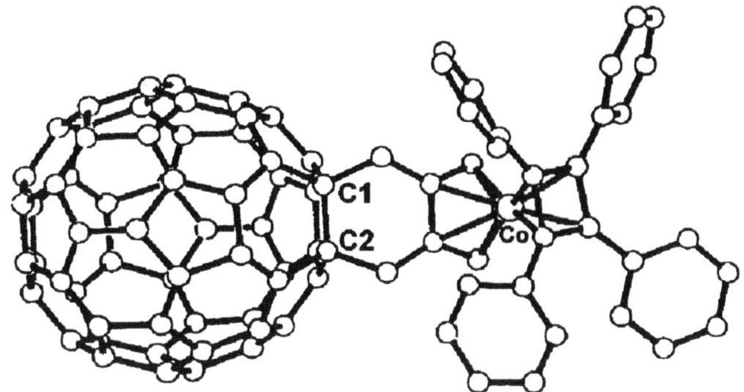

Fig. 20. X-Ray structure of C_{60}\{(η^5-bicyclo[3.2.0]hepta-1,3-dienyl)(η^4-tetraphenylcyclobutadiene)\}Co(I). C1-C2 = 1.59 Å. (Adapted from Ref. 23)

Following our examination, let us pass to the complex C_{60}\{(η^5-bicyclo[3.2.0] hepta-1,3-dienyl)(η^4-tetraphenylcyclobutadiene)\}cobalt(I), which, as illustrated in Figure 20 [23], constitutes another example of a metal fragment not directly bound to C_{60}.

Analysis of the cyclic voltammetric profile shown in Figure 21 indicates that the metal fragment, even if peripherically placed with respect to C_{60}, exerts the typical effect of making the sequential reductions of the fullerene ligand more difficult (by 0.15 V) relative to pure with respect to pure C_{60}.

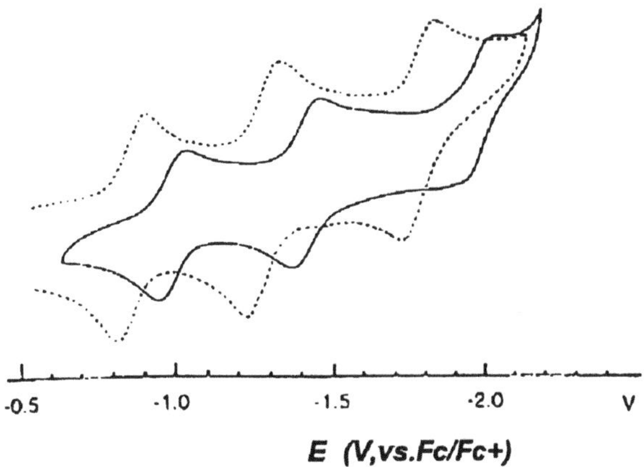

Fig. 21. Comparison between the cyclic voltammetric response of C_{60}\{(η^5-bicyclo[3.2.0] hepta-1,3-dienyl)(η^4-tetraphenylcyclobutadiene)\}Co(I) (——) and C_{60} (— —), in benzonitrile solution. (Adapted from Ref. 23)

A recent report on a series of complexes of formula $(\eta^2\text{-}C_{60})[M(NO)(PPh_3)_2]$ (M = Co, Rh, RuH) and $(\eta^2{:}\eta^2\text{-}C_{60})[Re_2H_8(PMe_3)_4]$ (Figure 22) makes it clear that electrochemical studies of metallofullerenes are sometimes not straightforward [11a].

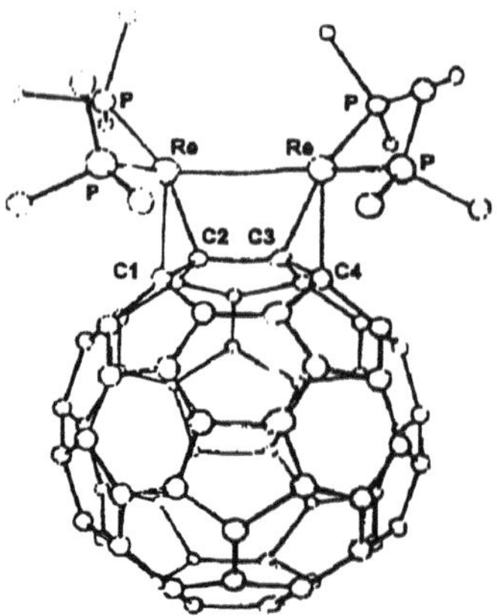

Fig. 22. X-Ray structure of $(\eta^2{:}\eta^2\text{-}C_{60})[Re_2H_8(PMe_3)_4]$. C1-C2 = 1.48 Å, Re1-Re2 = 2.89 Å, Re1-C1 = 2.23 Å, Re1-C2 = 2.19 Å. (Adapted from Ref. 11a)

As a matter of fact, as Figure 23 illustrates, the presence of free fullerene or fulleride ions (starred peaks) likely coming from equilibria setting up in solution complicates the redox patterns.

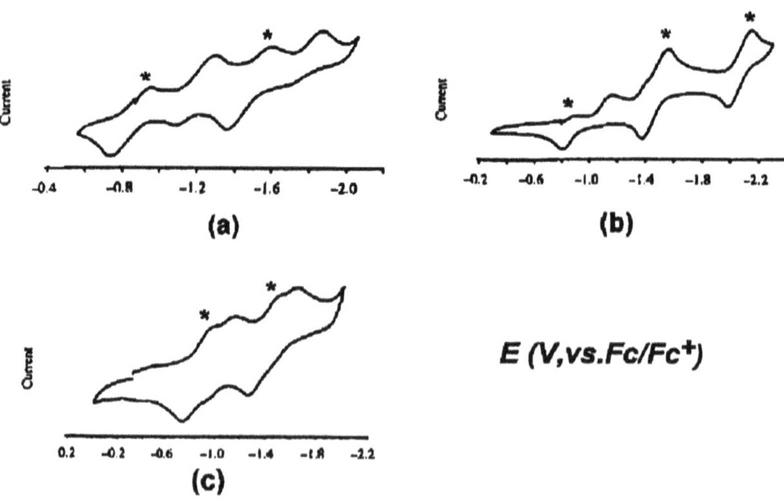

Fig. 23. Cyclic voltammograms recorded on THF solutions of: (a) $(C_{60})[Co(NO)(PPh_3)_2]$; (b) $(C_{60})[Rh(NO)(PPh_3)_2]$; (c) $(C_{60})[Re_2H_8(PMe_3)_4]$. Ag working electrode. T = -20°C. Scan rate 0.5 Vs^{-1}. (Adapted from Ref. 11a)

In all cases the reductions of the fullerene ligand are once again shifted (by about 0.3 V) towards more negative potential values with respect to free fullerene.

Particularly interesting from the electrochemical viewpoint is the adduct formed by C_{60} with tetraferrocenyl-[5]-cumulene, $(Fc)_2C{=}C{=}C{=}C{=}C{=}C(Fc)_2$ in that it offers the invaluable opportunity to couple the ferrocene donor units with the fullerene acceptor moiety [24]. Figure 24 shows the cyclic voltammetric profile of the fullerene adduct [b] together with that of free Fc_4C_6 [a]. The tetraferrocenyl-[5] -cumulene undergoes a first, ferrocene-based, four-electron oxidation ($E^{o\prime}$ = +0.23 V), followed by a cumulene-centered two-electron oxidation ($E^{o\prime}$ = +0.83 V), as well as a cumulene-centred two-electron reduction ($E^{o\prime}$ = –1.39 V). From a qualitative viewpoint, the overall picture given by the fullerene complex looks like the superposition of the response of the tetraferrocenyl-[5]-cumulene with the well-known sequential reductions of C_{60}. Indeed, we must take into account that the fullerene complex contains different redox-active subunities, *i.e.*, ferrocenes, cumulene, and fullerene.

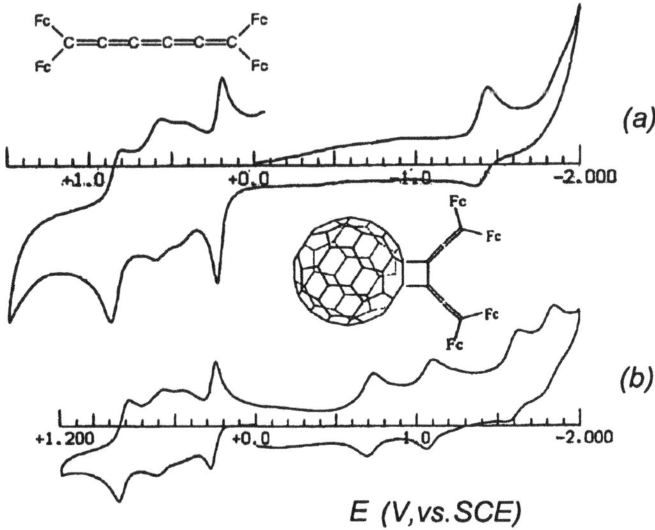

Fig. 24. Comparison between the cyclic voltammetric responses of Fc_4C_6 (a) and $(\eta^2\text{-}C_{60})[Fc_4C_6]$ (b) in CH_2Cl_2 solution. Pt working electrode. T = -10 °C. Scan rate 0.2 Vs^{-1}.

As deducible from Table 8: (i) the first ferrocene-based oxidation is shifted towards more positive potential values by 0.03 V with respect to that of free Fc_4C_6, indicating that the fullerene ligand acts, as expected, as the electron-acceptor unit; (ii) the cumulene based oxidation is substantially unaltered with respect to that of Fc_4C_6; (iii) the two first fullerene-centered reductions are shifted towards more negative potential values by 0.08 V with respect to those of free C_{60}. Less straightforward is to assign the nature of the third cathodic process ($E^{o\prime}$ = -1.56 V) as fullerene-centered or cumulene-centered step. Based on the relative peak-height, this third step has been taken as a cumulene-centered two-electron process, which hence results cathodically shifted by 0.17 V with respect to that of Fc_4C_6. On this basis, the fourth reduction (E°' = -1.90 V), which would correspond to the third reduction of the fullerene ligand, should be shifted towards more negative potential values by 0.38 V with respect to free C_{60}. From a speculative viewpoint the significant shifts of both steps are likely due to electrostatic repulsion arising from the addition of two electrons to the cumulene-fulleride dianion and to the addition of the third electron to the fulleride-cumulene tetraanion, respectively.

Table 8. Formal electrode potentials (V, *vs.* SCE) for the redox changes exhibited by complex $(C_{60})[Fc_4C_6]$ and related species, in dichloromethane solution, at -10 °C.

Complex	Cumulene oxidation	Ferrocene oxidation	Fullerene reductions				Cumulene reduction
	E°'	E°'	E°'1st	E°'2nd	E°'3rd	E°'4th	E°'
C_{60}	-	-	-0.63	-1.00	-1.45	-1.90	-
$(C_{60})[Fc_4C_6]$	+0.82	+0.26	-0.72	-1.08	-1.83	-	-1.56
Fc_4C_6	+0.83	+0.23	-	-	-	-	-1.39

Such behavior is somewhat reminiscent of that exhibited by the ferrocenyl fulleropyrrolidines illustrated in Scheme 5, the cyclic voltammetric behaviours of which are illustrated in Figure 25, and the relevant electrochemical data are compiled in Table 9 [25].

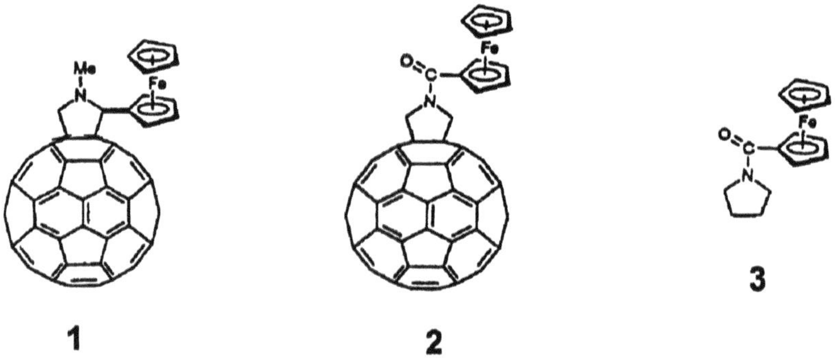

Scheme 5.

Table 9. Formal electrode potentials (V, *vs.* Fc/Fc+) for the redox changes exhibited by the ferrocenyl-fulleropyrrolidine adducts shown in Scheme 5, in 3:1 toluene-acetonitrile solution. T = -45 °C.

Complex	$E^{\circ\prime}_{+/0}$	$E^{\circ\prime}_{0/-}$	$E^{\circ\prime}_{-/2-}$	$E^{\circ\prime}_{2-/3-}$	$E^{\circ\prime}_{3-/4-}$	$E^{\circ\prime}_{4-/5-}$	$E^{\circ\prime}_{5-/6-}$
C_{60}	-	-0.94	-1.33	-1.83	-2.28	-2.74	-3.14
1	+0.05	-1.08	-1.47	-2.03	-2.44	-3.14	-
2	+0.16	-1.00	-1.38	-1.95	-2.36	-3.05	-
3	+0.12	-	-	-	-	-	-

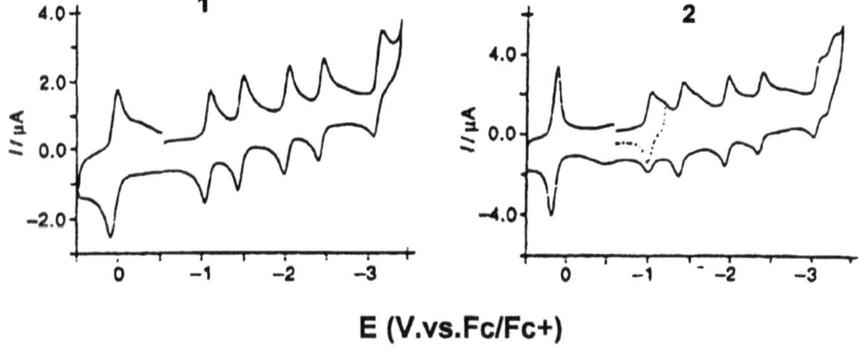

Fig. 25. Cyclic voltammetric responses recorded at a glassy-carbon electrode on 3:1 toluene-acetonitrile solutions of the ferrocenyl-fulleropyrrolidines shown in Scheme 5. (a) Complex **1**; (b) complex **2**. T = -45 °C. Scan rate 0.1 Vs^{-1}. (Adapted from Ref. 25)

To conclude the examination of these types of metallofullerenes, we like to mention that some electrochemical data on C_{60}:Ni-phthalocyanine [26a], C_{60}:Zn-porphyrin [26b], C_{60}:Cu-rotaxane [26c], and $[(2'\text{-}Pr^{i}\text{-}5'\text{-}Me\text{-}C_6H_8\text{-}O)P(Ph)(C_{60}H)]_2$ $PtCl_2$ [26d] adducts have been reported.

Let us now direct our attention to another class of structurally very spectacular exohedral metallofullerenes: those bearing metal-cluster fragments. X-ray structures of a few complexes have been reported, namely: $(\mu_3\text{-}\eta^2,\eta^2,\eta^2\text{-}C_{60})[Ru_3(CO)_9]$ [27a], $(\mu_3\text{-}\eta^2,\eta^2,\eta^2\text{-}C_{60})[Ru_5C(CO)_{11}(PPh_3)]$ [27b], $(\mu_3\text{-}\eta^2,\eta^2,\eta^2\text{-}C_{60})[Ru_5C(CO)_{10}(dppe)]$ [27c], $(\mu_3\text{-}\eta^2,\eta^2,\eta^2\text{-}C_{60})[Ru_5C(CO)_{10}(dppf)]$ (dppf = diphenylphosphinoferrocene) [27c], $(\mu_3\text{-}\eta^2,\eta^2,\eta^2\text{-}C_{60})[Ru_6C(CO)_{12}(dppm)]$ (dppm = diphenylphosphino-methane) [27b], $(\mu_3\text{-}\eta^2,\eta^2,\eta^2\text{-}C_{60})[Ru_5PtC(CO)_{11}(dppe)]$ [27c], but studies of their electrochemistry have not appeared in literature. In these derivatives the coordination of the metal cluster to the fullerene occurs through the donor-acceptor interaction between a C_{60} six-membered ring and the triangular face of the metal cluster.

Electrochemical studies have been reported on related osmium clusters adducts, where either (η^2-C_{60}) or (μ_3-η^2,η^2,η^2-C_{60}) coordinations have been found. As an example, Figure 26 shows the molecular structure of (η^2-C_{60})[$Os_3(CO)_{11}$] [28a] and (μ_3-η^2,η^2,η^2-C_{60})[$Os_3(CO)_8(PPh_3)$] [28b].

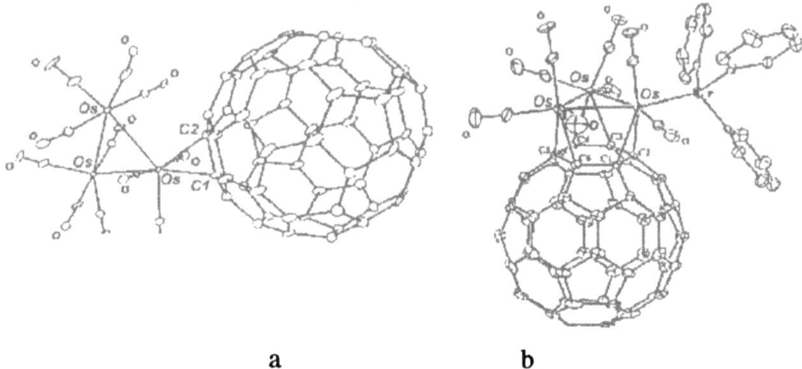

Fig. 26. X-Ray structures of: (a) (η^2-C_{60})[$Os_3(CO)_{11}$]. C1-C2 = 1.42 Å, averaged Os-Os distance = 2.88 Å, Os-C1 = 2.21 Å, Os-C2 = 2.26 Å; (b) (μ_3-η^2,η^2,η^2-C_{60})[$Os_3(CO)_8(PPh_3)$]. Averaged Os-Os distance = 2.92 Å; averaged Os-C(fullerene) distance = 2.28 Å; averaged C-C in the C_6 ring = 1.46 Å. (Adapted from Ref. 28a,b)

Figure 27 shows the cyclic voltammetric profiles exhibited by (η^2-C_{60})[$Os_3(CO)_{11}$], either at low and high scan rates [28c].

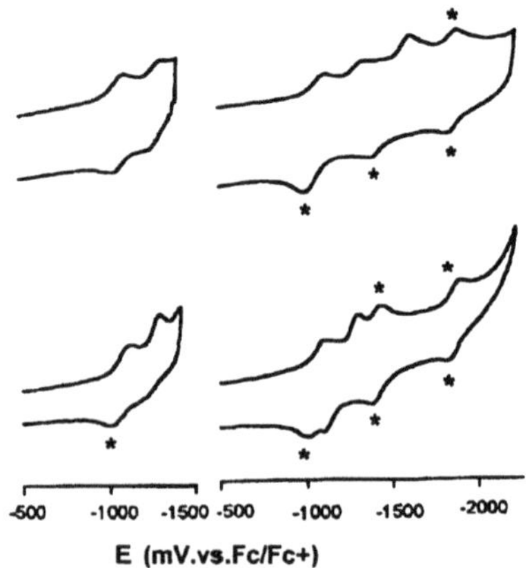

Fig. 27. Cyclic voltammetric responses of (η^2-C_{60})[$Os_3(CO)_{11}$] in 4:1 CH_2Cl_2-toluene solution. Pt working electrode. Scan rates: top, 20.48 Vs^{-1}; bottom, 0.1 Vs^{-1}. (Adapted from Ref. 28c)

As seen, the voltammograms at low scan rates put in evidence that the addition of the second electron causes decomplexation of the fullerene ligand (starred peaks); such chemical complication is prevented by scanning at high rates. The same behaviour holds for the phosphine substituted complexes $(\eta^2\text{-}C_{60})[Os_3(CO)_{10}(PPh_3)]$ and $(\eta^2\text{-}C_{60})[Os_3(CO)_9(PPh_3)_2]$. In these complexes the first electron addition occurs at potentials slightly more negative (from 0.03 V for the unsubstituted species to 0.14 V for the disubstituted species) than that of free C_{60}, but interestingly the second reduction of $(\eta^2\text{-}C_{60})[Os_3(CO)_{11}]$ occurs at potentials less negative by 0.10 V with respect to that of free C_{60}. It has been assumed that such second electron addition might be delocalised between the C_{60} moiety and the triosmium cluster.

Figure 28 shows the cyclic voltammograms of the series $(\mu_3\text{-}\eta^2,\eta^2,\eta^2\text{-}C_{60})[Os_3(CO)_{9\text{-}x}(PMe_3)_x]$ (x = 0 - 2) [28b].

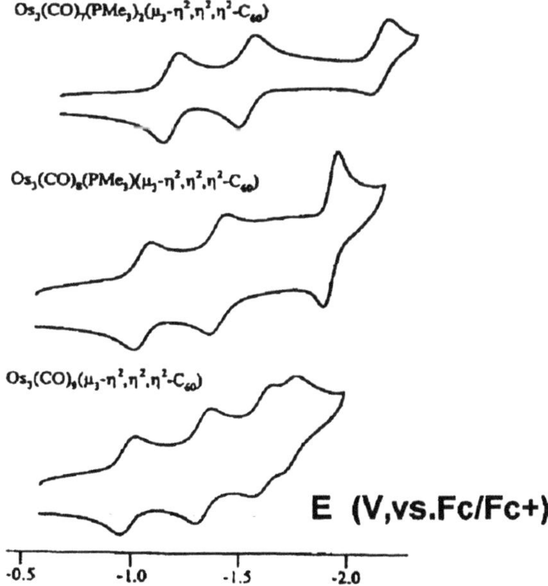

Fig. 28. Cyclic voltammograms of $(\mu_3\text{-}\eta^2,\eta^2,\eta^2\text{-}C_{60})[Os_3(CO)_{9\text{-}x}(PMe_3)_x]$ (x = 0 - 2) in 1,2-dichlorobenzene solution. Pt working electrode. Scan rate 0.05 Vs^{-1}. (Adapted from Ref. 28b)

In this case the complexes appear more stable towards electron additions. It is also important to point out that in this case, in particular for $(\mu_3\text{-}\eta^2,\eta^2,\eta^2\text{-}C_{60})[Os_3(CO)_9]$, all electron additions occur at potentials less negative by 0.10 V with respect to those of free C_{60}. Further supporting the preceding result, this finding has been interpreted as meaning that all electron additions to C_{60} are likely mediated by the triosmium cluster.

We will now examine the metallocomplexes of higher fullerenes, in particular those of C_{70}. As in the case of the C_{60} adducts, among the first derivatives prepared were the Vaska-type derivatives (η^2-C_{70}) [Ir(CO)Cl$(PPh_3)_2$] [29a], (η^2-C_{70}O)[Ir(CO)Cl$(PPh_3)_2$] [29b], and (η^2-C_{70}) [Ir(CO)Cl$(PPhMe_2)_2]_2$ [29c], as well as the platinum adduct (C_{70})[Pt$(PPh_3)_2]_4$ [29s], and the molybdenum complex (η^2-C_{70})[Mo$(CO)_3$(dppe)] [16b]. Unfortunately, no electrochemical studies have been performed on these complexes, nor for the cluster derivatives (μ_3-η^2,η^2,η^2-C_{70})[$Ru_3(CO)_9$] and (μ_3-η^2,η^2,η^2-C_{70})[$Ru_3(CO)_9]_2$ [30].

To the best of our knowledge, the first report on the electrochemical behaviour of C_{70}-metallocomplexes has dealt with (C_{70})[$Fe_2S_2(CO)_6$], an analog of the (C_{60})[$Fe_2S_2(CO)_6$] previously discussed [20b]. In 1,2-dichlorobenzene solution, the C_{70}-complex exhibits four one-electron reductions at potentials a few millivolts more negative than those of free C_{70}, Table 10 [20b], in contrast with the C_{60}–analogue, which was easier to reduce than C_{60}.

Table 10. Formal electrode potentials (V, *vs.* Ag/AgCl) for the redox changes exhibited by C_{70}[$Fe_2S_2(CO)_6$], in 1,2-dichlorobenzene solution.

Complex	$E^{\circ\prime}_{0/-}$	$E^{\circ\prime}_{-/2-}$	$E^{\circ\prime}_{2-/3-}$	$E^{\circ\prime}_{3-/4-}$
C_{70}[$Fe_2S_2(CO)_6$]	-0.49	-0.89	-1.26	-1.64
C_{70}	-0.49	-0.86	-1.25	-1.66

A brief electrochemical report appeared on (η^2-C_{70})[Pd$(PPh_3)_2$] [31], also an analog of the previously discussed (η^2-C_{60})[Pd$(PPh_3)_2$] [13c,15]. In this connection, Figure 29 shows the molecular structure of this C_{70} complex [32].

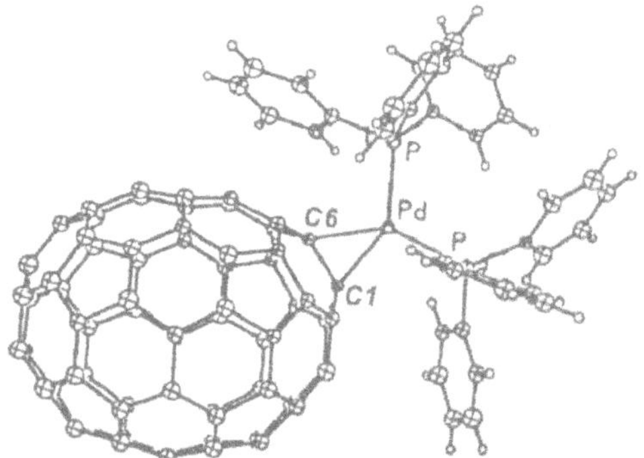

Fig. 29. X-Ray structure of (η^2-C_{70})[Pd$(PPh_3)_2$]. C1-C6 = 1.48 Å, Pd-C1 ≈ PdC2 = 2.11 Å. (Adapted from Ref. 32)

In toluene-acetonitrile solution (85:15 v/v) the first fullerene-centered reduction of the palladium derivative is cathodically shifted by 0.25 V with respect to free C_{70}. The same report indicates that in the platinum analogue $(\eta^2\text{-}C_{70})[Pt(PPh_3)_2]$ the cathodic shift is 0.26 V.

We examined the complex $(C_{70})[Mo(CO)_2(phen)(dbm)]$ [18c], which is also an analogue of the C_{60}-complex discussed above [18b]. As Figure 30 shows, with respect to free C_{70}, the presence of the Mo-fragment also induces a cathodic shift (by 0.10 V) of the first two C_{70}-centered one-electron reductions. However, the third cathodic wave appears as two, essentially overlapping, one-electron steps.

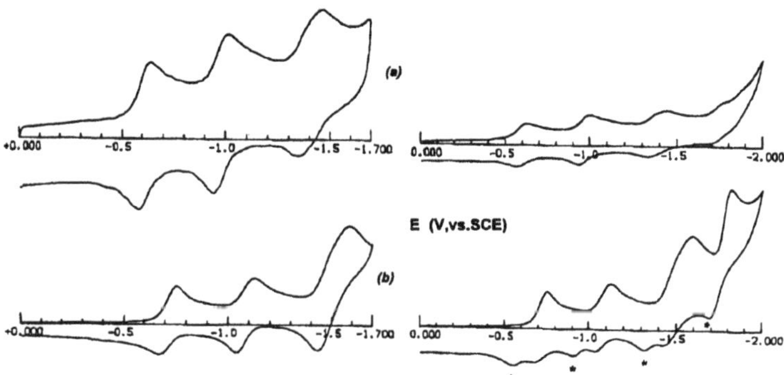

Fig. 30. Comparisons between the cyclic voltammetric responses of C_{70} (a) and $(C_{70})[W(CO)_2(phen)(dbm)]$ (b) in CH_2Cl_2 solution at different potential windows. Pt working electrode. T = -10 °C. Scan rates 0.2 Vs^{-1}. (Adapted from Ref. 18c)

Taking into account that it seems unlikely that the one-electron reduction of the metal fragment $Mo(CO)_2(phen)(dbm)$ might occur at potential values less negative either than that observed in $(C_{60})[Mo(CO)_2(phen)(dbm)]$ (E_p = -1.86 V) or than that of the precursor $Mo(CO)_2(phen)(dbm)_2$ itself ($E^{\circ\prime}$ = -1.67 V) [18b], we have assigned this third reversible wave to the almost concomitant third and fourth one-electron reduction of the C_{70} moiety, whereas the fourth reduction of the metallofullerene, which, as put in evidence by the starred peaks, induces C_{70} decomplexation, likely arises from the simple metal-centred EC step:

$$[M^0(\eta^2\text{-}C_{70})]^{4-} \xrightarrow{+e-} [M^{-1}(\eta^2\text{-}C_{70})]^{5-} \xrightarrow{k} [M]^- + [C_{70}]^{4-}$$

M = $Mo(CO)_2(phen)(dbm)$

This means that the reduction pathway of $(C_{70})[Mo(CO)_2(phen)(dbm)]$ is rather different from that of its C_{60} analogue [18b], in which the fourth reduction involved an ECE mechanism (see above). Table 11 compiles the relevant data, also in comparison with those of the related C_{60} complex. It is interesting to note that, as it happens for the couple $(\eta^2\text{-}C_{60})[Pt(PPh_3)_2]/(\eta^2\text{-}C_{70})[Pt(PPh_3)_2]$, the cathodic shift induced by the presence of $Mo(CO)_2(phen)(dbm)$ with respect to C_{70} is lower than that induced with respect to C_{60}, thus suggesting that the extent of breaking of the double bond conjugation exerted by the coordination of the metal fragment to C_{70} is smaller than that exerted by the same fragment with respect to C_{60}. Unfortunately, such a conclusion cannot be generalized since in the case of the couples $(C_{60})[(S_2)Fe_2(CO)_6]/(C_{70})[(S_2)Fe_2(CO)_6]$ and $(\eta^2\text{-}C_{60})[Pd(PPh_3)_2]/(\eta^2\text{-}C_{70})[Pd(PPh_3)_2]$ the cathodic shifts with respect to the corresponding fullerenes are higher in the C_{70}-complexes than in the C_{60}-complexes, thus preluding that the extent of double bond saturation is higher in the C_{70}-complexes. More data on complexes bearing the same metal fragment upon both C_{60} and C_{70} are hence needed to clarify this matter.

Table 11. Formal electrode potentials (V, *vs.* SCE) for the redox changes exhibited by the metallofullerenes $(C_{70})[Mo(CO)_2(phen)(dbm)]$ and $(C_{60})[Mo(CO)_2(phen)(dbm)]$ in dichloromethane solution. T = -10 °C.

Complex	$E^{\circ\prime}_{0/-}$	$E^{\circ\prime}_{-/2-}$	$E^{\circ\prime}_{2-/3-}$	$E^{\circ\prime}_{3-/4-}$	E_p
C_{70}	-0.61	-0.98	-1.41	-1.75	-
$(C_{70})[MoL]$	-0.71	-1.09	-1.51	-1.51	-1.82[a]
C_{60}	-0.63	-1.00	-1.45	-1.90	-
$(C_{60})[MoL]$	-0.77	-1.13	-1.60	-	-1.86[a]

[a] Peak potential value for irreversible processes

Finally, as far as the metallocomplexes of higher fullerenes are concerned, we are aware of only the Vaska-type complex $(\eta^2\text{-}C_{84})[Ir(CO)Cl(PPh_3)_2]$ [33] the electrochemistry of which has not been studied.

2. Endohedral metallofullerenes

Fullerene derivatives in which a metal(s) lies inside the spheroidal carbon cage are defined as *endohedral* metallofullerenes. Just after the discovery of fullerenes, it has been realized that the cavities of these new carbon assemblies, in particular those of the higher fullerenes, were sufficiently large to encapsulate metal atoms. Unfortunately, no crystal structure of these derivatives is yet available to validate definitively their existence, even if preliminary structural data from electron diffraction techniques [34] support the large body of theoretical and experimental evidences (mainly from mass and EPR spectroscopy).

A lot of endohedral metallofullerenes have been identified, ranging from monometallic [from $M@C_{60}$ (M = Ca, Sr, Ba) to $M@C_{90}$ (M = La)] to dimetallic [from $M_2@C_{74}$ (M = Sc) to $M_2@C_{84}$ (M = Sc, Y)] and trimetallic [from $M_3@C_{82}$ (M = Sc, Y, La) to $M_3@C_{112}$ (M = La)] complexes [6b,c], but the difficulties in their production on a macroscopic scale, as well as in their purification make their chemistry not easily accessible. On this basis it is well conceivable that electrochemical studies on these materials are still scanty. At the moment, the only electrochemical investigations concern those complexes in which group 3 elements are encaged inside C_{82} according to the general formula: $M@C_{82}$ (M = La, Sc, Y, Gd) and $Sc_3@C_{82}$, which can be produced at milligram levels. Figure 31 shows a schematic picture of the molecular structure of these endohedral complexes as derived from theoretical calculations [35]. Since inside the cavity the metals can occupy different positions (the less energetically favoured seems just the central position), for the sake of simplicity only one isomer is represented.

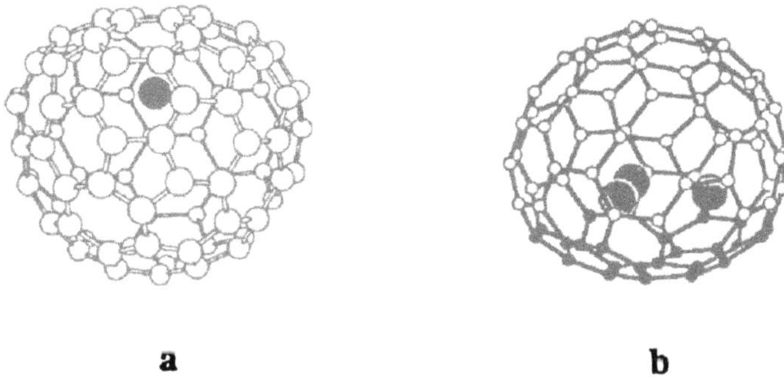

Fig. 31. Representation of one of the possible isomers of $M@C_{82}$ (M = lanthanide) (a) and $Sc_3@C_{82}$ (b). (Adapted from Ref. 35)

Figure 32 shows the cyclic voltammetric pattern exhibited by two isomers of $La@C_{82}$, in *o*-dichlorobenzene solution [36].

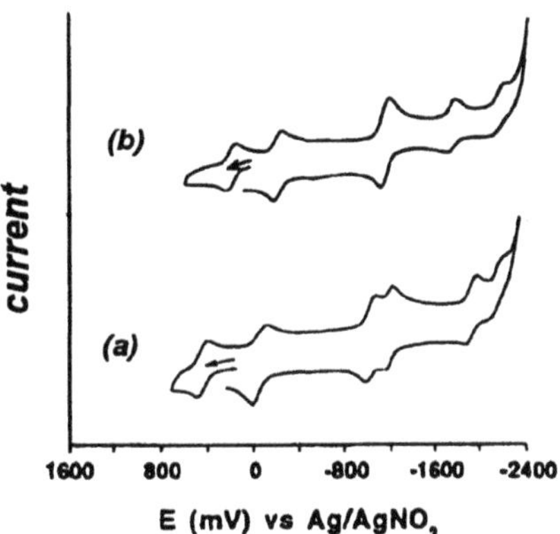

Fig. 32. Cyclic voltammetric responses recorded at a platinum electrode on 1,2-dichlorobenzene solutions of: (a) the major isomer of $La@C_{82}$; (b) the minor isomer of $La@C_{82}$. (Adapted from Refs. 36a,b)

As seen, they display a substantially similar redox pattern, both undergoing one oxidation and four/five reduction processes with features of chemical reversibility. From a qualitative viewpoint the only difference lies on the second reduction step, which in the major isomer appears as two slightly separated one-electron processes, whereas in the minor isomer it appears as a single two-electron process. Figure 33 shows that a similar electrochemical behaviour is exhibited by either $Y@C_{82}$ [37], or $Gd@\ C_{82}$ [38]. The relevant electrode potentials are compiled in Table 12.

Table 12. Formal electrode potentials (V, *vs.* Fc/Fc^{+}) for the reversible redox changes exhibited by the metallofullerenes $M@C_{82}$ in 1,2-dichlorobenzene solution.

Complex	$E^{o'}_{+/0}$	$E^{o'}_{0/-}$	$E^{o'}_{-/2-}$	$E^{o'}_{2-/3-}$	$E^{o'}_{3-/4-}$	$E^{o'}_{4-/5-}$	Ref.
$La@C_{82}$ [a]	+0.07	-0.42	-1.37	-1.53	-2.26	-2.5	36a
$La@C_{82}$ [b]	-0.07	-0.48	-1.41	-1.41	-2.01	-2.42	36b
$Y@C_{82}$	+0.10	-0.34	-1.34	-1.34	-2.20	-2.5	37
$Gd@C_{82}$	+0.09	-0.39	-1.3	-1.3	-2.3[c]	-2.3[c]	38
C_{82}	-	-0.69	-	-	-	-	38

[a] Major isomer; [b] minor isomer; [c] irreversible process.

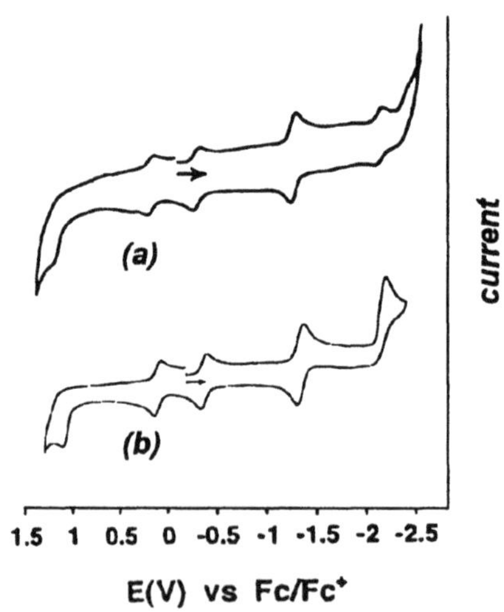

Fig. 33. Cyclic voltammetric responses recorded at a platinum electrode on 1,2-dichlorobenzene solutions of: (a) $Y@C_{82}$; (b) $Gd@C_{82}$. (Adapted from Refs. 37, 38)

A problem arises regarding the oxidation state of the encapsulated metal. It is commonly accepted that in these complexes charge transfer takes place from the encaged metal atoms to the fullerene hosts, but it is still debated if they must be formally described as $M^{3+}@C_{82}^{3-}$ or $M^{2+}@C_{82}^{2-}$ [39], even if a theoretical analysis assigns the oxidation states of $M^{2+}@C_{82}^{2-}$ to M = Sc and Y, and $M^{3+}@C_{82}^{3-}$ to M = La. Such a charge distribution should essentially depend on the relative position that the metal atom occupies, due to its dimensions, inside the cavity [35a]. In this connection, it must be said that none of the above mentioned oxidation states seem to agree with the above described electrochemical profiles. As a matter of fact, taking into account that C_{82} does not exhibit reversible oxidation processes [40], if we confidently assume that all the reversible processes illustrated in Figures 32 and 33 are fullerene-centered, we must conclude that in solution these complexes would have to be formulated as: $M^{+}@C_{82}^{-}$.

The same contrasting conclusions must be drawn for $Sc_3@C_{82}$. Figure 34 compares the square wave voltammogram of $Sc_3@C_{82}$ with that of C_{82}, in pyridine solution [41]. As a matter of fact, the trimetallocomplex exhibits two (assumed) one-electron oxidations and two one-electron reductions, whereas free fullerene exhibits four one-electron reductions, Table 13.

Once again, assigning the redox sequence exhibited by the metallocomplex as due to fullerene-centered processes, it would have to be concluded that in solution the complex possesses the oxidation state: $Sc_3^{2+}@C_{82}^{2-}$ (in spite of the fact that the same authors assign, rather surprisingly, the oxidation state: $Sc_3^{4+}@C_{82}^{4-}$).

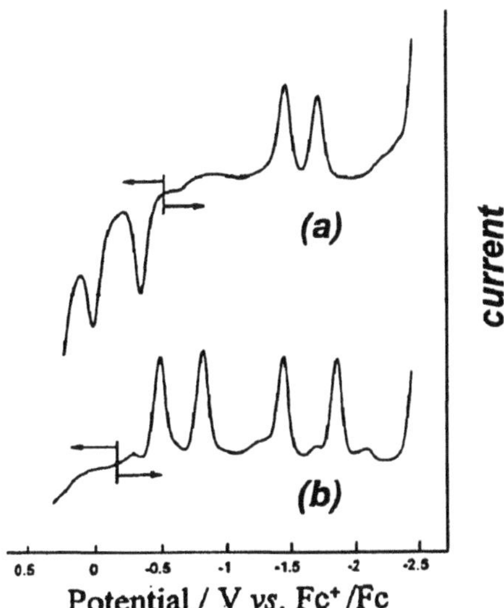

Fig. 34. Square wave voltammograms recorded at a glassy carbon electrode on pyridine solutions of: (a) $Sc_3@C_{82}$; (b) C_{82}. (Adapted from Ref. 41)

Table 13. Formal electrode potentials (V, *vs.* Fc/Fc^+) for the reversible redox changes exhibited by $Sc_3@C_{82}$ in pyridine solution.

Complex	$E^{o'}_{2+/+}$	$E^{o'}_{+/0}$	$E^{o'}_{0/-}$	$E^{o'}_{-/2-}$	$E^{o'}_{2-/3-}$	$E^{o'}_{3-/4-}$
$Sc_3@C_{82}$	-0.04	-0.32	-1.42	-1.84	-	-
C_{82}	-	-	-0.47	-0.80	-1.42	-1.67

References

1. F.Diederich and Y.Rubin, *Angew.Chem.Int.Ed.Engl.*, *31*, 1101 (1992).
2. (a) Q.Xie, E.Pérez-Cordero, and L.Echegoyen, *J.Am.Chem.Soc.*, *114*, 3978 (1992); (b) C.Boudon, J.-P.Gisselbrecht, M.Gross, A.Herrmann, M.Rüttimann, J.Crassous, F.Cardullo, L.Echegoyen, and F.Diederich, *J.Am.Chem.Soc.*, *120*, 7860 (1998), and references therein; (c) P.B.Burbank, J.R.Gibson, H.C.Dorn, and M.R.Anderson, *J.Electroanal.Chem.*, *417*, 1 (1996); (d) M.S.Meier, T.F.Guarr, J.P.Selegue, and V.K.Vance, *J.Chem.Soc., Chem.Commun.*, *63* (1993).
3. J.Heinze, *Angew.Chem.Int.Ed.Engl.*, *23*, 831 (1984).
4. (a) R.C.Haddon, L.E.Brus, and K.Raghavachari, *Chem.Phys.Letters*, *125*, 459 (1986); (b) P.W.Fowler and J.Woolrich, *Chem.Phys.Letters*, *127*, 78 (1986).
5. (a) M.Fedurco, M.M.Olmstead, and W.R.Fawcett, *Inorg.Chem*, *34*, 390 (1995); (b) W.C.Wan, X.Liu, G.M.Sweeney, and W.E.Broderick, *J.Am.Chem.Soc.*, *117*, 9580 (1995); (c) P.Paul, Z.Xie, R.Bau, P.D.W.Boyd, and C.A.Reed, *J.Am.Chem.Soc.*, *116*, 4145 (1994); (d) K.Himmel and M.Jansen, *Inorg.Chem.*, *37*, 3437 (1998); (e) K.Himmel and M.Jansen, *Eur.J.Inorg.Chem.*, 1183 (1998).
6. (a) P.J.Fagan, J.C.Calabrese, and B.Malone, *Acc.Chem.Res.*, *25*, 134-142 (1992); (b) J.R.Bowser, *Adv.Organometal.Chem.*, *36*, 57-94 (1994); (c) W.Šliwa, *Transition Met.Chem.*, *21*, 583-592 (1996); (d) A.H.H.Stephens and M.L.H.Green, *Adv.Organometal.Chem.*, *44*, 1-43 (1997); (e) P.L.Boulas, M.Gómez-Kaifer, and L.Echegoyen, *Angew.Chem.Int.Ed.Engl.*, *37*, 217-247 (1998); (f) A.L.Balch and M.M.Olmstead, *Chem.Rev.*, *98*, 2123-2165 (1998).
7. J.M.Hawkins, A.Meyer, T.A.Lewis, S.Loren, F.J.Hollander, *Science*, *252*, 312 (1991).
8. (a) A.L.Balch, V.J.Catalano, and J.W.Lee, *Inorg.Chem.*, *30*, 3980 (1991); (b) A.L.Balch, V.J.Catalano, J.W.Lee, and M.M.Olmstead, *J.Am.Chem.Soc.*, *114*, 5455 (1992); (c) A.L.Balch, D.A.Costa, J.W.Lee, B.C.Noll, and M.M.Olmstead, *Inorg.Chem.*, *33*, 2071 (1994); (d) A.L.Balch, D.A.Costa, B.C.Noll, and M.M.Olmstead, *Inorg.Chem.*, *35*, 458 (1996); (e) A.L.Balch, D.A.Costa, B.C.Noll, and M.M.Olmstead, *J.Am.Chem.Soc.*, *117*, 8926 (1995).
9. (a) A.L.Balch, J.W.Lee, B.C.Noll, and M.M.Olmstead, *J.Am.Chem.Soc.*, *114*, 10984 (1992); (b) A.L.Balch, J.W.Lee, B.C.Noll, and M.M.Olmstead, *Inorg.Chem.*, *33*, 5238 (1994).
10. (a) A.L.Balch, J.W.Lee, B.C.Noll, and M.M.Olmstead, *Inorg.Chem.*, *32*, 3577 (1993); (b) Y.Ishii, H.Hoshi, Y.Hamada, and M.Hidai, *Chem. Letters*, 801 (1994); (c) M.Soimasuo, T.T.Pakkanen, M.Ahlgrén, and T.A.Pakkanen, *Polyhedron*, *17*, 2073

(1998); (d) M.Rasinkangas, T.T.Pakkanen, T.A.Pakkanen, M.Ahlgrén, and J.Rouvinen, *J.Am.Chem.Soc.*, *115*, 4901 (1993).

11. (a) A.N.Chernega, M.L.H.Green, J.Haggitt, and A.H.H.Stephens, *J.Chem.Soc., Dalton Trans.*, 755 (1998); (b) I.J.Mavunkal, Y.Chi, S.-M.Peng, and G.-H.Lee, *Organometallics*, *14*, 4454 (1995).

12. (a) M.-J.Arce, A.L.Viado, S.I.Khan, and Y.Rubin, *Organometallics*, *15*, 4340 (1996); (b) M.Sawamura, H.Iikura, and E.Nakamura, *J.Am.Chem.Soc.*, *118*, 12850 (1996); (c) T. Kusukawa and W. Ando, *J.Organomet.Chem.*, *561*, 109 (1998).

13. (a) S.A.Lerke, B.A.Parkinson, D.H.Evans, and P.J.Fagan, *J.Am.Chem.Soc.*, *114*, 7807 (1992); (b) P.J.Fagan, J.C.Calabrese, B.Malone, *Science*, *252*, 1160 (1991); (c) V.V.Bashilov, P.V.Petrovskii, V.I.Sokolov, S.V.Lindeman, I.A.Guzey, and Y.I.Struchkov, *Organometallics*, *12*, 991 (1993); (d) P.J.Fagan, J.C.Calabrese, and B.Malone, *J.Am.Chem.Soc.*, *113*, 9408 (1991).

14. (a) R.S.Koefod, M.F.Hudgens, and J.R.Shapley, *J.Am.Chem.Soc.*, *113*, 8957 (1991); (b) R.S.Koefod, C.Xu, W.Lu, J.R.Shapley, M.G.Hill, and K.R.Mann, *J.Phys.Chem.*, *96*, 2928 (1992); (c) J.R.Shapley, *personal communication.*

15. T.V.Magdesieva, V.V.Bashilov, S.I.Gorelsky, V.I.Sokolov, and K.P.Butin, *Russ.Chem.Bull.*, *43*, 2034 (1994).

16. (a) J.R.Shapley, Y.Du, H.-F.Hsu, and J.J.Way, *Fullerenes: Recent advances in the Chemistry and Physics of fullerenes and related Materials*, K.M.Kadish, R.S.Ruoff, eds., The Electrochemical Society: Pennington. N.J. 1994, 24, 1255; (b) H.F.Hsu, Y.Du, T.E.Albrecht-Schmitt, S.R.Wilson, and J.R.Shapley, *Organometallics*, *17*, 1756 (1998).

17. (a) L.-C.Song, Y.-H.Zhu, and Q.-M.Hu, *Polyhedron*, *16*, 2141 (1997); (b) P.Zanello, F.Laschi, M.Fontani, L.-C.Song, and Y.-H.Zhu, *J.Organomet.Chem.*, *593-594*, 7 (2000).

18. (a) K.Tang, S.Zheng, X.Jin, H.Zeng, Z.Gu, X.Zhou, and Y.Tang, *J.Chem.Soc., Dalton Trans.*, 3585 (1997); (b) P.Zanello, F.Laschi, M.Fontani, C.Mealli, A.Ienco, K.Tang, X.Jin, and L. Li, *J.Chem.Soc., Dalton Trans.*, 965 (1999); (c) P.Zanello, F.Laschi, A.Cinquantini, M.Fontani, K.Tang, X.Jin, and L. Li, *Eur. J. Inorg.Chem., in the press.*

19. H.Song. K.Lee, J.T.Park, I.-H.Suh, *J.Organomet.Chem.*, *584*, 361 (1999).

20. (a) M.D.Westmeyer, C.P.Galloway, and T.B.Rauchfuss, *Inorg.Chem.*, *33*, 4615 (1994); (b) M.D.Westmeyer, T.B.Rauchfuss, and A.K.Verma, *Inorg.Chem.*, *35*, 7140 (1996).

21. D.Armspach, E.C.Constable, F.Diederich, C.E.Housecroft, and J.-F.Nierengarten, *Chem.Eur.J.*, *4*, 723 (1998).

22. L.Shu, S.Pyo, J.Rivera, L.Echegoyen, *Inorg.Chim.Acta*, *292*, 34 (1999).

23. M.Iyoda, F.Sultana, S.Sasaki, and H.Butenschön, *Tetrahedron Letters*, *36*, 579 (1995).

24. B.Bildstein, M.Schweiger, H.Angleitner, H.Kopacka, M.Mitterböck, K.Wurst, K.-H.Ongania, M.Fontani, and P.Zanello, *Organometallics*, *18*, 4286 (1999).

25. M.Maggini, A.Karlsson, G.Scorrano, G.Sandonà, G.Farnia, and M.Prato, *J.Chem.Soc., Chem.Commun.*, 589 (1994)

26. (a) T.G.Linssen, K.Dürr, M.Hanack, and A.Hirsch, *J.Chem.Soc., Chem. Commun.*, 103 (1995); (b) E.Dietel, A.Hirsch, E.Eichorn, A.Rieker, S.Hackbarth, and B.Röder,

Chem.Commun., 1981 (1998); N.Armaroli, F.Diederich, C.O.Dietrich-Buchecker, L.Flamigni, G.Marconi, J.-F.Nierengarten, and J.-P.Sauvage, *Chem.Eur.J.*, *4*, 406 (1998); (d) S.Yamago, M.Yanagawa, H.Mukai, and E.Nakamura, *Tetrahedron*, *14*, 5091 (1996).

27. (a) H.-F.Hsu and J.R.Shapley, *J.Am.Chem.Soc.*, *118*, 9192 (1996); (b) K.Lee, H.-F.Hsu, and J.R.Shapley, *Organometallics*, *16*, 3876 (1997); (c) K.Lee and J.R.Shapley, *Organometallics*, *17*, 3020 (1998).

28. (a) J.T.Park, H.Song, J.-J.Cho, M.-K.Chung, J.-H.Lee, and I.-H.Suh, *Organometallics*, *17*, 227 (1998); (b) H.Song, K.Lee, J.T.Park, and M.-G.Choi, *Organometallics*, *17*, 4477 (1998); (c) J.T.Park, J.-J.Cho, H.Song, C.-S.Jun, Y.Son, and J.Kwak, *Inorg.Chem.*, *36*, 2698 (1997).

29. (a) A.L.Balch, V.J.Catalano, J.W.Lee, M.M.Olmstead, and S.R.Parkin, *J.Am.Chem.Soc.*, *113*, 8953 (1991); (b) A.L.Balch, D.A.Costa, and M.M.Olmstead, *Chem.Commun.*, 2449 (1996); (c) A.L.Balch, J.W.Lee, and M.M.Olmstead, *Angew.Chem.Int. Ed.Engl.*, *31*, 1356 (1992); (d) A.L.Balch, L.Hao, and M.M.Olmstead, *Angew.Chem.Int. Ed.Engl.*, *35*, 188 (1996).

30. H.-F.Hsu, S.R.Wilson, and J.R.Shapley, *Chem.Commun.*, 1125 (1997).

31. T.V.Magdesieva, V.V.Bashilov, D.N.Kravchuk, P.V.Petrovskii, V.I.Sokolov, and K.P.Butin, *Fullerenes: Recent advances in the Chemistry and Physics of fullerenes and related Materials.* K.M.Kadish, R.S.Ruoff, eds., The Electrochemical Society: Pennington. N.J. 1997, 14, 209.

32. M.M.Olmstead, L.Hao, and A.L.Balch, *J.Organomet.Chem.*, *578*, 85 (1999).

33. A.L.Balch, A.S.Ginwalla, J.W.Lee, B.C.Noll, and M.M.Olmstead, *J.Am.Chem.Soc.*, *116*, 2227 (1994).

34. R.Beyers, C.H.Kiang, R.D.Johnson, J.R.Salem, M.S. de Vries, C.S.Yannoni, D.S.Bethune, H.C.Dorn, P.Burbank, K.Harich, and S.Stevenson, *Nature*, *370*, 196 (1994).

35. (a) S.Nagase and K.Kobayashi, *Chem.Phys.Letters*, *214*, 57 (199; (b) J.R.Ungerer and T.Hughbanks, *J.Am.Chem.Soc.*, *115*, 2054 (1993).

36. (a) T.Suzuki, Y.Maruyama, T.Kato, K.Kikuchi, and Y.Achiba, *J.Am.Chem.Soc.*, *115*, 11006 (1993); (b) K.Yamamoto, H.Funasaka, T.Takahashi, T.Suzuki, and Y.Maruyama, *J.Phys.Chem.*, *98*, 12831 (1994).

37. K.Kikuchi, Y.Nakao, S.Suzuki, and Y.Achiba, *J.Am.Chem.Soc.*, *116*, 9367 (1994).

38. T.Akasaka, S.Nagase, K.Kobayashi, T.Suzuki, T.Kato, K.Yamamoto, H.Funasaka, andT.Takahashi, *J.Chem.Soc., Chem.Commun.*, 1343 (1995).

39. S.Nagase, K.Kobayashi, T.Kato, and Y.Achiba, *Chem.Phys.Letters*, *201*, 475 (1993), and references therein.

40. P.M.Burbank, J.R.Gibson, H.C.Dorn, and M.R.Anderson, *J.Electroanal.Chem.*, *417*, 1 (1996).

41. M.R.Anderson, H.C.Dorn, S.Stevenson, P.M.Burbank, and J.R.Gibson, *J.Am.Chem.Soc.*, *119*, 437 (1997).

Polyoxometalates and Coordination Polymers

Christian Robl

Institut für Anorganische und Analytische Chemie der Friedrich-Schiller-Universität Jena,
August-Bebel-Str. 6-8, D-07743 Jena , Fed. Rep. of Germany
e-mail: crr@rz.uni-jena.de

Abstract. New polyoxometalate anions with unusual chemical composition and structural features could be prepared by carefully tuning the synthetic procedures. Polyoxometalate anions containing a high amount of Te or Se or residues of acetic acid, fumaric acid, succinic acid, diglycolic acid or thio-diacetic acid as ligands have been characterized by X-ray structure analysis. Coordination polymers with one-dimensionally, two-dimensionally and three-dimensionally infinite structural characteristics could be prepared employing anions of dihydroxy-p-benzoquinones, benzene carboxylic acids, methylene diphosphonic acid, and phosphono-carboxylic acids. Coordination polymers with chain-like or layer-like structure carrying negative excess charge as well as compounds with an open framework structure have been studied.

I. Polyoxometalates

I.1. Introduction

Various polyoxometalate anions, the derivatives of the so-called heteropoly acids, with the general formula $[A_xB_yO_z]^{n-}$ (A=Si, P, Se, Te etc., B=V, Nb, Mo etc.), are known in great number, the first of them having been described in the early decades of the 19th century [1, 2]. Nevertheless, these large family of versatile chemical compounds attracts great interest still today [3, 4].

A few types of polyoxometalate anions are favoured due to their outstanding stability and are known as, e.g. Keggin, Anderson-Evans, Dexter-Silverton and Dawson structures [3]. However, it has to be assumed that further anion types occur during the various intermediate steps of condensation which can be isolated by carefully tuning the synthetic procedure. Furthermore a large number of novel polyoxometalate anions will become accessible, if organic acids are used as condensation reagents [5 – 11].

I.2. $[Te_6Mo_{12}O_{60}]^{12-}$ and $[Se_2MoO_8]_n^{2n-}$

Most classical types of heteropolyanions are characterized by a rather low stoichiometric content of hetero-elements. E.g. the Te to Mo ratio in the Anderson-Evans type anion $[TeMo_6O_{24}]^{6-}$ is 1:6 [12]. We were able to prepare the novel tellurium-rich cyclic heteropolyanion $[Te_6Mo_{12}O_{60}]^{12-}$, which has been isolated as $(NH_4)_{12}[Te_6Mo_{12}O_{60}]\cdot 8H_2O$ [11], representing a derivative of the hypothetical cyclic hexatelluric acid $[TeO(OH)_4]_6$. A single crystal analysis showed a puckered ring composed of six corner-sharing TeO_6 octahedra to be the dominant structural feature of the $[Te_6Mo_{12}O_{60}]^{12-}$ anion. A pair of face-sharing, considerably distorted, MoO_6 octahedra is attached to each TeO_6 octahedron via common corners and common edges, respectively (Fig. 1).

Crystal data for $(NH_4)_{12}[Te_6Mo_{12}O_{60}]\cdot 8H_2O$: trigonal, space group R$\bar{3}$ (no. 148), a=2452.3(3), c=993.4(1) pm, V=5173.8$\cdot 10^6$ pm^3.

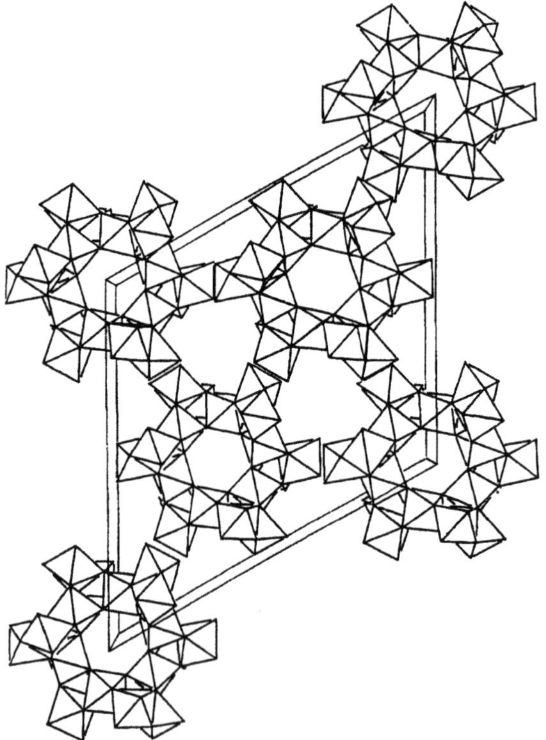

Fig. 1. The tellurium-rich cyclic polyoxometalate anion $[Te_6Mo_{12}O_{60}]^{12-}$ consists of a centrosymmetric ring made up by 6 corner-sharing TeO_6 octahedra. Six pairs of face-sharing MoO_6 octahedra are attached to this ring. There is no continuous -O-Mo-O-Mo-O- bond sequence.

The chain-like and highly selenium-rich polyoxometalate anion $[Se_2^{IV}MoO_8]_n^{2n-}$ has been prepared in aqueous solution applying a high excess of SeO_3^{2-}. $K_2[Se_2MoO_8]\cdot 3H_2O$ has been isolated and characterized by X-ray structure analysis

[13]. Distorted MoO_6 coordination octahedra are linked by two crystallographically independent SeO_3 pseudo-tetrahedra to yield a sinusoidally corrugated chain running parallel to [100] with the amplitude of corrugation parallel to [010] (Fig. 2). This chain-like anion is chiral and polar and lacks a continuous -O-Mo-O-Mo-O- bond sequence.

Crystal data for $K_2[Se_2MoO_8]\cdot 3H_2O$: monoclinic, space group $P2_1/a$ (no. 14), a=932.5(2), b=1233.0(2), c=1058.0(1) pm, ß=112.31(1)°, V=1125.4·10^6 pm^3.

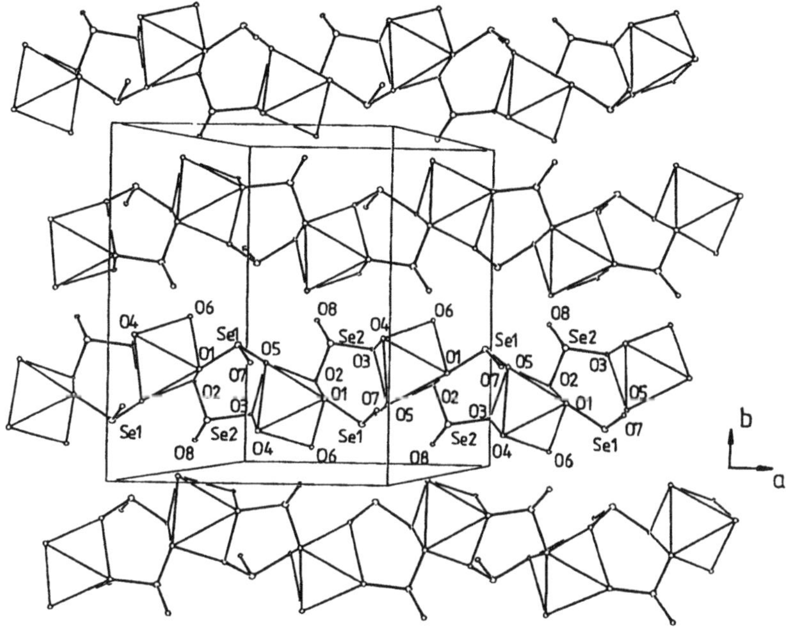

Fig. 2. The selenium-rich chain-like anion consists of MoO_6 octahedra linked by Se-centered pseudo-tetrahedra [13].

I.3. $[XY_nMo_6O_{21+4n}(OH)_{6-2n}]^{8-}$ ($X=Se^{IV}$, $Y=S^{VI}$, Se^{VI}) and related anions

A six-membered octahedra ring belting a pyramidal SeO_3 moiety is the prominent structural feature of the cyclic polar polyoxometalate anion $[Se^{IV}S_3{}^{VI}Mo_6O_{33}]^{8-}$ that has been characterized by X-ray structure analysis of $K_8[SeS_3Mo_6O_{33}]\cdot 5,5H_2O$ [14]. The octahedra ring consists of three pairs of edge-sharing MoO_6 coordination octahedra. These octahedra pairs share common corners yielding a cyclic arrangement of six octahedra connected alternatively by edges and corners, respectively. Two edge-sharing octahedra at one time are capped by a SO_4 tetrahedron sharing one common corner with each of the two octahedra.

The anions $[Se^{IV}Se_3{}^{VI}Mo_6O_{33}]^{8-}$ (Fig. 3) and $[Se^{IV}Se_2{}^{VI}Mo_2O_{29}(OH)_2]^{8-}$ (Fig. 4) have been prepared as well and X-ray structure analysis [7, 8] has proven them to posses a structure analoguous to $[SeS_3Mo_6O_{33}]^{8-}$. These have to be considered as members of a new versatile family of polar cyclic heteropolyanions with the general formula $[XY_nMo_6O_{21+4n}(OH)_{6-2n}]^{8-}$ (n=0, 1, 2, 3; X=Se^{IV}, Y=S^{VI}, Se^{VI}) and are derivatives of the $[Se^{IV}Mo_6O_{21}(OH)_6]^{8-}$ anion, which has bot been isolated yet. Crystal data for $K_8[SeS_3Mo_6O_{33}]\cdot 5.5H_2O$: triclinic, space group P-1 (no. 2), a=1028.5(2), b=1144.5(2), c=1677.6(2) pm, α=95.46(1), ß=96.12(1), γ=103.60(1)°, V=1893.7·10^6 pm^3.

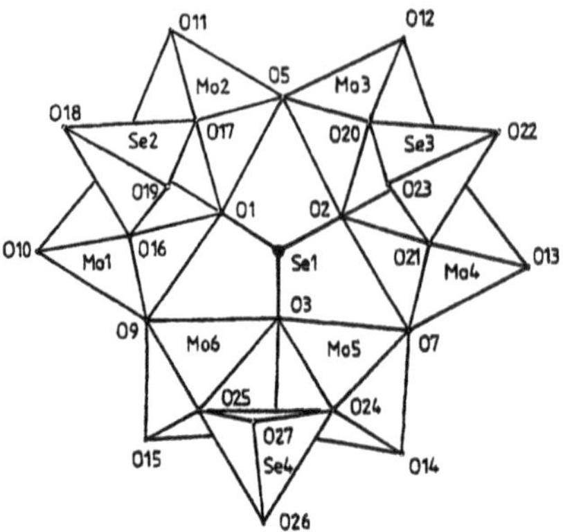

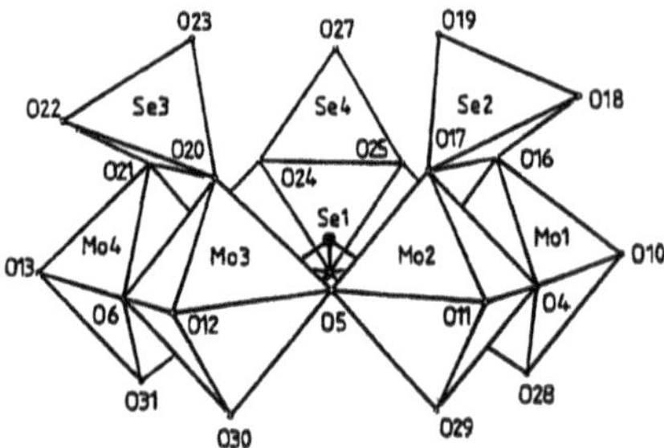

Fig. 3. The novel cyclic polar heteropolyanion $[Se^{IV}Se_3{}^{VI}Mo_6O_{33}]^{8-}$ bearing a pyramidal SeO_3 group in its centre and featuring a ring of six MoO_6 coordination octahedra linked by common corners and edges, respectively, to which three SeO_4 tetrahedra are attached [8].

Further derivatives of $[Se^{IV}Mo_6O_{21}(OH)_6]^{8-}$ can be prepared by carrying out condensation reactions with various inorganic or organic acids. By use of selenous acid the $[Se_4{}^{IV}Mo_6O_{30}]^{8-}$ anion could be isolated as $Li_8[Se_4{}^{IV}Mo_6O_{30}]\cdot 18H_2O$ [7, 8]. Two $[Se_4{}^{IV}Mo_6O_{30}]^{8-}$ anions related by a centre of symmetry are linked by Li^+ cations to yield a large cluster with moderate negative excess charge of $\{Li_{10}[Se_4{}^{IV}Mo_6O_{30}]_2(H_2O)_{20}\}^{6-}$ composition (Fig. 5). The Li^+ cations in the centre of this cluster are tetrahedrally coordinated by three water molecules and one oxygen atom stemming from a SeO_3 moiety of the $[Se_4{}^{IV}Mo_6O_{30}]^{8-}$ anion. Further Li^+ cations are closely bound to the octahedra ring by three oxygen atoms of the anion and two water molecules yielding a distorted trigonally bipyramidal coordination sphere. Additional Li^+ cations in the vicinity of the anionic cluster compensate for the negativ excess charge.

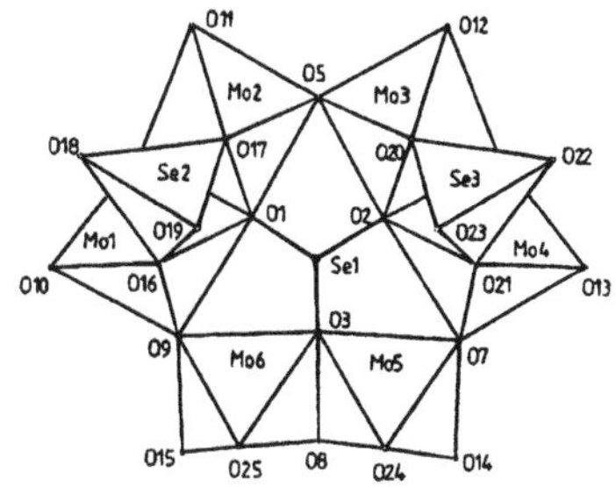

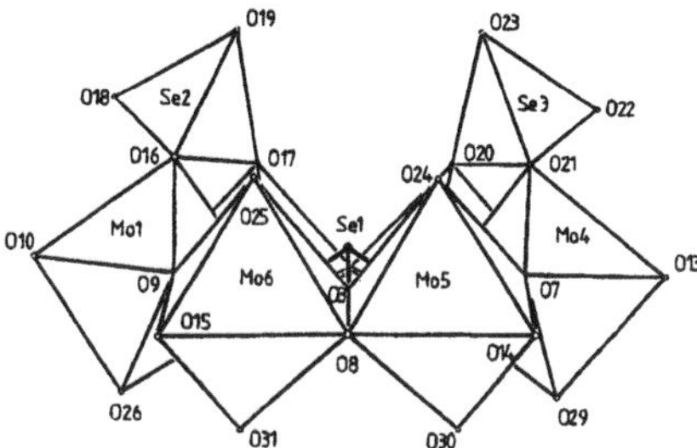

Fig. 4. The $[Se^{IV}Se_2{}^{VI}Mo_6O_{29}(OH)_2]^{8-}$ anion: a lacunary derivative of $[Se^{IV}Se_3{}^{VI}Mo_6O_{33}]^{8-}$ [8].

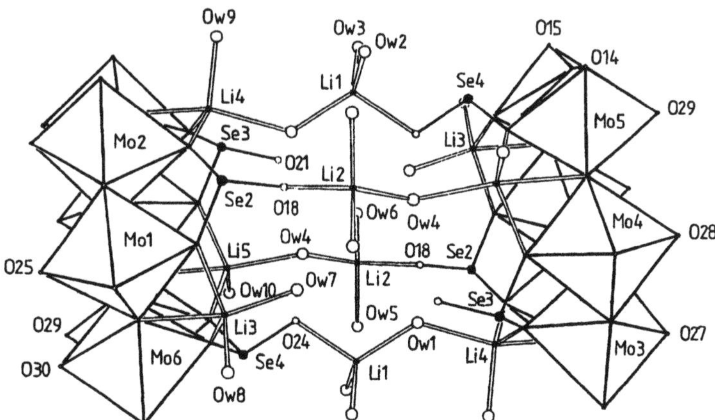

Fig. 5. The $\{Li_{10}[Se_4{}^{IV}Mo_6O_{30}]_2(H_2O)_{20}\}^{6-}$ cluster in the crystal structure of $Li_8[Se_4{}^{IV}Mo_6O_{30}]\cdot 18H_2O$ [8].

Condensation with acetic acid yields the polar $[Se^{IV}Mo_6O_{21}(OOCCH_3)_3]^{5-}$ anion (Fig. 6) verified by X-ray structure analysis of $Na_5[Se^{IV}Mo_6O_{21}(OOCCH_3)_3]\cdot 13H_2O$ [7, 8]. A related anion with a methylene diphosphonate ligand having one PO_3 group at the place of the SeO_3 pseudo tetrahedron and the second PO_3 group as a tetrahedral building block connected to a pair of edge sharing octahedra has been reported recently [15].

Crystal data for $Li_8[Se_4Mo_6O_{30}]\cdot 18H_2O$: triclinic, space group P-1 (no. 2), a=1054.57(15), b=1300.2(2), c=1656.3(2) pm, α=80.425(11), ß=85.969(12), γ=73.321(11)°, $V=2144.6\cdot 10^6$ pm^3.
$Na_5[Se^{IV}Mo_6O_{21}(OOCCH_3)_3]\cdot 13H_2O$: monoclinic. Space group C2/c (no. 15), a=2521.2(8), b=1952.4(6), c=1765.7(6) pm, ß=90.68(3)°, $V=8691\cdot 10^6$ pm^3.

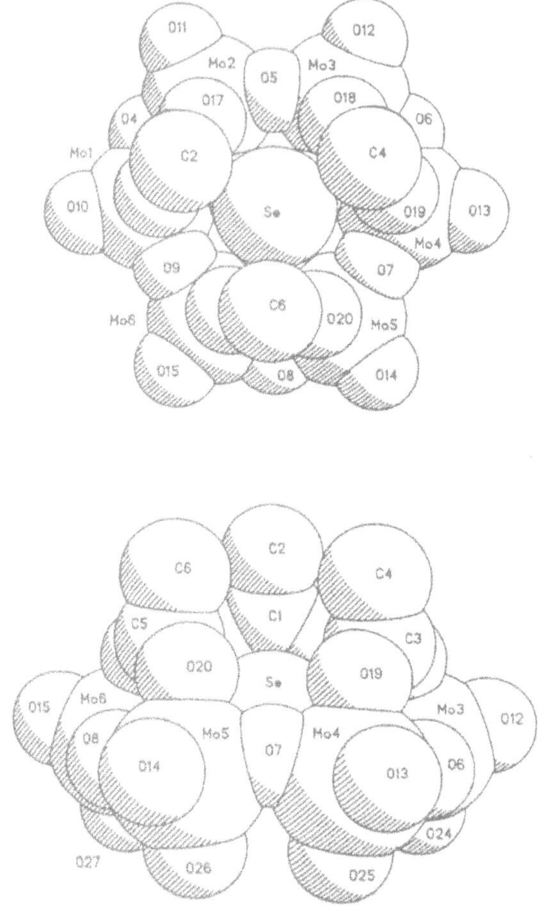

Fig. 6. A space-filling model of the polar $[Se^{IV}Mo_6O_{21}(OOCCH_3)_3]^{5-}$ anion [8].

Further organic acids can be used as well, e. g. fumaric acid or succinic acid. The polar dimeric anion $[Se_2^{IV}Mo_{12}O_{42}(C_{12}H_6O_{12})]^{10-}$ results from the condensation reaction with fumaric acid (Fig. 7) [7, 8]. It resembles the Li-rich cluster $\{Li_{10}[Se_4^{IV}Mo_6O_{30}]_2(H_2O)_{20}\}^{6-}$ (Fig. 5) encountered in the crystal structure of $Li_8[Se_4^{IV}Mo_6O_{30}]\cdot 18H_2O$. Guanidinium cations and H_3O^+ ions serve as counter ions. Crystal data: $(CH_6N_3)_8(H_3O)_2[Se_2Mo_{12}O_{42}(C_{12}H_6O_{12})]\cdot 18H_2O$, monoclinic, space group C2/c (no. 15), a=1321.4(5), b=2279.2(6), c=3236.4(10) pm, ß=101.71(3)°, $V=9544\cdot 10^6$ pm^3.

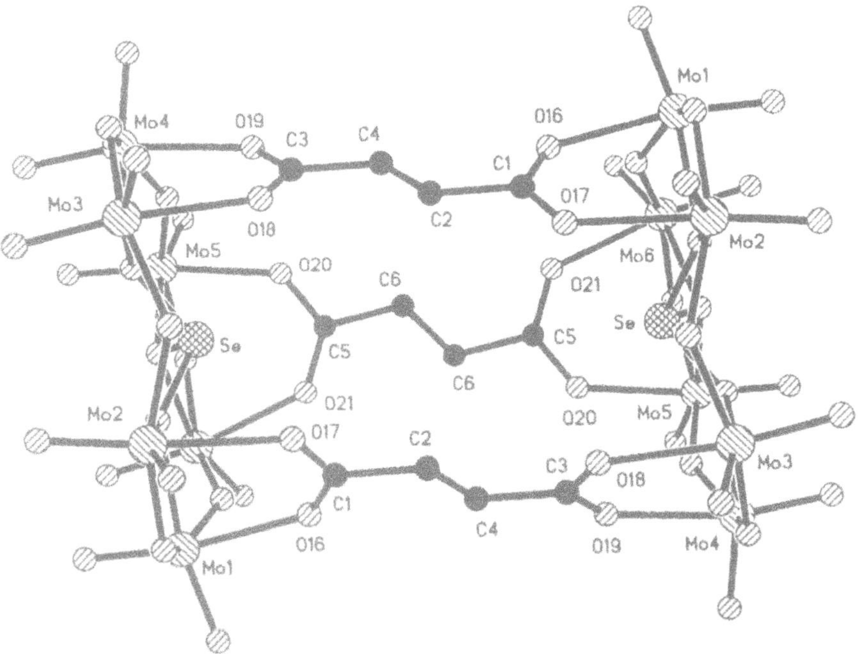

Fig. 7. The polar dimeric anion $[Se_2^{IV}Mo_{12}O_{42}(C_{12}H_6O_{12})]^{10-}$ with twofold symmetry [8].

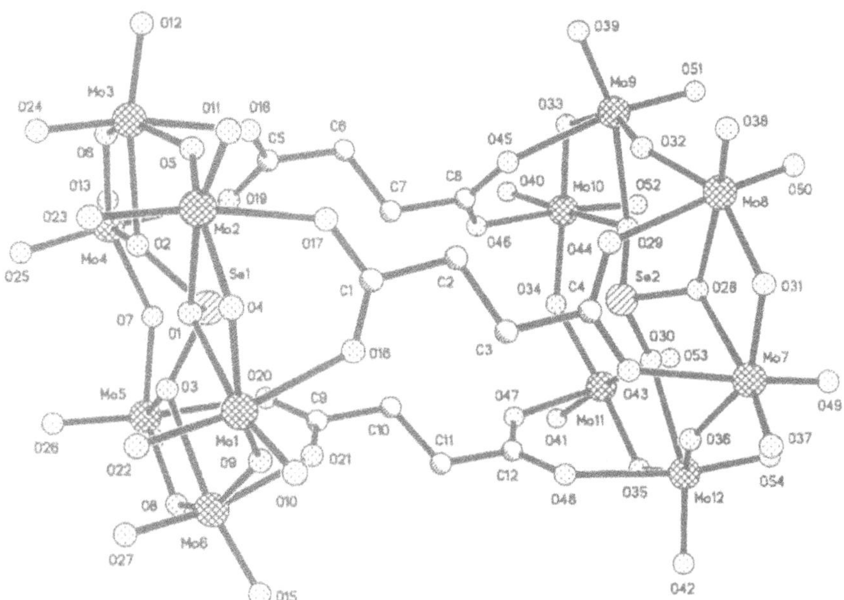

Fig. 8. The $[Se_2^{IV}Mo_{12}O_{42}(C_{12}H_{12}O_{12})]^{10-}$ anion.

Condensation with succinic acid yields the quite similar dimeric anion $[Se_2^{IV}Mo_{12}O_{42}(C_{12}H_{12}O_{12})]^{10-}$ (Fig. 8). Crystal data: triclinic, space group P-1 (no. 2), a=1119.76(11), b=1601.7(2), c=2314.3(2) pm, α=86.614(8), ß=77.351(7), γ=71.871(7)°, $V=3848.7 \cdot 10^6$ pm^3.
The reaction of MoO_3, Li_2CO_3, SeO_2, HOOC-CH_2-O-CH_2-COOH and $(CH_3)_4NCl$ yielded colourless, needle-like crystals of $Li_{2,5}[(CH_3)_4N]_{3,5}[SeMo_6O_{31}C_8H_8]\cdot 9H_2O$. Crystal data: monoclinic, space group P2/c (no. 13), a=1312.6(2), b=1285.7(2), c=3378.2(5) pm, ß=92.424(9)°, $V=5696 \cdot 10^6$ pm^3.

There are three crystallographically independent residues of diglycolic acid. Two of them are situated on twofold axes and serve to link neighbouring six-membered rings of Mo-centered octahedra (Fig. 9). The third diglycolic acid residue is bound to one of the three pairs of edge-sharing octahedra and linked to tetrahedrally coordinated Li^+ via an oxygen atom of the carboxylate group [7, 8].

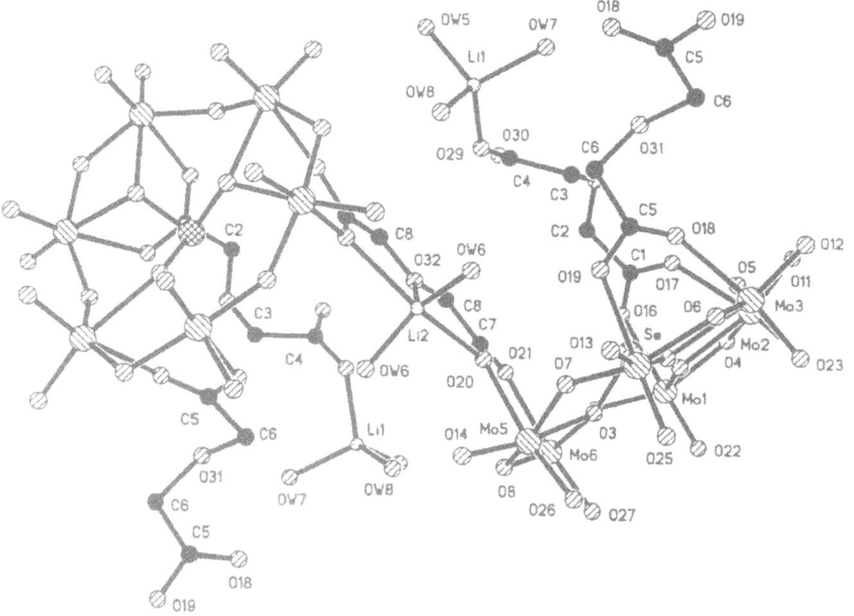

Fig. 9. A section of the chain-like polyanion $[Se^{IV}Mo_6O_{31}C_8H_8]_n^{6n-}$ with four and five-coordinated Li^+ cations [8].

Thus a one dimensionally infinite anion with negative excess charge is the prominent feature of the crystal structure of $Li_{2,5}[(CH_3)_4N]_{3,5}[Se\ Mo_6O_{31}C_8H_8]\cdot 9H_2O$ (Fig. 10).

Further Li^+ cations are attached to the polyanion via ether oxygen atoms of the diglycolic acid residues (Fig. 9). These Li^+ cations (Li(2)) are coordinated in a distorted trigonally bipyramidal manner by two water molecules, two carboxylate oxygen atoms and the ether oxygen atom (O(32)).

Hence Li^+ cations serve to lower the negative excess charge of the $[SeMo_6O_{31}C_8H_8]_n^{6n-}$ polyanions.

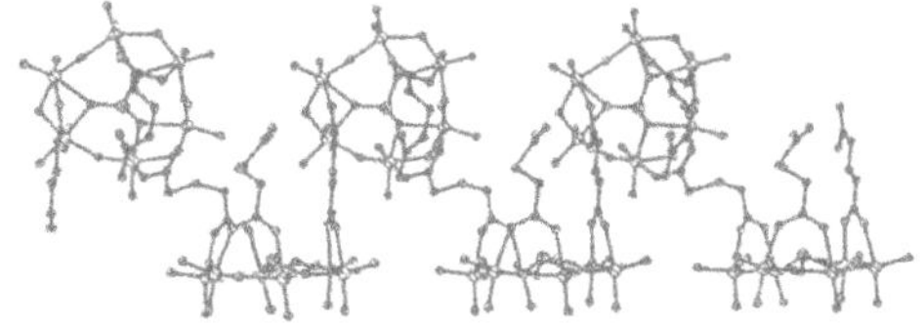

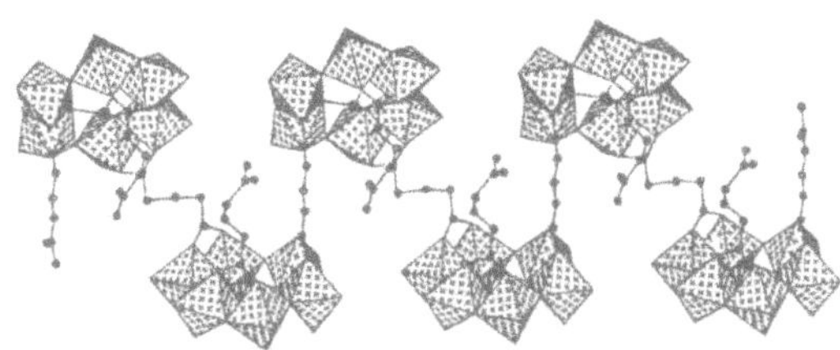

Fig. 10. The one-dimensionally infinite anion $[Se^{IV}Mo_6O_{31}C_8H_8]_n^{6n-}$ [8].

Two different products of the reaction with thio-diacetic acid could be identified. A monomeric anion (Fig. 11) in $(NH_4)_4Li_2[SeMo_6O_{29}S_2C_8H_8]\cdot 7H_2O$ and a cyclic tetrameric anion $[Se_4Mo_{24}O_{108}S_4C_{16}H_{24}]^{24-}$ which has been verified by a preliminary crystal structure analysis [7, 8].

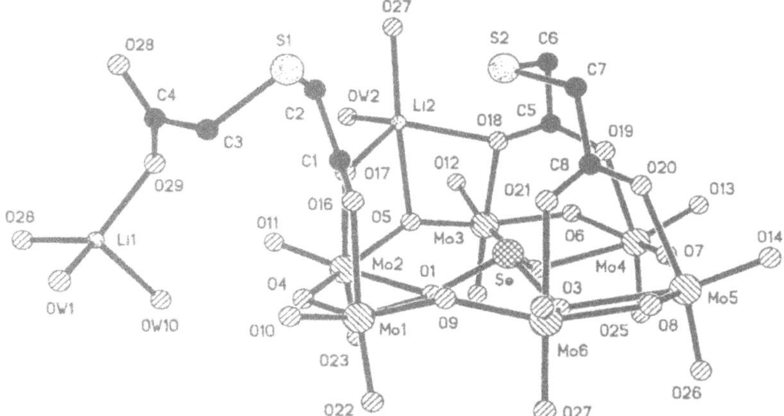

Fig. 11. The reaction with thio-diacetic acid yields the monomeric $[Se^{IV}Mo_6O_{29}S_2C_8H_8]^{6-}$ anion [8].

Crystal data for $(NH_4)_4Li_2[SeMo_6O_{29}S_2C_8H_8]\cdot 7H_2O$: triclinic, space group P-1 (no. 2), a=938.5(4), b=1305.0(3), c=1765.0(5) pm, α=69.300(10), ß=84.90(2), γ=87.080(10)°, V=2013.7$\cdot 10^6$ pm^3.

One thio-diacetic acid residue is bound to two pairs of edge-sharing octahedra (Fig. 11), thus it cannot link neighbouring anions. A further thio-diacetic acid residue is linked to tetrahedrally coordinated Li^+. Five-coordinated Li^+ cations are bound tightly to the anions and link adjacent anions to infinite strands. Neighbouring strands are connected by four-coordinated Li^+ cations yielding a double chain-like pattern as the prominent structural feature (Fig. 12).

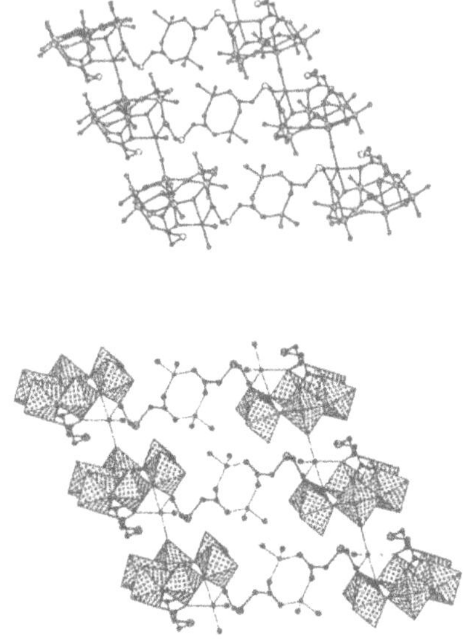

Fig. 12. A double chain-like pattern is the prominent structural feature in $(NH_4)_4Li_2[SeMo_6O_{29}S_2C_8H_8]\cdot 7H_2O$ [8].

II. Coordination Polymers

II.1. Introduction

Coordination polymers are capable of forming one-, two- or three-dimensionally infinite arrangements. Hence there is a situation similar to the structural chemistry of silicates, which is famous for chain-like and layer-like structural features carrying negative excess charge and for open framework structures. Anionic ligands like the anions of various dihydroxy-p-benzoquinones, benzene carboxylic acids, phosphono-carboxylic acids and diphosphonic acids are appropriate building blocks for the preparation of coordination polymers with interesting structural properties. More than a decade ago we started to prepare systematically coordination polymers with various structural characteristics (cf. e.g. [16 – 20]) which was a nearly unnoticed field of research at that time. Today coordination polymers have succeeded in attracting the interest of a great number of chemists in the entire world [21 – 24].

II.2. One-dimensionally Infinite Coordination Polymers

II.2.1. $Cu_{1,5}[OOC\text{-}CH_2\text{-}PO_3] \cdot 5H_2O$

There are two crystallographically unique Cu^{2+} sites. Cu(1) is linked to one water molecule (O(w1)) and oxygen atoms (O(1), O(3), 2 x O(4)) belonging to the anion of phosphono-acetic acid. Thus a chain-like polymeric anion extending parallel to [100] with $\{[Cu(H_2O)(OOC\text{-}CH_2\text{-}PO_3)]^-\}_n$ composition is made up (Fig. 13) [25]. The negative excess charge of this anionic coordination polymer is compensated for by $Cu(H_2O)_6^{2+}$ ions intercalated between the polyanionic chains and situated on crystallographic centres of symmetry. Hence $Cu_{1,5}[OOC\text{-}CH_2\text{-}PO_3] \cdot 5H_2O$ has to be regarded as the Cu^{2+} loaded form of a crystalline cation exchanger. The theoretical exchange capacity is 5.6 mval/g.

Crystal data for $Cu_{1,5}[OOC\text{-}CH_2\text{-}PO_3] \cdot 5H_2O$: triclinic, space group P-1 (no. 2), a=608.2(1), b=800.1(1), c=1083.6(1) pm, α=94.98(1), ß=105.71(1), γ=109.84(1)°, V=468.08·10^6 pm^3.

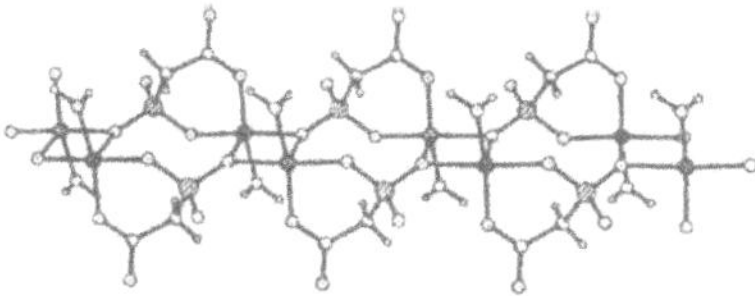

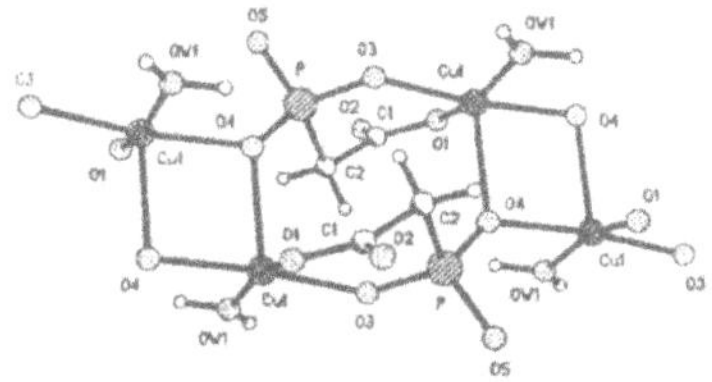

Fig. 13. The polymeric chain-like anion $\{[Cu(H_2O)(OOC\text{-}CH_2\text{-}PO_3)]^-\}_n$. The negative excess charge is compensated for by $Cu(H_2O)_6^{2+}$ cations [25].

II.2.2. $Co_2[C_6H_2(COO)_4] \cdot 18H_2O$ and $Co_3[C_6(COO)_6] \cdot 18H_2O$

Co(2) and $C_6H_2(COO)_4^{4-}$ anions are connected to infinite chains extending in the [101] direction (Fig. 14) [26, 27]. These chains are stacked in ...AAA... sequence in the [010] direction possessing $\{[Co(H_2O)_4C_6H_2(COO)_4]^{2-}\}_n$ stoichiometry and excessive negative charge, thus representing an infinite polyanion. This negative excesss charge is compensated for by $Co(H_2O)_6^{2+}$ octahedra intercalated between neighbouring chains. Thus $Co_2[C_6H_2(COO)_4] \cdot 18H_2O$ can be regarded as a crystalline cation exchanger. The theoretical exchange capacity for the $\{[Co(H_2O)_4C_6H_2(COO)_4]^{2-}\}_n$ polyanions is 5.2 mval/g with a linear excess charge density of 0.18 e/10^2 pm.

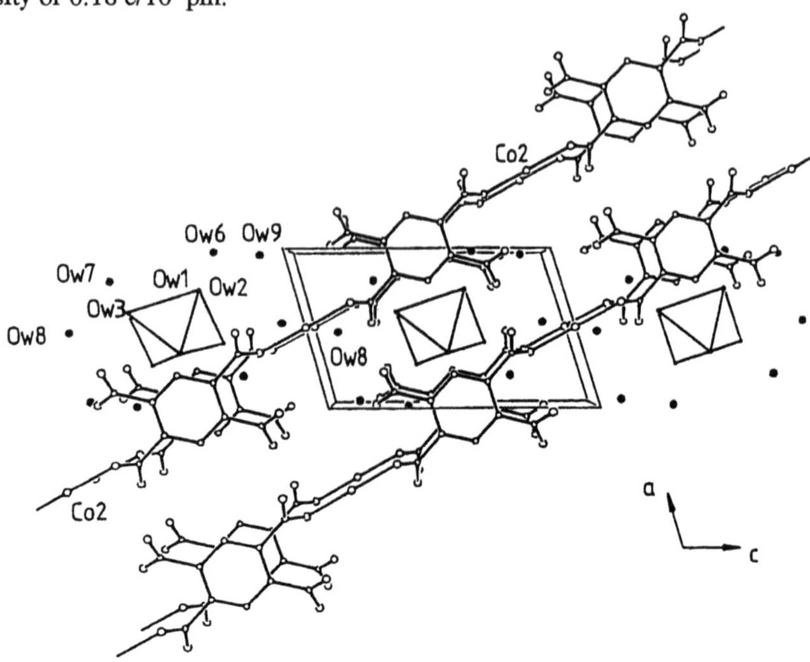

Fig. 14. Polymeric anionic chains of $\{[Co(H_2O)_4C_6H_2(COO)_4]^{2-}\}_n$ composition in $Co_2[C_6H_2(COO)_4]\cdot 18H_2O$. These chains with negative excess charge are made up by Co(2) and pyromellitate tetraanions. Water molecules bound to Co(2) are not shown. $Co(H_2O)_6^{2+}$ octahedra formed by Co(1), 2x O(w1), 2x O(w2), and 2x O(w3) are intercalated between neighbouring chains. Water molecules not coordinated to Co^{2+} are represented as black balls [26].

A similar crystal structure was found in tricobaltmellitate-octadecahydrate $Co_3[C_6(COO)_6]\cdot 18H_2O$ [28]. Chain-like polyanions of $\{[Co(H_2O)_2C_6(COO)_6]^{4-}\}_n$ stoichiometry are the characteristic feature in the crystal structure of this compound (Fig. 15) possessing, however, a linear excess charge density twice as high as that of the polyanions in cobaltpyromellitate-octadecahydrate. The theoretical exchange capacity is 9.3 mval/g. The negative excess charge is compensated for by Co^{2+} (Co(1)) which is merely bound by the Co(1)-O(2) bond to the polyanion. Water molecules not shown in Fig. 15 complete the coordination spheres of Co(1) and Co(2) to yield slightly distorted octahedra. Crystal data for $Co_2[C_6H_2(COO)_4]\cdot 18H_2O$: triclinic, space group P-1 (no. 2), a=686.4(1), b=1000.0(2), c=1093.2(2) pm, α=93.00(2), ß=104.86(1), γ=103.59(1)°, V=699.9·10^6 pm^3.

Crystal data for $Co_3[C_6(COO)_6]\cdot 18H_2O$: orthorhombic, space group Pbca (no. 61), a=852.7(2), b=2015.6(3), c=1712.3(3) pm, V=2943.0·10^6 pm^3.

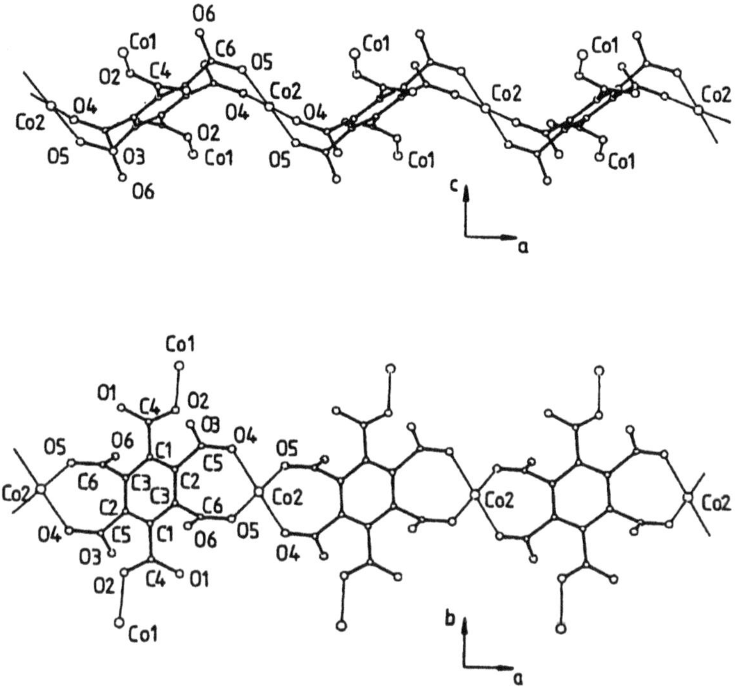

Fig. 15. Corrugated anionic chains with $\{[Co(H_2O)_2C_6(COO)_6]^{4-}\}_n$ composition extending parallel to [100] are the prominent structural feature in Co-mellitate ($Co_3[C_6(COO)_6] \cdot 18H_2O$). The plane formed by Co(2), O(4) and O(5) is tilted by 80° to the plane of the C_6-ring [28].

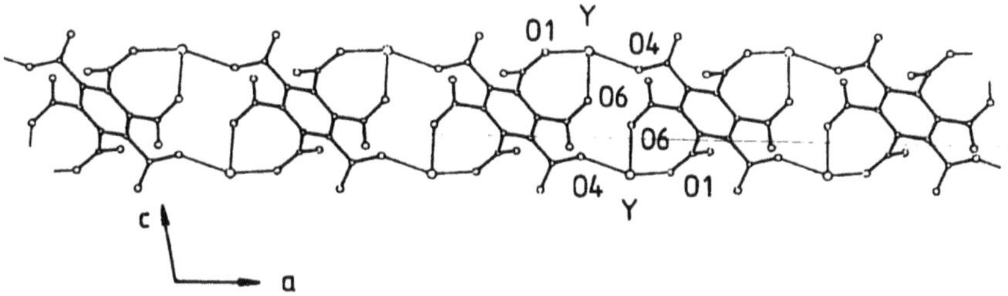

Fig. 16. The electrically neutral $\{[Y(H_2O)_5]_2[C_6(COO)_6]\}_n$ chain in Y-mellitate ($Y_2[C_6(COO)_6] \cdot 14H_2O$) viewed from the [010] direction. (Water molecules have been omitted.)[29].

II.2.3. $Y_2[C_6(COO)_6] \cdot 14H_2O$

Y^{3+} is coordinated by five water molecules and three oxygen atoms stemming from the $C_6(COO)_6^{6-}$ anion [29]. Y^{3+} cations related by a centre of symmetry connect

$C_6(COO)_6^{6-}$ anions to yield an electrically neutral chain with $\{[Y(H_2O)_5]_2[C_6(COO)_6]\}_n$ composition extending parallel to [100] (Fig. 16). Hydrogen bonds connect adjacent chains.
Crystal data for $Y_2[C_6(COO)_6]\cdot 14H_2O$: monoclinic, space group $P2_1/n$ (no. 14), a=847.5(1), b=923.4(2), c=1632.0(3) pm, ß=100.33(1)°, $V=1256.5\cdot 10^6$ pm^3.

II. 2.4. $Na_2Zn_2[C_6H_3(COO)_3]_2\cdot 11H_2O$

Double chains made up by Zn^{2+} cations and trimesinate anions (benzene-1,3,5-tricarboxylic acid = trimesic acid) are the prominent structural feature in the crystal structure of $Na_2Zn_2[C_6H_3(COO)_3]\cdot 11H_2O$. The chains are parallel to [010] with pairs of anions related by a centre of symmetry (Fig. 17) [30].Zn^{2+} is coordinated

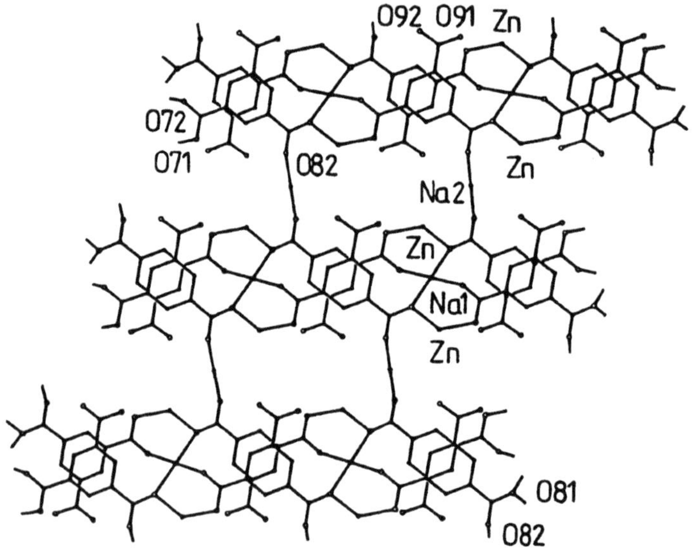

Fig. 17. Anionic double chains of $\{Na(H_2O)_2[Zn(H_2O)C_6H_3(COO)_3]_2^-\}_n$ composition are linked by $Na(H_2O)_4^+$ groups to yield an electrically neutral layer-like arrangement in the crystal structure of $Na_2Zn_2[C_6H_3(COO)_3]_2\cdot 11H_2O$ [30].

by two oxygen atoms belonging to carboxylate groups and three water molecules in a distorted trigonally bipyramidal manner. Na^+ cations (Na(1)) situated on centres of symmetry and coordinated octahedrally by two water molecules and four carboxylate oxygen atoms are bound in the central part of the double chain-like polyanions. Further Na^+ cations (Na(2)) occupying crystallographic centres of symmetry as well and being intercalated between neighbouring double chains connect the polyanions to yield a layer-like structure. These Na^+ cations carry four water molecules and two carboxylate oxygen atoms in the octahedral coordination sphere. Taking water molecules into account the stoichiometry of the anionic double chains can be given as $\{Na(H_2O)_2[Zn(H_2O)C_6H_3(COO)_3]_2^-\}_n$. Further water molecules not linked to Na^+ or Zn^{2+} serve as space filling particles. Crystal data for $Na_2Zn_2[C_6H_3(COO)_3]_2\cdot 11H_2O$: triclinic, space group P-1 (no. 2), a=710.0(1), b=979.3(2), c=1120.2(2) pm, α=66.94(2), ß=73.86(1), γ=84.76(1)°, V=688.3·10^6 pm^3.

II.3. Two-dimensionally Infinite Coordination Polymers

II.3.1. $Ca[C_6(i\text{-}C_3H_7)_2O_4H]_2 \cdot 2/3H_2O$

Ca^{2+} cations and $C_6(i\text{-}C_3H_7)_2O_4H^-$ anions form protonated puckered layers extending parallel to (001), which are stacked in ...ABCABC... sequence in the crystal structure (Fig. 18) [31]. Hence this coordination polymer corresponds to the H^+ loaded form of a weakly acidic cation exchanger with layer structure. Layers carrying negative excess charge are commonly characterised by their reciprocal excess charge densities.

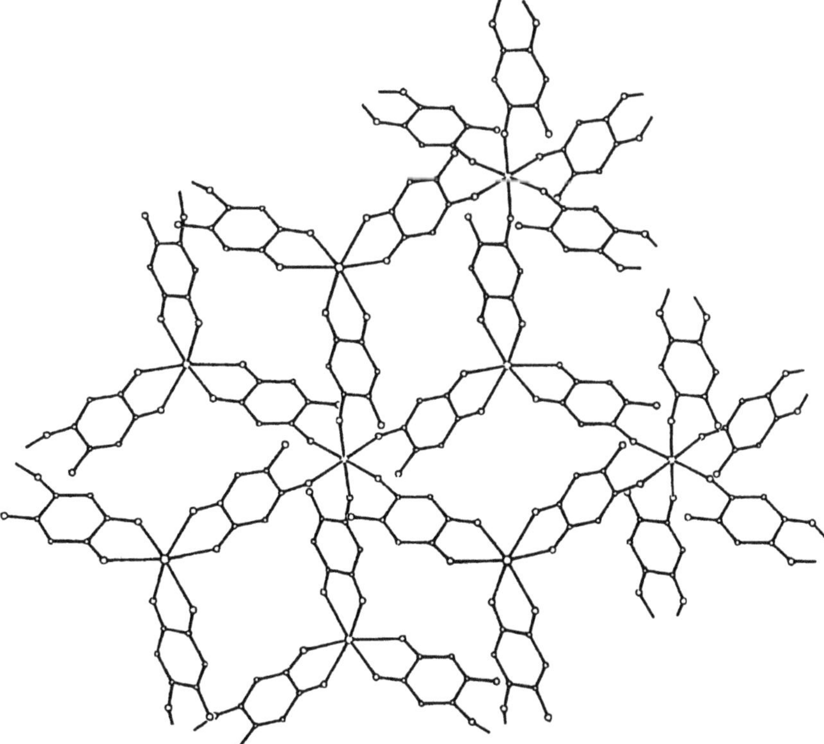

Fig. 18. A protonated layer of $\{Ca[C_6(i\text{-}C_3H_7)_2O_4H]_2\}_n$ composition. (Water molecules, H-atoms and C_3H_7-groups have been omitted.)[31].

E.g. layered silicates like montmorillonite or beidellite have a reciprocal excess charge density between 41 and 75 Å^2 /equivalent. Micas with higher excess charge density like margarite or muskovite reach 12 and 24 Å^2 /equivalent, respectively. Deprotonated $\{Ca[C_6(i\text{-}C_3H_7)_2O_4]_2 \cdot 2/3H_2O^{2-}\}_n$ layers posses a theoretical exchange capacity of 4.01 mval/g with a reciprocal excess charge density of 28.7 $Å^2$ /equivalent. Crystal data for $Ca[C_6(i\text{-}C_3H_7)_2O_4H]_2 \cdot 2/3H_2O$: trigonal, space group R$\bar{3}$ (no. 148), a=b=1408.9(2), c=3524.2(6) pm, V=6058.3·10^6 pm^3.

II.3.2. $Cu[HOOC\text{-}CH_2\text{-}CH_2\text{-}PO_3] \cdot 2H_2O$

A layer-like pattern extending parallel to (001) is made up by PO_3 groups and Cu-centered coordination polyhedra linked by common corners (Fig. 19) [25]. Carboxylic acid residues are bound to each side of the layers. The –COOH group is not involved in the coordination of Cu^{2+}. The water of crystallization is bound to Cu^{2+}. Hence the layers posses $Cu(H_2O)_2[HOOC\text{-}CH_2\text{-}CH_2\text{-}PO_3]$ stoichiometry.

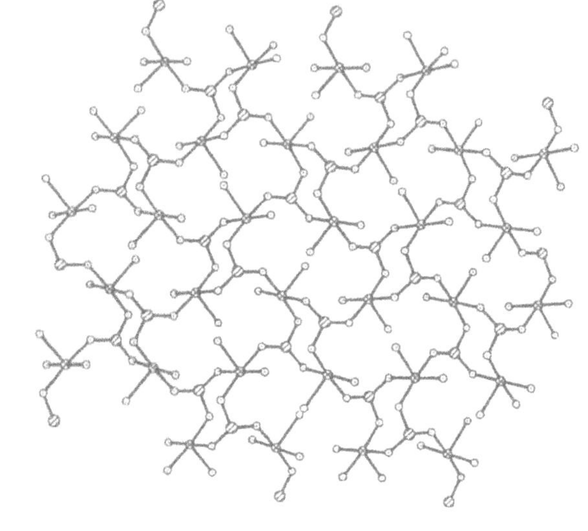

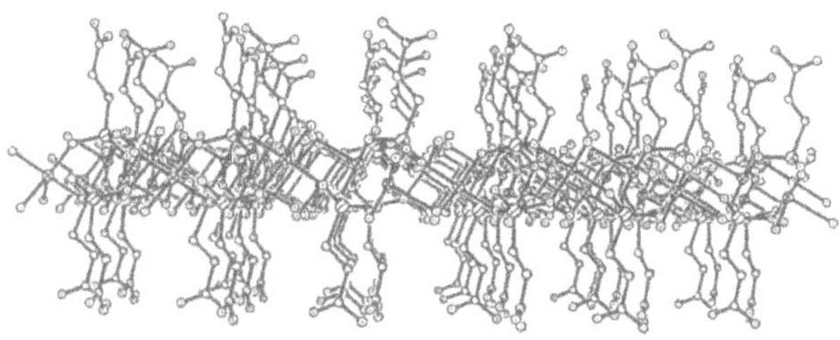

Fig. 19. A protonated layer of $\{Cu(H_2O)_2[HOOC\text{-}CH_2\text{-}CH_2\text{-}PO_3]\}_n$ composition. Upper part: PO_3 groups and Cu-centered coordination polyhedra linked by common corners make up a layer-like pattern. Lower part: The layers carry carboxylic acid residues on each side [25].

Consequently we deal with a weakly acidic cation exchanger material with layer structure in its protonated form. The theoretical exchange capacity equals 4.0 mval/g. The excess charge density is rather high and similar to that known from mica minerals like muskovite.

Crystal data for $Cu[HOOC-CH_2-CH_2-PO_3]\cdot 2H_2O$: orthorhombic, space group Pbca (no. 61), a=821.5(2), b=919.0(9), c=2102.3(2) pm, $V=1569.7\cdot 10^6 pm^3$.

II.3.3. $Zn_2[C_6H_2(COO)_4]\cdot 7H_2O$

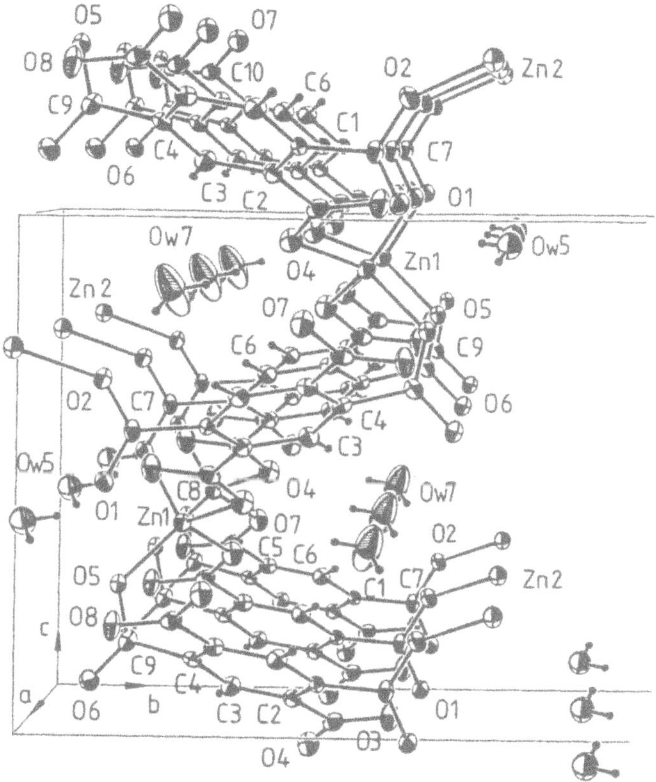

Fig. 20. Zinc pyromellitate heptahydrate $Zn_2[C_6H_2(COO)_4]\cdot 7H_2O$: the negative excess charge of the $\{Zn[C_6H_2(COO)_4]^{2-}\}_n$ layer is compensated for by $Zn(H_2O)_5^{2+}$ ions [32].

The crystal structure of zinc pyromellitate heptahydrate comprises features of ion exchangers, layer-like and zeolitic structures. There are two unique Zn^{2+} sites. Zn(1) is coordinates tetrahedrally by oxygen atoms belonging to the carboxylate groups of the pyromellitate anion [32] (Fig. 20). Thus a corrugated layer of $\{Zn[C_6H_2(COO)_4]^{2-}\}_n$ composition is established. The negative excess charge is compensated for by $Zn(H_2O)_5^{2+}$ (Zn(2)) which is linked to the layer by a Zn-O bond involving an oxygen atom of a carboxylate group. Thus yielding an octahedral coordination sphere. Further water molecules not bound to Zn^{2+} are accomodated between adjacent $\{Zn[C_6H_2(COO)_4]^{2-}\}_n$ layers. One half of them is situated in channel-like folds resembling the situation known from zeolites. According to these structural properties $Zn(H_2O)_5[ZnC_6H_2(COO)_4]\cdot 2H_2O$ is an appropriate formula. Crystal data for $Zn_2[C_6H_2(COO)_4]\cdot 7H_2O$: monoclinic, space group $P2_1/n$

(no. 14), a=593.1(1), b=2359.1(3), c=1168.7(1) pm, ß=96.42(1)°, V=1624.97·10^6 pm^3.

II.3.4. $Zn[O(CH_2COO)_2]$·1/4H_2O

Anions of diglycolic acid and Zn^{2+} cations form the two-dimensionally infinite coordination polymer shown in Fig. 21 [33]. Zn^{2+} is five coordinated in a trigonally bipyramidal fashion. The electrically neutral $Zn[O(CH_2COO)_2]$ layers are parallel to (001) possessing rectangular meshes of different size. Water molecules not bound to Zn^{2+} occupy the centres of the largest meshes. Since the $Zn[O(CH_2COO)_2]$ layers are stacked in ...ABAB... sequence along [001] every wide mesh is followed by a narrow one. Therefore there is no appropriate channel-like diffusion pathway for the removal or uptake of water molecules. Consequently this coordination polymer is more related to an inclusion compound rather than to common layered compounds.

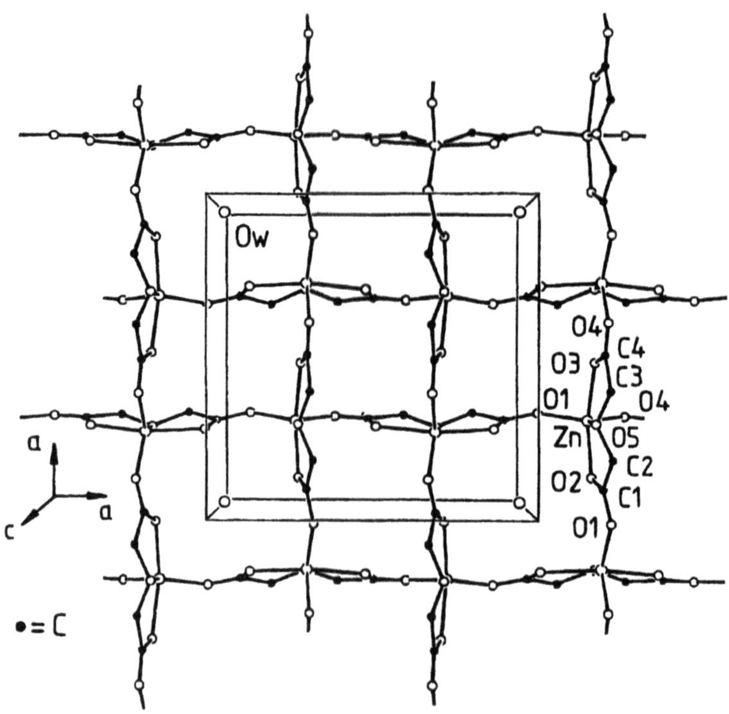

Fig. 21. Water molecules are accommodated between $\{Zn[O(CH_2COO)_2]\}_n$ layers [33].Crystal data for $Zn[O(CH_2COO)_2]$·1/4H_2O: tetragonal, space group I-4 (no. 82), a=990.06(11), c=1352.0(2) pm, V=1325.2·10^6 pm^3.

II.4. Three-dimensionally Infinite Coordination Polymers

II.4.1. $Cd_{3-x}Na_{2x}[C_6(COO)_6]\cdot 9.5H_2O$ (x=0.56)

Cd^{2+} (Cd(1), Cd(2), Cd(3)) and mellitate anions make up a three-dimensional coordination polymer with channel-like voids parallel to [001] (Fig. 22, 23) [33]. Cd(2) is eight-coordinated by oxygen atoms belonging to carboxylate groups.
Cd(1) and Cd(3) are octahedrally coordinated by water molecules and carboxylate oxygen atoms. The Cd(3) site is partly occupied by Na^+ . The empiral formula of this coordination polymer is in rough approximation $Cd_{2.5}Na[C_6(COO)_6]\cdot 9.5H_2O$. Consequently there is negative excess charge on the three-dimensional framework which is compensated for by Na^+ on the Na(2) site (Fig. 23). Na(2) is octahedrally coordinated by water molecules yielding strands of octahedra parallel to [001] linked by common corners in trans-position. The channel-like voids are alternatively filled in a checker board-like manner by strands of Na-centered octahedra and additional water molecules.

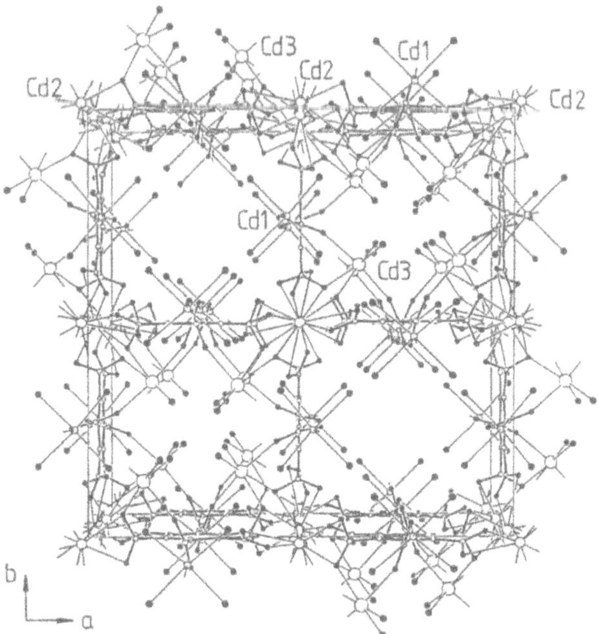

Fig. 22. An open framework with channels extending parallel to [001] characterizes the crystal structure of $Cd_{3-x}Na_{2x}[C_6(COO)_6]\cdot 9.5\ H_2O$ [33].

Crystal data for $Cd_{3-x}Na_{2x}[C_6(COO)_6]\cdot 9.5H_2O$ (x=0.56): tetragonal, space group $P4_2/nbc$ (no. 133), a=1957.2(2), c=1160.91(12) pm, V=4447.0$\cdot 10^6$ pm^3.

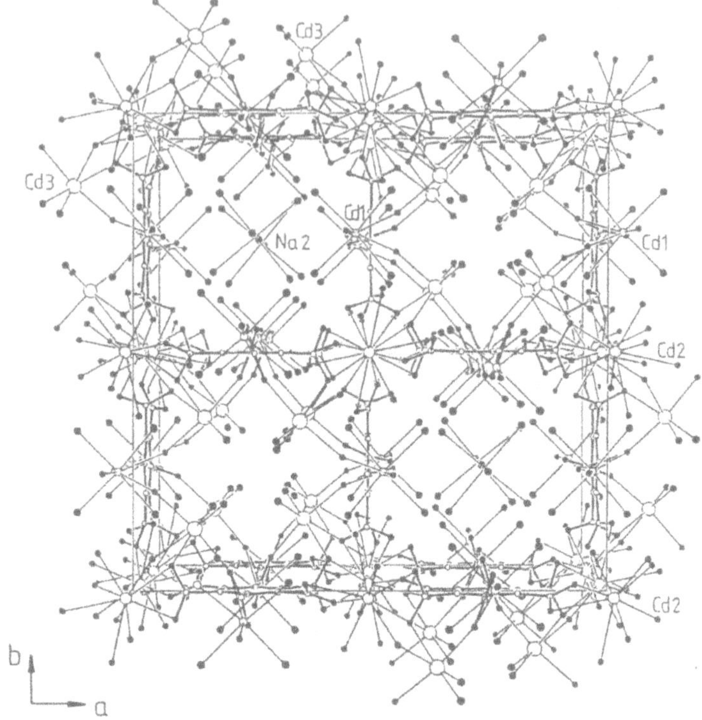

Fig. 23. The cannel-like voids are filled by water molecules and hydrated Na^{+} ions, respectively, in a checker board-like manner [33].

II.4.2. $Na_2Zn[C_6H_2(COO)_4]\cdot 9H_2O$

There are two crystallographically independent Na^{+} ions (Na(1), Na(2)), which are coordinated in a distorted octahedral manner by four water molecules and two oxygen atoms stemming from carboxylate groups [34]. The Na^{+} coordination octahedra share a common face, thus establishing a short Na(1)-Na(2) contact of 324.1(2) pm. Zn^{2+} is exclusively coordinated to oxygen atoms of carboxylate groups forming a moderately distorted tetrahedron, that shares a common corner with the octahedron around Na(1). The $C_6H_2(COO)_4^{4-}$ anions connect these coordination polyhedra yielding a three-dimensional zeolite-like framework with large channels extending along [100] (Fig. 24) and accomodating water molecules associated to clusters by intermolecular hydrogen bonds (Fig. 25). The water cluster is primarily made up by an eight membered centrosymmetric ring. Further water molecules are attached to this ring and hydrogen bonds to oxygen atoms of the pyromellitate anions link the water cluster with the zeolitic framework.
Crystal data for $Na_2Zn[C_6H_2(COO)_4]\cdot 9H_2O$: monoclinic, space group $P2_1/n$ (no. 14), a=1092.6(1), b=1391.7(1), c=1312.2(1) pm, ß=102.14(1)°, V=1950.67$\cdot 10^{6}$ pm^{3}.

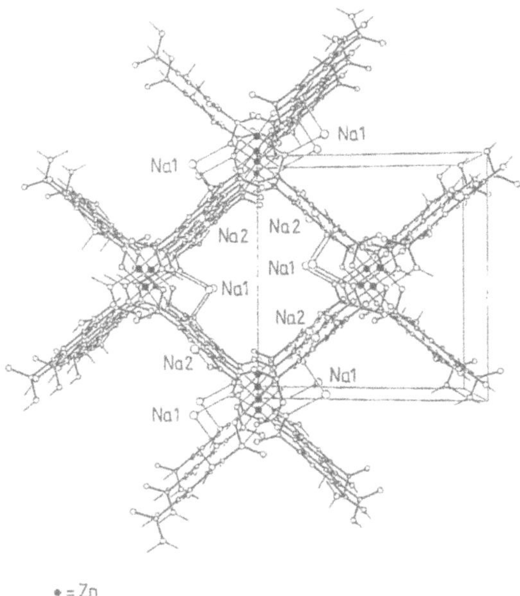

Fig. 24. The zeolite-like framework of $Na_2Zn[C_6H_2(COO)_4]{\cdot}9H_2O$ viewed from [100]. Water molecules have been omitted. Coordinative bonds are shown as thin lines [34].

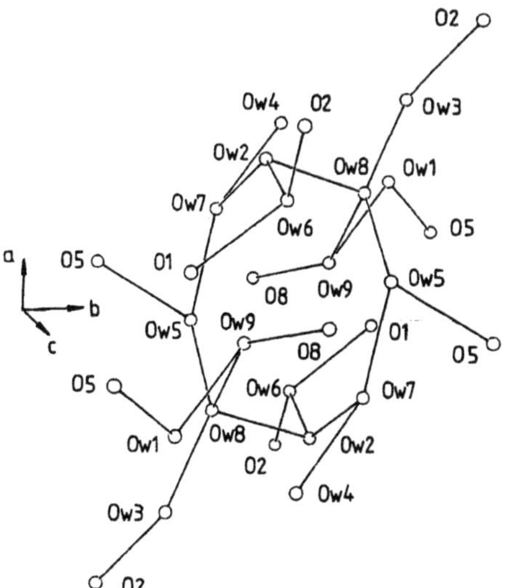

Fig. 25. The centrosymmetric water cluster. O(1), O(2), O(5) and O(8) belong to the three-dimensional framework [34].

II.4.3. $Cu_2[O_3P\text{-}CH_2\text{-}CH_2\text{-}PO_3]\cdot 3H_2O$

A highly symmetric open framework structure is the characteristic feature of $Cu_2[O_3P\text{-}CH_2\text{-}CH_2\text{-}PO_3]\cdot 3H_2O$ (Fig. 26) [25]. Most atoms are situated on crystallographic mirror planes of space group C2/m. A view along the [001] direction shows primarily layer-like patterns parallel to (100) made up by PO_3 groups and Cu^{2+} which are coordinated in a square planar fashion. $\text{-}CH_2\text{-}CH_2\text{-}$ moieties serve to link those layers in the [100] direction, thus yielding a three-dimensional coordination polymer. Fig. 26 shows a space-filling model of the $Cu_2[O_3P\text{-}CH_2\text{-}CH_2\text{-}PO_3]$ coordination polymer with channel-like voids extending parallel to [001]. The diameter of these channels is approx. 3Å x 4.5 Å. The channels are filled by water molecules.

Crystal data for $Cu_2[O_3P\text{-}CH_2\text{-}CH_2\text{-}PO_3]\cdot 3H_2O$: monoclinic, space group C2/m (no. 12), a=1483.6(2), b=668.44(8), c=436.30(6) pm, ß=93.28(2)°, $V=431.96\cdot 10^6$ pm^3.

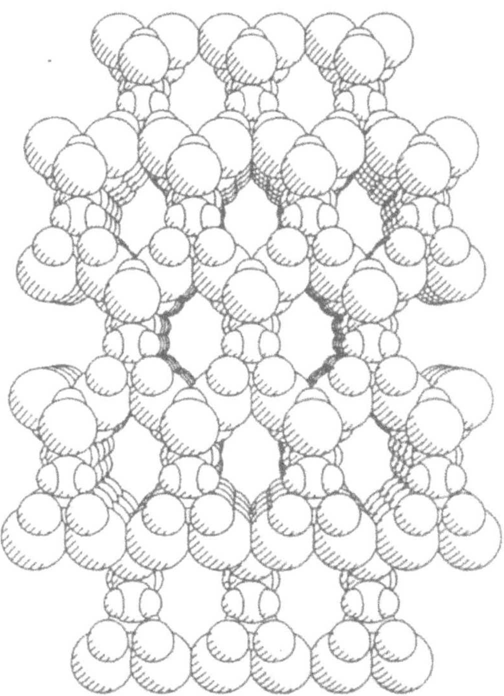

Fig. 26. The highly symmetric open framework of $Cu_2[O_3P\text{-}CH_2\text{-}CH_2\text{-}PO_3]\cdot 3H_2O$. Water molecules have been omitted [25].

II.4.4. $Cu_{1.5}[OOC\text{-}CH_2\text{-}CH_2\text{-}PO_3]\cdot 4H_2O$

$Cu_{1.5}[OOC\text{-}CH_2\text{-}CH_2\text{-}PO_3]\cdot 4H_2O$ is a further three-dimensional coordination polymer featuring channel-like voids filled by water molecules [25]. PO_3 groups, -COO groups and Cu^{2+} make up column-like strands parallel to [100] (Fig. 27). The $\text{-}CH_2\text{-}CH_2\text{-}$ moieties establish connections between those columns in the [011] and

[01-1] directions (Fig. 28). The diameter of the channels which are occupied by water molecules is approx. 6Å x 8Å.

Crystal data for $Cu_{1.5}[OOC-CH_2-CH_2-PO_3]\cdot 4H_2O$: monoclinic, space group $P2_1/n$ (no. 14), a=504.44(12), b=1236.7(3), c=1546.2(3) pm, ß=91.28(2)°, $V=964.1.10^6$ pm^3.

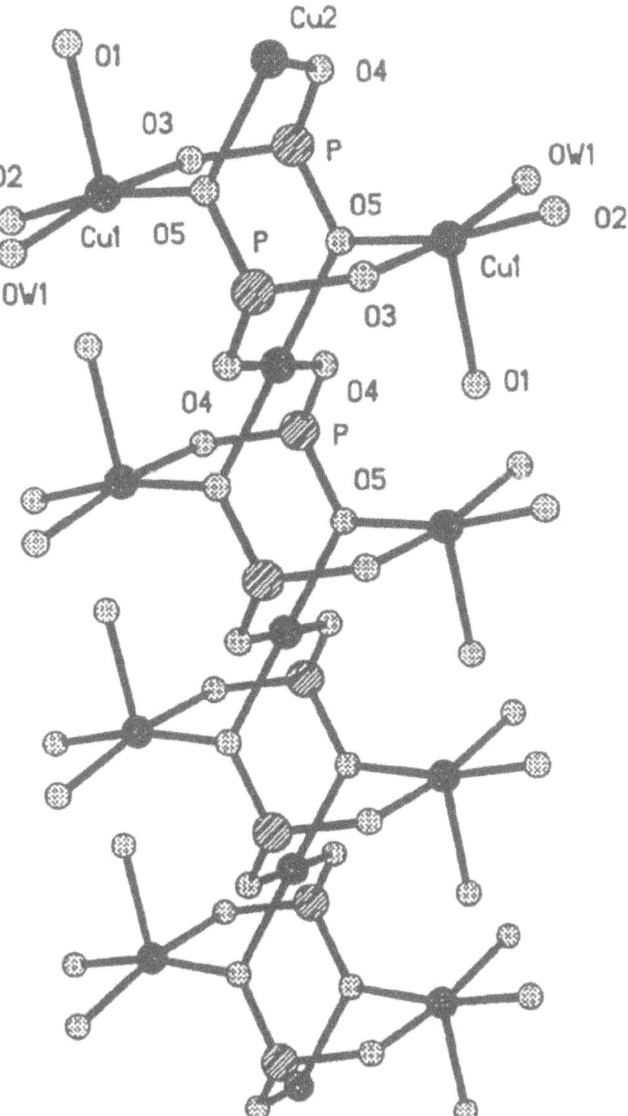

Fig. 27. Columns made up by Cu^{2+} , PO_3 groups and COO groups extend parallel to [100] in the crystal structure of $Cu_{1.5}[OOC-CH_2-CH_2-PO_3]\cdot 4H_2O$ [25].

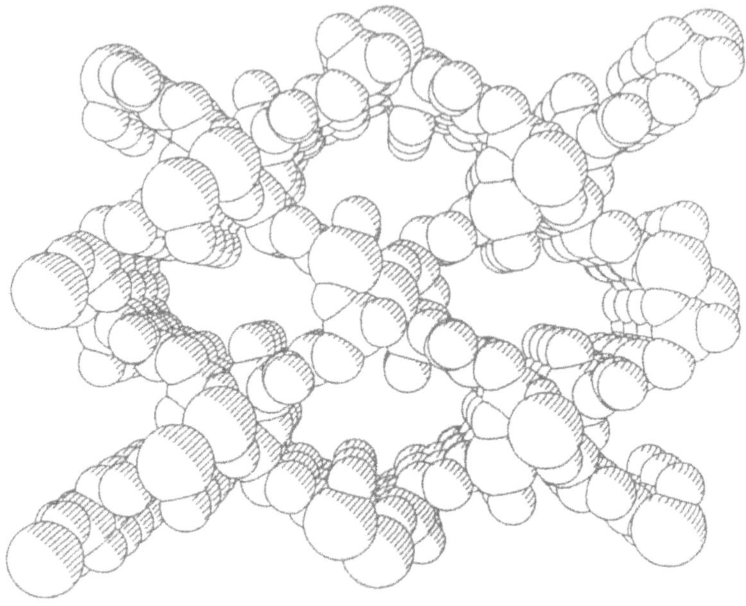

Fig. 28. A space-filling model of the open framework coordination polymer with channels extending parallel to [100] in $Cu_{1.5}[OOC\text{-}CH_2CH_2\text{-}PO_3]\cdot 4H_2O$. Water molecules occupying the channels have been omitted [25].

Acknowledgements

Valuable support by Fonds der Chemischen Industrie and Deutsche Forschungsgemeinschaft is gratefully acknowledged.

References:

[1] J. Berzelius, *Pogg. Ann.* **1826** *6*, 369, 380.
[2] L. Svanberg, H. J. Struve, *J. Prakt. Chem.* **1848** *44*, 257, 291.
[3] M. T. Pope, A. Müller, *Angew. Chem.* **1991** *103*, 56.
[4] B. Krebs, R. Klein in M. T. Pope and A. Müller (ed.) polyoxometalates: From Platonic Solids to Anti-retroviral Activity. Kluwer Academic Publishers, Dordrecht, The Netherlands **1994**, p. 41.
[5] B. Krebs, I. Paulat-Böschen, *Acta Crystallogr.* **1982** *B38*, 1710.
[6] A. Müller, R. Rohlfing, E. Krickemeyer, H. Bögge, *Angew. Chem., Int. Ed. Engl.* **1993** *32*, 909.
[7] C. Robl in **1995** International Chemical Congress of Pacific Basin Societies, ISBN 0-8412-3251-2

[8] K. Haake, Dissertation in preparation.
[9] T. Yamase, M. Inoue, H. Naruke, K. Fukaya, *Chem. Lett.* **1999**, 563.
[10] M. I. Khan, Q. Chen, J. Zubieta, D. V. Gosborn, *Inorg. Chem.* **1992** *31*, 1558
[11] C. Robl, M. Frost, *J. Chem. Soc., Chem. Commun.* **1992**, 248.
[12] H. T. Evans, *J. Amer. Chem. Soc.* **1948** *70*, 1291.
[13] C. Robl, K. Haake, *J. Chem. Soc., Chem. Commun.* **1992**, 1786.
[14] C. Robl, K. Haake, *J. Chem. Soc., Chem. Commun.* **1993**, 397.
[15] U. Kortz, M. T. Pope, *Inorg. Chem.* **1995** *34*, 2160
[16] C. Robl, *Mater, Res. Bull.* **1987** 22, 1483.
[17] C. Robl, *Z. Naturforsch.* **1987** *42b*, 972.
[18] C. Robl, W. F. Kuhs, *J. Solid State Chem.* **1988** *74*, 21.
[19] C. Robl, *Z. Naturforsch.* **1988** *43b*, 993.
[20] C. Robl, *Z. anorg. allg. Chem.* **1988** *566*, 144.
[21] Q. Gao, N. Guillou, M. Nogues, A. K. Cheetham, G. Férey, *Chem. Mater.* **1999** *11*, 2937
[22] T. J. Barton, L. M. Bull, W. G. Klemperer, D. A. Loy, B. McEnaney, M. Misono, P. A. Monson, G. Pez, G. W. Scherer, J. C. Vartuli, O. M. Yaghi, *Chem. Mater.* **1999** *11*, 2633.
[23] C. Livage, C. Egger, G. Férey, *Chem. Mater.***1999** *11*, 1546.
[24] P. J. Hagrman, D. Hagrman, J. Zubieta, *Angew. Chem., Int. Ed. Engl.* **1999** *38*, 2638.
[25] M. Arnold, Dissertation, Univ. Jena **1998**.
[26] C. Robl, S. Hentschel, *Mater. Res. Bull.* **1991** *26*, 1355.
[27] D. Poleti, L. Karanovic, *Acta Crystallogr.* **1989** *C45*, 1716.
[28] C. Robl, S. Hentschel, *Z. Naturforsch.* **1991** *46b*, 1188.
[29] C. Robl, S. Hentschel, *Z. Naturforsch.* **1992** *47b*, 1561.
[30] C. Robl, *Z. anorg. allg. Chem.* **1988** *561*, 57.
[31] C. Robl, G. M. Sheldrick, *Z. Naturforsch.* **1988** *43b*, 733.
[32] C. Robl, *Z. anorg. allg. Chem.* **1987** *554*, 79.
[33] S. Hentschel, Dissertation, Univ. München **1993**.
[34] C. Robl, *Mater. Res. Bull.* **1992** *27*, 99.

Modelling Interpretation of the Kinetics of Metabolic Processes

S. Bastianoni, C. Bonechi, A. Gastaldelli, S. Martini and C. Rossi

Dipartimento di Scienze Chimiche e dei Biosistemi, Università di Siena, Pian dei Mantellini 44, 53100 Siena, Italy
E-mail: rossi@unisi.it

Abstract.

In vivo NMR spectroscopy and sugars selectively enriched with ^{13}C were used to follow the step-by-step metabolic kinetics of microorganisms. A new modelling approach is discussed for the analysis of microbial dynamics. It is based on structurally non linear compartmental models and on the dynamics of the substrate and product. This combined approach was tested with the fermentation of sugars by bacteria and yeasts. The models were fitted with experimental data to obtain the values of the kinetic constants related to the investigated metabolic path.

Introduction

The kinetic behaviour of complex organisms, like cell cultures, are often difficult to analyse. The cell is the fundamental unit of higher organisms and even in unicellular species it still contains subcellular structures with specialized functions. Different approaches have been used to investigate cell organization and functions. These include study of cell cycles, uptake processes, energy metabolism, communications with other cells, storage of genetic information and so forth. Molecular studies of cell organization have revealed numerous important biological roles and functions. However, methods of studying a second level of cell biomolecular organization need to be developed. Classical methods of biological investigation have revealed the general trends of metabolic pathways. The information that can be obtained by traditional experimental approaches is not sufficient when we want to clarify the relationships between the various components of a system. For example, microbe metabolism is generally analyzed using Michaelis-Menten dynamics to describe experimental data. This type of analysis only gives information on the speed of the reaction and the rate of degradation of the substrate, which is insufficient when we need information about interactions between cells, precursors and products.

Although much data are now available on the solution structures of different biological components, such as proteins [1,2], nucleic acids [3] and membranes, little is known about the details of cell organization and the role of each constituent in modulating the chemical properties of bio-structures. The biological significance of each bio-constituent can be completely understood only by considering the complex network of interactions of micro and macro components that take place within the system. Tracer techniques and mathematical modelling, applied to pharmacokinetics and physiology, can be used to study microbial metabolic processes [4].

Interpretation of experimental results obtained by organized complex systems like cell organisms requires an appropriate approach. In the past, theoretical models have been proposed [5-7] and used for the elucidation of metabolic steps and for the calculation of kinetic parameters. Considering the cellular metabolic reactions resulting from activation, inhibition and feed-back activities, the development of new theoretical models is of great importance for the biomolecular sciences. Such models must deal with a large number of interactions and must be flexible enough to adapt to existing approaches and experimental results. When a comparison between different approaches is required, the following properties must be analyzed and considered:

i) the model should require the lowest number of parameters to be calculated;
ii) the difference between the theoretical and the experimental behavior (residuals) should be reduced at the minimum value at any stage of the process. Residuals must be randomly distributed around zero. Regions where great discrepancy between the theoretical and experimental values must be avoided;
iii) the model must be flexible enough to allow the addition or elimination of individual parts in order to fit specific metabolic activities;
iv) the model should refer to parameters having precise biological meanings.

In order to evaluate which could be the best available model in the interpretation of biological events, a careful comparison of theoretical approaches is required.

The aim of this paper is to show the possibility of developing flexible compartmental models to describe the microorganism metabolisms. This approach provides several new points of view on the kinetics of biological reactions:

i) the metabolization process is described using concise symbols for energy and matter fluxes within the system and all the main events related to the sugar metabolism that occur in the cell culture;
ii) all the parameters describing the state of the system are treated as kinetic parameters and non-linear equations are developed to describe their evolution in time;
iii) specific kinetic constants can be calculated, each one correlated with a specific biological path and meaning.

This approach could be an alternative to the Michaelis-Menten approach for describing enzyme kinetics. The models developed also contain biological information indicating the strategies used by organisms to exploit sources of energy and matter.

In this study we analyse the metabolism of the bacteria (*Klebsiella planticola* G11 and ATCC 33531) and yeast *Saccharomyces cerevisiae* KL-144A by ^{13}C-NMR "in vivo" spectroscopy using ^{13}C sugar substrate enrichments.

Because of its "non-invasive" nature, NMR spectroscopy can be useful in kinetic studies in which several samplings are required during metabolic activity [4,8,9]. This technique also reveals the metabolic route of enriched carbon-13 nuclei from substrate to the end-product.

Methods

Spectroscopic Measurements

^{13}C-NMR spectra were recorded with a Varian XL-200 spectrometer operating at 200.085 MHz and 50.288 MHz for proton and carbon respectively. Carbon spectra were recorded under broad-band proton decoupling conditions using a low power MLEV-16 pulse in order to avoid sample temperature effects. 10 mm coaxial tubes, containing 99.75% D_2O in the outer part, were used for the NMR measurements. The ^{13}C-NMR were recorded in blocks of 10 minutes until the end of the fermentation process unless otherwise specified.

Bacterial and Yeast Cultures

a) *Klebsiella planticola* G11

Wild strain G11 of *K. planticola* was isolated from a corn field and characterized in a previous study [9]. The culture medium consisted of 5.25 g/l KH_2PO_4, 6.85 g/l K_2HPO_4, 5 g/l $NaHCO_3$, 0.1 g/l $MgSO_4$, 0.1 g/l NaCl, 0.2 g/l $(NH_4)_2SO_4$,), 0.3 g/l urea, 0.02 g/l $CaCl_2$, 0.2 g/l yeast extract and 10 g/l xylose unless otherwise stated. The following trace elements were also present: Fe, Cu, Co, Mo, Mn. *Klebsiella planticola* G11 was grown in flasks at 35 $^{\circ}$C in a nitrogen atmosphere. The pH of the medium was adjusted to 7.5. Growth was followed by spectrophotometric optical density (O.D.) measurements at 660 nm. A unitary value of O.D. was verified corresponding to 0.53 g/l of dry weight of biomass. Inocula for in vivo microbatch ^{13}C-NMR experiments were prepared by growing a single agar colony overnight in the medium with 10 g/l of D-xylose. A fraction of this culture was diluted 2:100 in the medium for 2-3 further duplication cycles. The cells were then collected by centrifugation and used as the inoculum for NMR measurements. The initial O.D. value of the cell culture was 0.5. The ^{13}C enriched sugar substrates [2-^{13}C]glucose and [1-^{13}C]xylose were obtained from Cambridge Isotope Laboratories and used without any further purification. The two substrates were enriched with carbon-13 at two different positions, in order to avoid spectral line superposition and to simplify the measurements of NMR parameters, e.g. intensities and chemical shifts.

b) *Klebsiella planticola* ATTCC 33531

K. planticola ATCC 33531, previously isolated and identified [10] is grown at 35°C under nitrogen atmosphere using the same culture medium reported for *K. planticola* G11. The *K. planticola* ATCC 33531 cell cultures for NMR measurements were prepared following the procedure used for the *K. planticola* G11.

c) *Saccharomyces cerevisiae* KL-144A

Saccharomyces cerevisiae strain KL-144A was grown at 31°C in a liquid medium containing 6 g yeast extract, 0.5 g L-cysteine HCl, 5.6 g KH_2PO_4, 7 g K_2HPO_4, 1.0 g/l $NaHCO_3$, 1.5 g $(NH_4)_2SO_4$, 0.15 g $MgCl_2.6H_2O$, 0.01 g $FeSO_4.6H_2O$, 3 g/l sodium citrate.$2H_2O$. A density of 2x109 cell/ml was used for each sample analyzed. A pH of 6.5 was used in all determinations and kept constant during the fermentation process. The initial glucose concentrations were 85 and 200 g/l. The number of transients necessary for a spectrum with a high signal to noise ratio was reduced by adding 5 g/l of [1-^{13}C] 90% enriched glucose (from Stohler Isotopic Chemicals) to each sample.

Mathematical Analysis

The mathematical program used to calculate the optimal parameters of the model is MLAB (Modelling LABoratory) [11]. This program, originally developed at the National Institute of Health, has shown very good flexibility and adaptability in solving simulation and modelling problems.

Compartmental Models in Biological Systems

The use of compartmental analysis implies the adoption of a phenomenological and macroscopic viewpoint in modelling physical chemical processes. This type of approach is assuming a growing importance in quantitative studies of metabolic processes, in pharmacokinetics, in ecological studies and so on. In literature different meanings for compartmental models can be found. We use this term meaning that a compartment is a quantity of matter or energy that is homogeneous and distinguishable from the rest of the system: "homogeneous" because we assume there are no internal differences, relatively to the aspects of the phenomena we are dealing with; "distinguishable" because a compartment should have characteristics that are different, in some aspects, from those of the other compartments.

The choice of the compartments depends on the aims of the study for which the model is developed, as well as on the information available. Two compartments may differ for site, physical or chemical characteristics. The quantities in the compartments may vary in time for transport phenomena, or for variation in the physical chemical state of the matter or energy of the compartment.

In general terms the process of modelling can be divided into two main categories: models of processes and models of data. The former are built up from models of the basic process or mechanisms at work in a system; the latter are chosen in order to fit particular sets of data without referring to the process that produces the data themselves [12]. The linear, or polynomial, regression models, widely used in statistical analysis, are examples of models of data. Compartmental models are usually meant to be models of processes, even though sometimes one can find a compartmental model that fits the data set very well but the interpretation and the meaning of the compartments and flows are not obvious.

Making a model of process for a biological system requires a large background research. The compartmental model should be such that its structure and flows have a precise meaning in terms of the assumed process and system.

Energy system diagrams

Energy system diagrams are here used to portray the models we used for the description of microrganisms metabolism. This language was developed by H.T. Odum in the early nineteen-seventies [13] as a very general tool with possible applications in any kind of systems. It has proven to be a rather simple and flexible way to visualize the structure and dynamics of the systems discussed.

The models we will use in the following paragraphs contain only few of the symbols, but for the sake of completeness we present all the symbols and a brief explanation in Figure 1. This type of diagrams was used because they can be seen as pictorial mathematical equations which include energy constraints [14,15]. Despite this, their meaning and patterns of performance may be read without technical knowledge. There is a univocal correspondence passing from a diagram to a set of equations; in general it is possible to draw more than one diagram starting from a set of equations.

In Figure 1, the symbols for *producer* and *consumer* are only used to show the role of a system component, but their presence has no mathematical translation. Usually inside these two symbols others are present, explaining how that component behaves. The main symbols are those of *flow* and *storage*. The *storage* represents what in modellistic terms is a state variable, a measurable (or estimable) quantity. The *flows* are in general energy flows (but sometimes also matter) and are used for the connection of the storages. They indicate that energy is transformed from a state to another or transferred from a place to another, depending on the meaning of the storages.

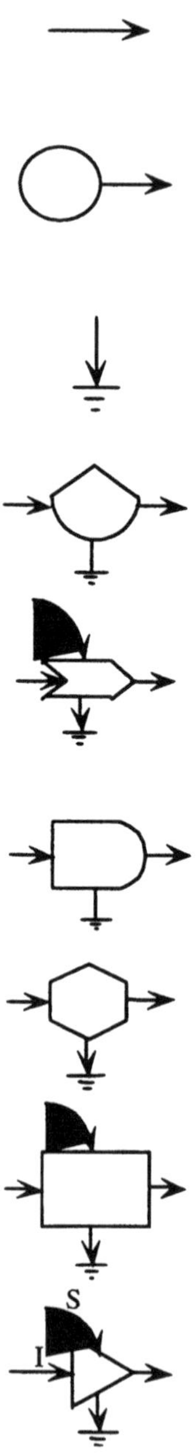

Energy circuit . A pathway whose flow is proportional to the quantity in tl storage or source upstream.

Source . Outside source of energy; a forcing function.

Heat sink . Dispersion of potential energy into heat that accompanies all rea transformation processes or storages.

Storage . A compartment where energy is stored within the system; the qua is a balance of inflows and outflows; a state variable.

Interaction . Interactive intersection of two pathways; it indicates control a for one flow to another or limiting factors.

Producer . Unit that collects low quality energy and stores it in compartmei of the system (e.g. photosynthesis).

Consumer . Unit that transforms energy quality, stores it and feeds it back.

Box . Miscellaneous symbol to use for whatever unit or function is labeled.

S

I

Amplifier . A unit that delivers an output in proportion to input I as long as energy source S is sufficient.

Fig. 1. Energy System Diagrams Symbols.

Usually one or more *source* symbols are present in a model, indicating a forcing function, an external input that influence the system but it is not part of it. Sources are drawn outside the box representing the system under study.

When a flow from a storage is controlled only by other factors, and not by the quantity stored, the *amplifier* symbol is used. For example, as long as water is available, the quantity of water outflowing from a tap is regulated by the tap itself, not by the quantity of water stored. This symbol is used also for zero order chemical reactions (see Figure 2). Remaining in the chemical realm, a first order kinetic reaction is represented by an outflow from a storage. This in fact is linear if there is no other symbol (except for other storages) at the end (Figure 2).

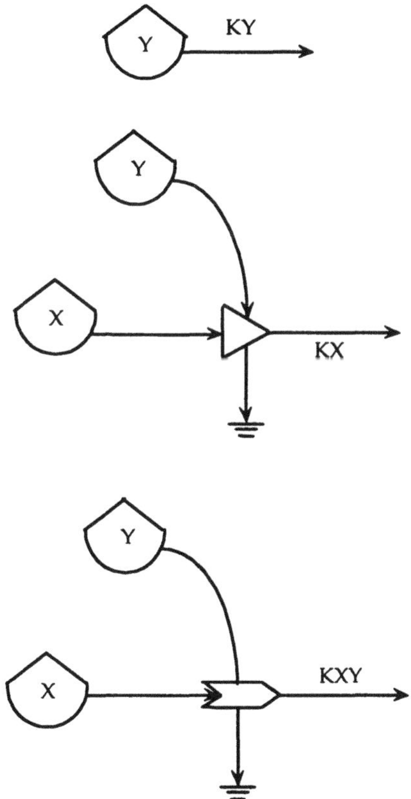

Fig. 2. Diagram of the Lotka-Volterra model.

For second order reactions the symbol of *interaction* is used. This symbol has a much wider use, that, is whenever two or more types of energy interact to produce another type. Mathematically, this means that in an interaction all inflows and outflows to this symbol depend on all the quantities that interact (see Figure 2). A self interaction indicates an autocatalytic loop: a storage interacts with a source (or another storage) and receives a benefit according to the quantity present in the storage and to the availability of the source.

Other symbols are the the *switch*, that is used when a flow is present only in some determined conditions; the *money transaction*, in which matter and/or energy are exchanged with money; and the heat sink, that accounts for the energy dispersed as low temperature heat in all the transformations.

In Figure 3 a diagram of the well known Lotka-Volterra model, in its simplest version is represented:

$$X' = aX - bXY$$

$$Y' = cXY - dY$$

where the symbol ' indicates the derivative with respect to time, X and Y are the quantities of preys and predators, respectively, and a, b, c and d are parameters of the equation. The main factor in this model is the presence of the interaction that benefits the predators (and disadvantages the prey) proportionally to the number of preys and predators in the system, that is, proportionally to the probability of encounters between preys and predators.

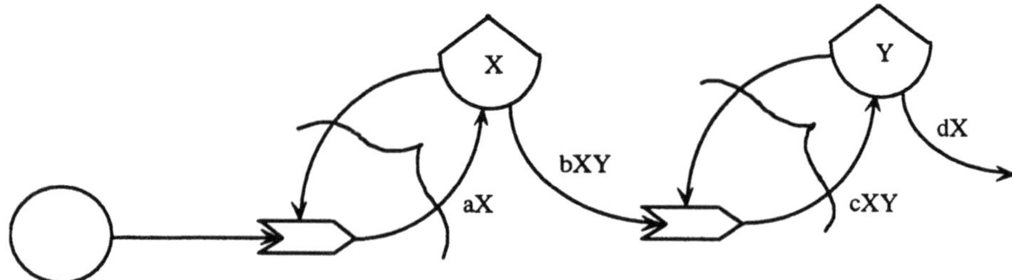

Fig. 3. Diagram of the Lotka-Volterra model.

Metabolic Modelling

i) Xylose degradation by *K. planticola*

Both *K. planticola* G11 and ATCC 33531 can use xylose as energy and carbon source. The xylose metabolic pathway by *K. Planticola* is shown in Figure 4. Here we would compare three different approaches utilized for the analysis of the metabolism of sugar by a bacterium cell culture. Our purpose is to identify a model which allows for the best fit with the experimental results and considers a limited number of parameters all related to an identified biological function.

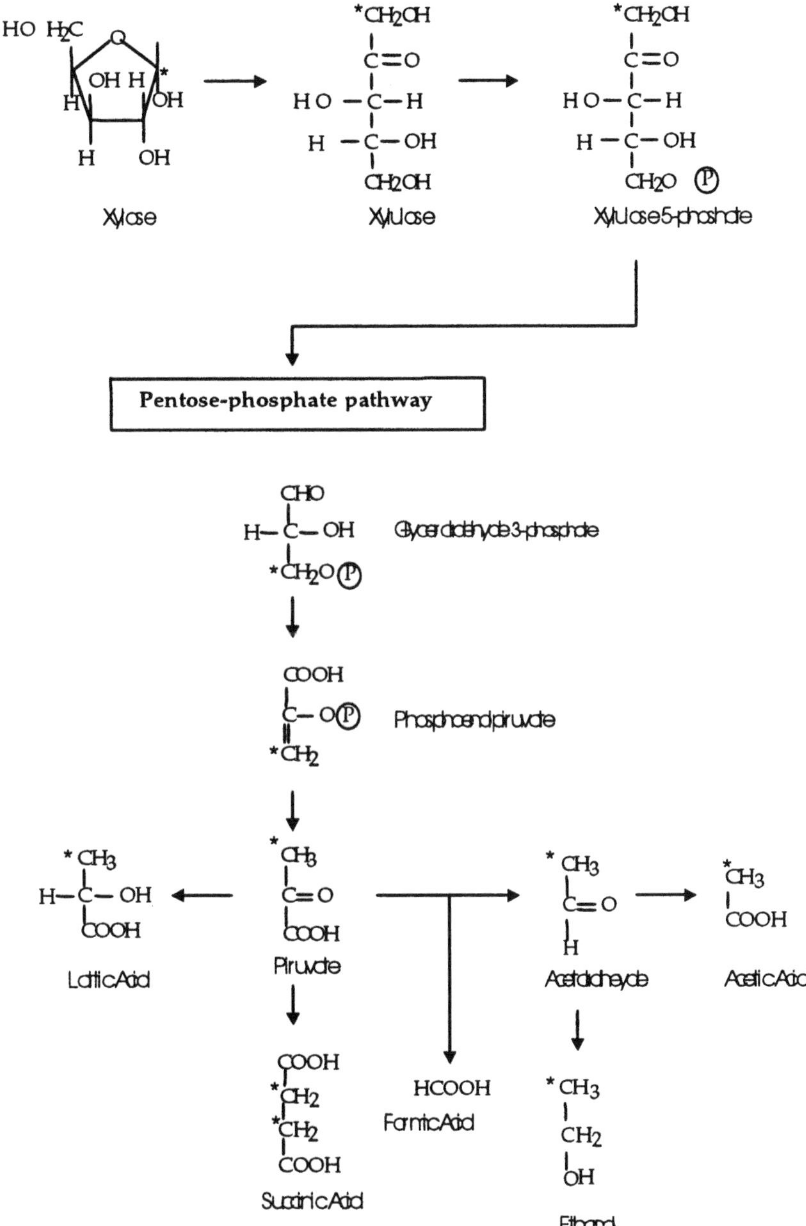

Fig. 4. Xylose metabolic pathway by *K. planticola*.

Three different models were compared: a model based on a pure exponential behavior [16-18], a model which considers the metabolization process based on the classical Michaelis-Menten kinetic analysis [19,20] and finally the model developed to describe energy and matter fluxes that pass through complex systems and based on the language of the *Energy System Diagrams* introduced by H.T. Odum [21]. The three theoretical models were used to fit a set of experimental results related to the xylose metabolization by both *Klebsiella planticola* G11 and ATCC 33531 strains. The experimental substrate consumption and the end-product formation were detected by "in-vivo" NMR spectroscopy using selective carbon-13 enrichment of the xylose substrate.

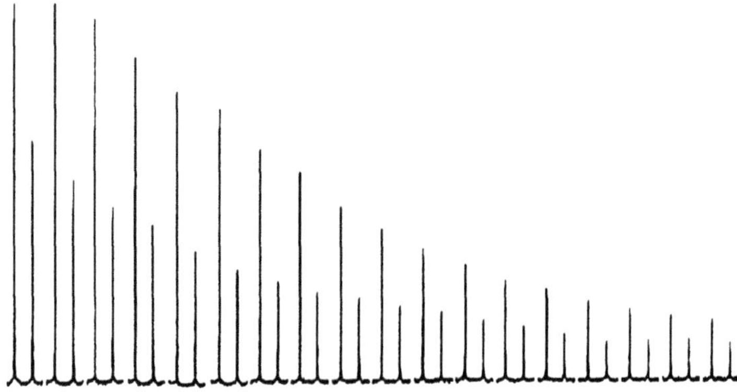

Fig. 5. ^{13}C-NMR spectra of xylose in relation to time.

Figure 5 shows the ^{13}C-NMR spectra which describe the sugar metabolization process. The NMR approach provides a step-by-step description of how the ^{13}C-labeled atom is transferred along the metabolic pathway and precise information on the amount of substrate consumed and ethanol yielded.

In Figure 6 the experimental xylose concentrations observed at different times in a culture of *K. planticola* G11 are reported.

We utilized as a first model, to analyse the data, a model based on an exponential fitting with governing equations for the xylose (X(t)) and ethanol (E(t)) in the form:

$$X(t) = XMAX \cdot \exp\{-KX_1 \cdot t\} + XMIN \qquad [1]$$

$$E(t) = EMAX \cdot (1 - \exp\{-KE_1 \cdot t\}) \qquad [2]$$

where XMAX is the maximum value in the concentration of xylose; XMIN is the concentration at which the degradation of xylose is restricted; EMAX is the plateau concentration of ethanol; KX_1 and KE_1 are the exponents of the xylose and ethanol exponential dynamics.

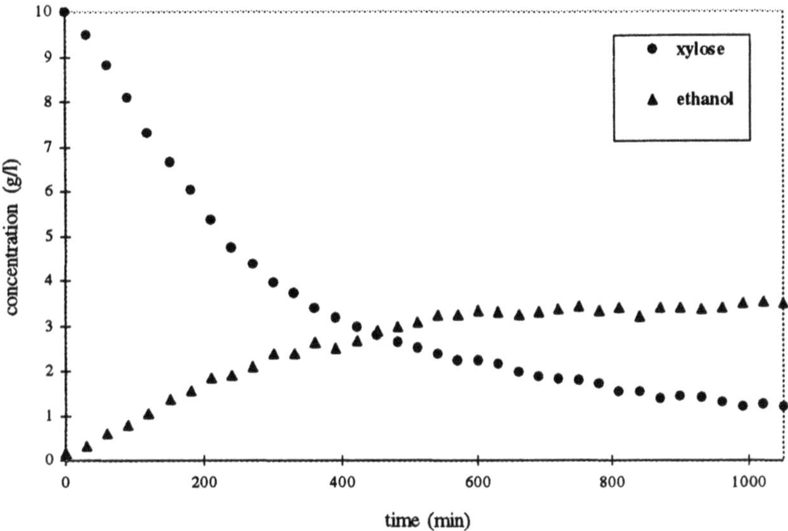

Fig. 6. Experimental xylose concentrations observed at different times in a culture of *K. planticola* G11.

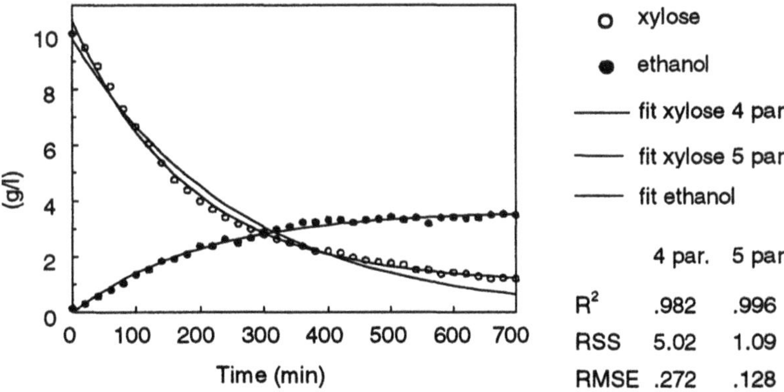

Fig. 7. The results of the fitting procedure in the case of the exponential model (four and five parameters). Data for xylose and ethanol were collected in a NMR experiment where the initial concentration of xylose was 10 g/l, as described in Materials and Methods. Values of R^2, residual sum of squares (RSS), and root-mean-square error (RMSE) for these two models are used to compare the results.

Two versions of the model (equation [1] above) are considered, containing either five or four parameters. In the second case, the fit was carried out constraining

XMIN to a zero value; the exponentials and ethanol plateau were then estimated. In Figure 7 the results of the fitting procedure in the two cases are shown. The model with four parameters is not able to follow the experimental curve; the introduction of the parameter XMIN makes the value of R^2 rise until 0.996 with a Residual Sum of Squares (RSS) from 5.02 to 1.09 (Figure 7).

The Michaelis-Menten model was used in its differential version [19] to study the degradation dynamics of the sugar while the ethanol dynamics were fitted with the traditional Michaelis-Menten curve (growth and saturation):

$$\frac{dX}{dt} = -VX\max\frac{X}{KX_2 + X} \quad [3]$$

$$E(t) = VE\max\frac{t}{KE_2 + t} \quad [4]$$

where VEmax is the plateau level of the ethanol, VXmax is the initial degradation rate, and KX_2 and KE_2 are the constant of the Michaelis-Menten dynamics. This model (see Figure 8) has four parameters and gives results that are markedly inferior to the exponential model. The use of the best fitting procedure is difficult as a result of the distance between the theoretical curve and the experimental data with regards to the xylose dynamics. The obtained parameters have an error with a coefficient of variation of nearly 100%. The description of the ethanol production data is slightly better even if R^2 has smaller and RSS has greater values than the ones obtained using the first model.

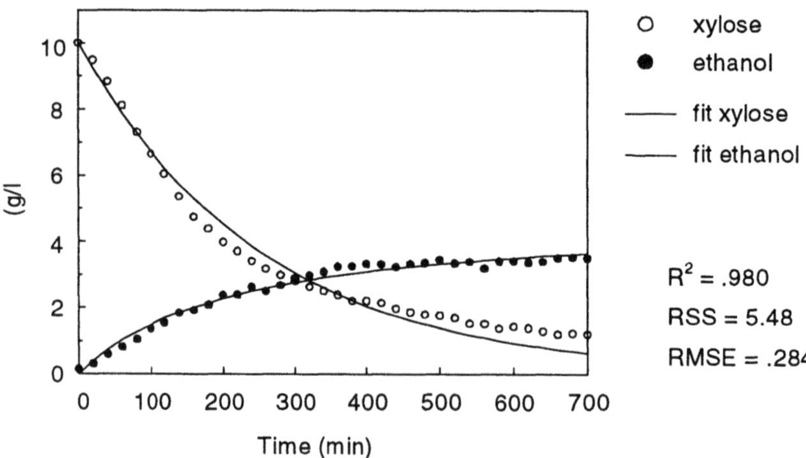

Fig. 8. The Michaelis-Menten model compared with the data set reported in Figure 7.

The third model is based upon the energy system diagrams introduced by H.T. Odum to describe energy fluxes [13,15]. This type of model could be described as compartmental where the fundamental components are identified and their relationships modelled using kinetic equations that are usually non-linear.

The model is composed by four storages (see Figure 9): xylose, that is the (limited) source of energy; active cells of *Klebsiella planticola;* ethanol, that is the main product of the process and bacteria cells, that are inhibited by the end-product.

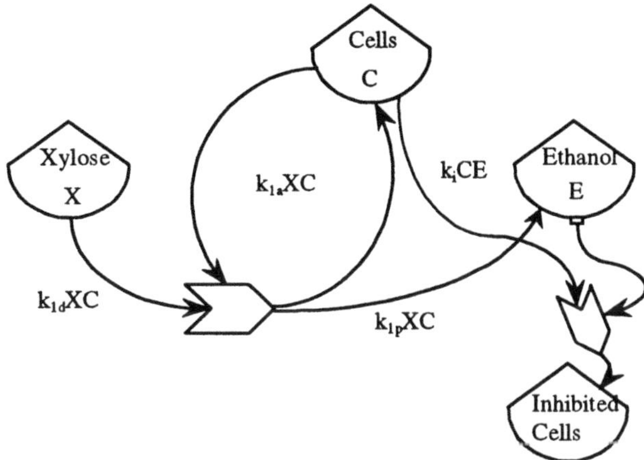

Fig. 9. Energy system diagram of the conversion of xylose to ethanol by *K. planticola.*

The dynamics of the sugar metabolization is assumed to be the result of an autocatalytic process that depends on the concentration of substrate and on the number of active cells. The model takes into account of the fact that the presence of sugar substrates promotes an energy flow from the sugar to the active cells. The energy flow is used for increasing the concentration of active cells. Ethanol acts as a controller of the quantity of active cells, shown in Figure 9 as an outflow from the active to the inhibited cell tank. The interaction between cells and ethanol does not imply an outflow of ethanol and this is graphically expressed by the box under the storage of ethanol.

In the present case the xylose degradation rate becomes irrelevant when the sugar concentration approaches low values (1 g/l). The resulting model is represented in Figure 9. The systems can be described by the following equations, in which sugar ([X(t)]) degradation, ethanol ([E(t)]) production and cell ([C(t)]) activation are linked together as follow:

$$\frac{d[X]}{dt} = -K_{1d}[X][C] \qquad [5]$$

$$\frac{d[C]}{dt} = K_{1a}[X][C] - K_i[C][E] \qquad [6]$$

$$\frac{d[E]}{dt} = K_{1p}[X][C] \qquad [7]$$

where the four parameters are the coefficients relating the concentration of xylose [X], cells [C] and ethanol [E] to the degradation of sugar (k_{1d}), activation (k_{1a}) and inhibition (k_i) of cells and production of ethanol (k_{1p}). These parameters have dimensions: for k_{1d} and k_{1p} min^{-1}; for k_{1a} and k_i $lg^{-1}min^{-1}$. A fourth equation, the one relative to inhibited cells (I) has to be written, even if it is not important for the simulation:

$$\frac{dI}{dt} = K_i[C][E] \qquad [8]$$

The result of the fitting procedure is shown in Figure 10. This model with its four parameters can now be compared with the previous ones. Such a comparison demonstrates that the compartmental model fit is always superior to the other two, both in terms of precision of parameter estimates and RSS. Our compartmental model behaves better across the complete degradation dynamics. Moreover our model describes the interaction between cells and substrate and cells and end-product, while in the other cases the dynamics of xylose and ethanol are separately described.

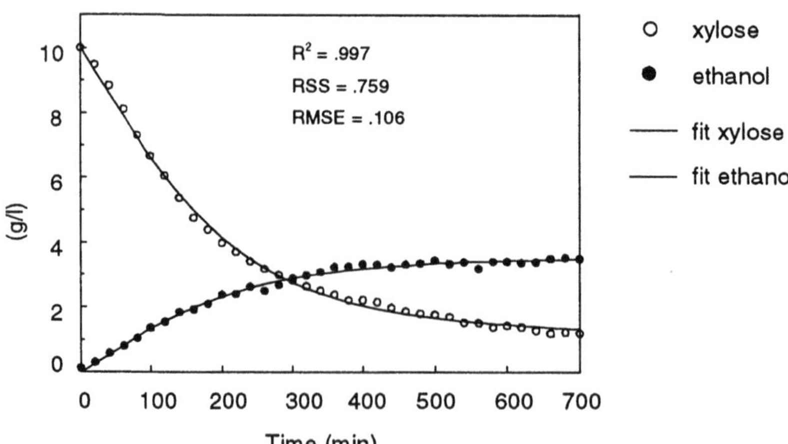

Fig. 10. The result of the fitting procedure of the model reported in Figure 9.

In conclusion, it appears that the Michaelis-Menten dynamic model is not adequate for the description of this system, even if its parameters have a biological significance. The exponential model with five parameters gives a good response but its use is purely mathematically descriptive. Our model shows a superior performance from the modellistic view and contains parameters that have a precise biological significance. Additionally, the degradation dynamics of the substrate, the activation and inhibition processes and the ethanol production are strictly coupled. This model allows us to evaluate the real level of the cellular activity, the value of which is not correlated only to the concentration of biomass and is difficult to measure directly. From the viewpoint of our model we can now compare the performances of the two strains of *Klebsiella planticola*, the wild G11 and the selected ATCC 33531. The values of the parameters after the best fit procedure and the respective coefficients of variation are presented in Table 1. We obtained very similar values for each of the parameters. Nonetheless, we can say that the strain ATCC 33531 has a more active metabolic process, having a higher rate of degradation of xylose (k_{1d}), a higher level of activation of the cells (k_{1a}), and rate of production of ethanol (k_{1p}), while the parameter that accounts for the inhibition (k_i) is lower. We believe that this model can constitute a basis for a deeper understanding of microorganism dynamics, permitting to identify variations in individual aspects of the degradation processes. This modelling approach has allowed us to reveal a more comprehensive comparison of two microorganisms which are important in the sugar to ethanol transformation.

Table 1. Comparison of the values of the parameters between two strains of *K. planticola* .

Kinetic Constants	Optimal Values (G11)	CV (%)	Optimal Values (ATCC 33531)	CV (%)
k_{1d}	9.145×10^{-4}	3	1.038×10^{-3}	8.1
k_{1p}	1.829×10^{-4}	3	1.964×10^{-4}	8.4
k_{1a}	3.294×10^{-4}	10	4.399×10^{-4}	12.2
k_i	2.042×10^{-3}	7	1.551×10^{-3}	8.0

ii) Simultaneous glucose and xylose degradation by *K. planticola*

The previously studied wild strain of *Klebsiella planticola*, G11, was used to follow the simultaneous metabolization of glucose and xylose [9]. Data were obtained by Nuclear Magnetic Resonance (NMR) spectroscopy using substrates, selectively enriched with carbon-13 in different positions, in order to avoid spectral line superposition: [2-^{13}C]glucose and [1-^{13}C]xylose.

When grown in a medium containing both glucose and xylose, *Klebsiella planticola* strain G11 shows non-diauxic behaviour, i.e. it can metabolize glucose and other sugar substrates simultaneously. The kinetics of glucose and xylose degradation as well as ethanol production are shown in Figure 11. In order to study the kinetics of the metabolization process we used an approach based on the compartmental modelling approach. In this case the metabolic kinetics are complicated by the simultaneous degradation of two substrates. The two processes apparently do not interfere with each other: the kinetic pathways can be considered separate and studied independently.

In order to build a model for the present system and to develop a set of kinetic equations, we first considered the various compartments: the glucose "storage", the xylose "storage" and the "storages" of active cells, inhibited cells and ethanol.

Cell metabolism was then described as the overlap of two kinetic processes: glucose fermentation, which depends on the number of active cells and xylose fermentation, an "autocatalytic" process related to the number of active cells and to xylose concentration [22].

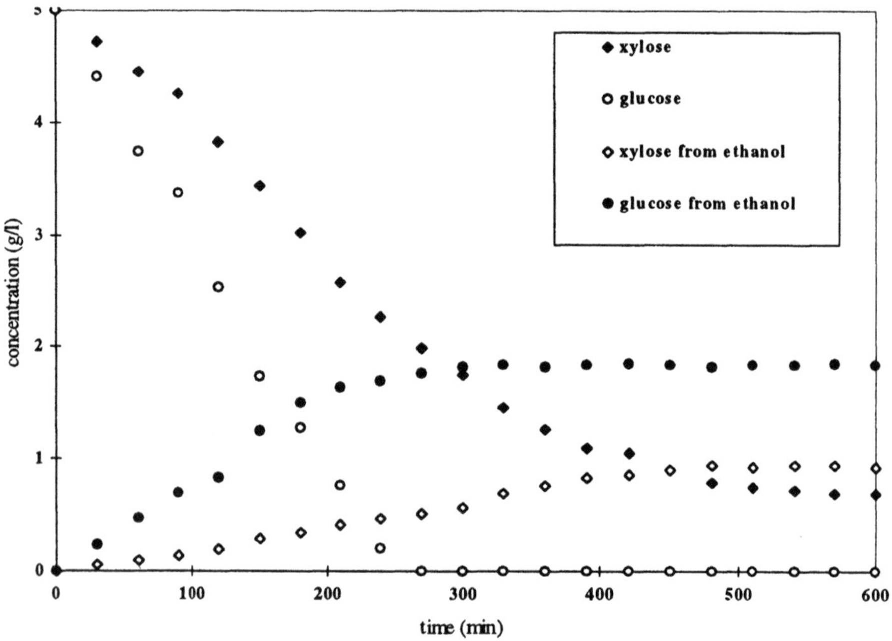

Fig. 11. Experimental data collected by ^{13}C-NMR of the non-diauxic degradation of glucose and xylose by *K. planticola* G11.

Figure 12 shows the model we developed. The set of non-linear equations derived from the model is:

$$\frac{d[X]}{dt} = -k_{1d}[X][C_x] \qquad [9]$$

$$\frac{d[G]}{dt} = -k_{2d}[C_G] \qquad [10]$$

$$\frac{d[C_X]}{dt} = k_{1a}[X][C_X] - k_{1i}[C_X][E] \qquad [11]$$

$$\frac{d[C_G]}{dt} = k_{2a}[C_G] - k_{2i}[C_G][E] \qquad [12]$$

$$\frac{d[E]}{dt} = \frac{d[E_X]}{dt} + \frac{d[E_G]}{dt} = k_{1p}[C_X][X] + k_{2p}[C_G] \qquad [13]$$

where [G], [X] and [E] are the concentrations of glucose, xylose and ethanol, respectively (measured in g/l). [E] is the sum of $[E_X]$ and $[E_G]$, that are the amount of ethanol produced by xylose and glucose fermentation respectively. Both the ethanol productions can be followed simultaneously due to the different

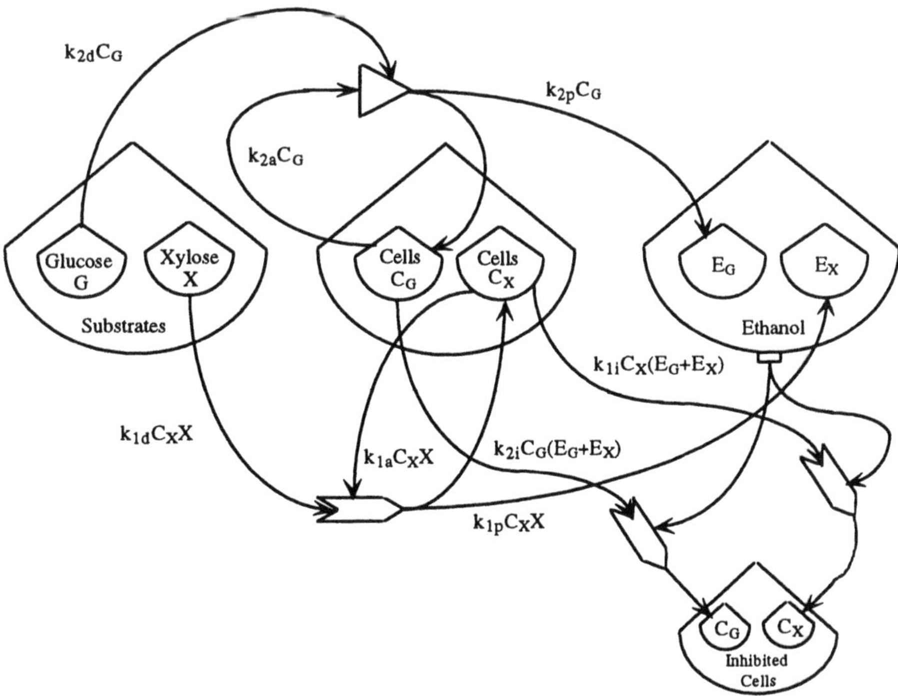

Fig. 12. Diagram of the interactions of sugars and bacteria cells. The result is the activation of the cells and the production of ethanol. The latter inhibits the cellular activity.

enrichment positions of ^{13}C in the substrates. On the basis of both the experimental results and the numerical integration of the differential equations (9-13), the kinetic constant parameters k_{α_j} ((j = 1, 2 ; α = d, a, i, p) can be determined. In equations (9-13) the subscripts 1, 2 refer to xylose and glucose metabolization pathways and the subscripts d, a, i and p refer to the degradation, the activation, the inhibition and the production processes respectively.

In this model xylose and glucose fermentation pathways are assumed to be independent: the two quantities C_X and C_G are related to the number of active cells present in the medium. The C_X and C_G activities are considered, as they would refer to different populations, with their different degradation activity, inhibition and production performances.

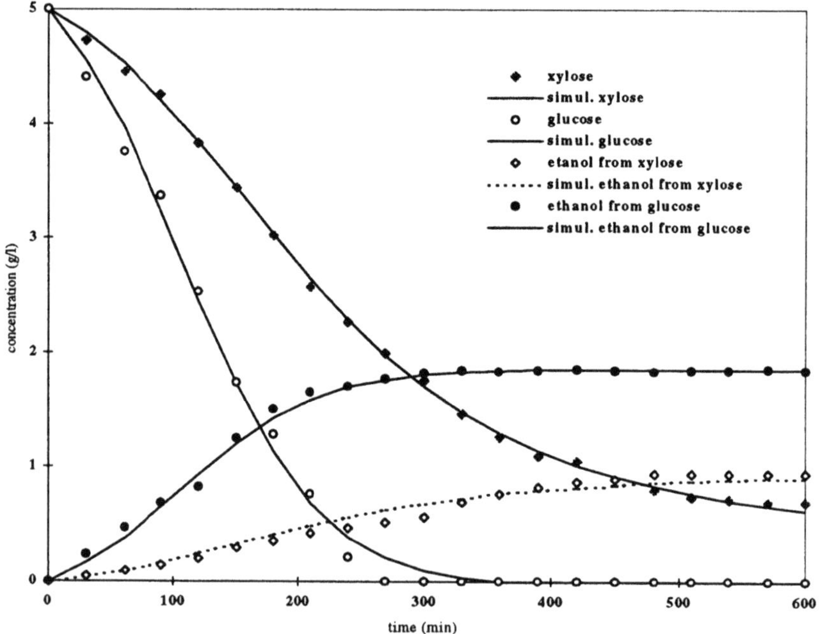

Fig. 13. Result of the fitting procedure. R^2=0.9975, RMSE=0.0689.

The validity of the model and the set of kinetic equations was checked against the experimental results. A non-linear least-square estimation procedure implemented in the MLAB program (using the Marquardt-Levemberg method) was used to calculate the kinetic constants k_{α_j}.

The results of the fitting procedure follow the experimental data very closely. Figure 13 shows the experimental and calculated curves together with the R^2 and the root-mean-square error (RMSE). The kinetic constants and their coefficients of variation (CV) are reported in Table 2.

Table 2. Values of the Kinetic Parameters and of the Coefficients of Variation (CV).

Kinetic Constants	Value	CV (%)
k_{1d}	1.19×10^{-3}	8.1
k_{1p}	2.49×10^{-4}	8.4
k_{1i}	1.92×10^{-3}	12.2
k_{1a}	2.15×10^{-3}	8.0
k_{2d}	1.23×10^{-2}	6.2
k_{2p}	4.55×10^{-3}	6.3
k_{2i}	1.25×10^{-2}	5.9
k_{2a}	1.26×10^{-2}	7.8

As shown by the model, the glucose and xylose metabolic pathways are dominated by two different kinetics. Glucose, the main nutrient, is metabolized by an equal distribution strategy among the cell population until the resource is totally consumed. Xylose, a supplementary energy and carbon source, becomes important when its concentration in the media justifies activation of the enzymatic systems necessary for its uptake and metabolization. These different sugar metabolization strategies result in the different kinetic equations required to describe sugar metabolism.

iii) Glucose yeast metabolism

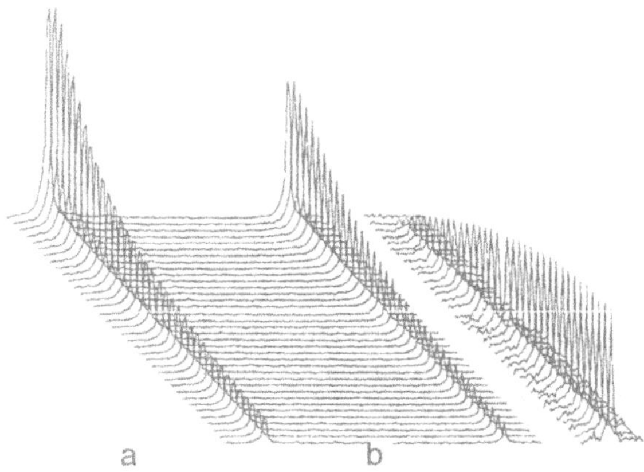

Fig. 14. ^{13}C-NMR spectra describing the glucose metabolization process (a) and ethanol production (b).

The metabolism of the yeast *Saccharomyces cerevisiae* was also investigated using the compartmental model [21,22].

Two samples containing 85 and 200 g/l of sugar substrate respectively, were analyzed using in vivo ^{13}C-NMR spectroscopy. In this range of concentrations, the cell culture behaved in a homogeneous manner.

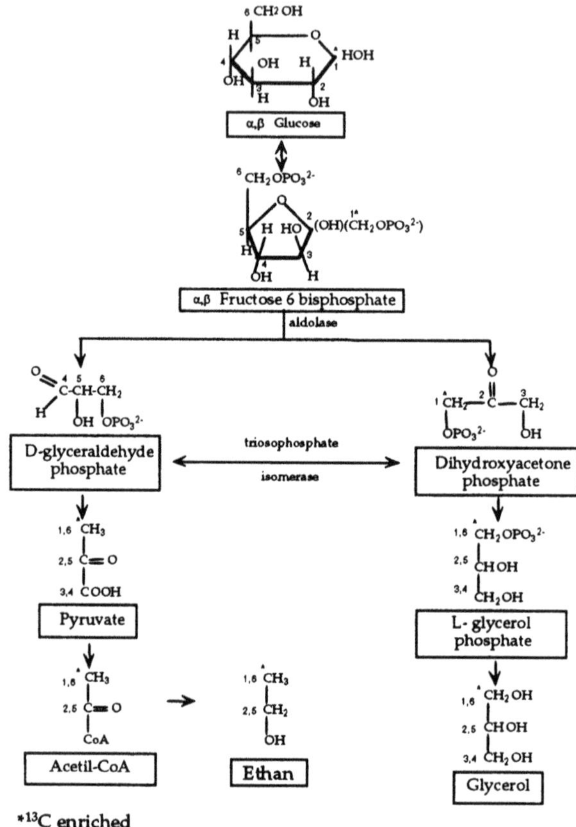

Fig. 15. Glucose metabolic pathway by *S. cerevisiae*.

Figure 14 shows the ^{13}C-NMR spectra which describe the sugar metabolization process (A) and ethanol production (B). It emerges that the [1-^{13}C] glucose labeled carbon is transferred to the methyl carbon of the ethanol as suggested by the analysis of the metabolic pathway (Figure 15).

The model we used to describe this process (Figure 16) consists of storages of: glucose, the substrate; ethanol, the main metabolic product; active cells, that transform glucose; "inhibited cells", the yeast cells inhibited by ethanol as its concentration increases. The model considers only ethanol production, neglecting glycerol (1-2 % the total end products), which we assumed not able to effect the metabolic process.

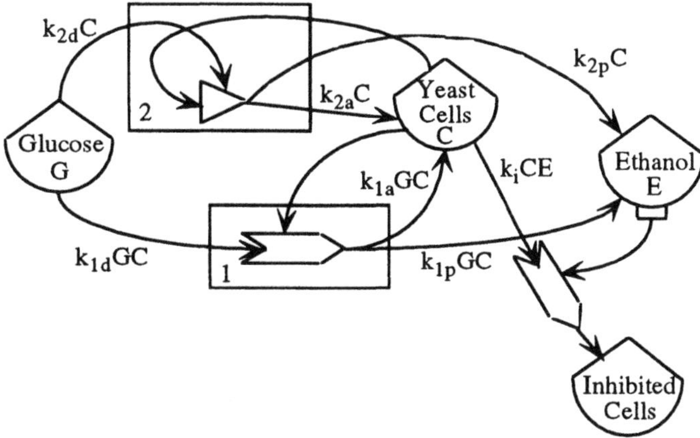

Fig. 16. Diagram of the model of the fermentation process.

Glucose consumption is described as the overlap of two kinetic processes: an autocatalytic one (box 1), and the other dependent only on the active cell concentration (box 2). Yeast activity and fermentation can be modelled as flows from the interactions between glucose and the active cells, meaning that part of the glucose is used by the yeast for the production of ethanol, and the remainder for feeding cell activity. Inhibition of the ability of yeast to convert glucose, due to the presence of ethanol, has been modelled as an outflow from the storage of active to the inhibited cells. This is proportional to the ethanol concentration which acts as controller (which is the meaning of the box under the storage symbol).

The differential equations derived from the model are:

$$\frac{d[G]}{dt} = -K_{1d}[G][C] \qquad [14]$$

$$\frac{d[C]}{dt} = K_{1a}[G][C] - K_i[C][E] \qquad [15]$$

$$\frac{d[E]}{dt} = K_{1p}[G][C] \qquad [16]$$

where G is glucose concentration (g/l), C is an index related to the metabolic efficiency of the active cells, and E is ethanol concentration (g/l). k_{1d}, k_{1a} and k_{1p} are respectively, in the autocatalytic pathway, the kinetic constants of the glucose degradation, of the activation of the yeast cells and of the ethanol production; k_{2d}, k_{2a} and k_{2p} have the same meanings in the pathway dependent only on the cell activity; k_i is the kinetic constant of the ethanol inhibition affecting cell activation. It is evident that the autocatalytic part is relatively more important at the beginning of the transformation process, in relation to the quantity of glucose in the system. The part that is proportional to the number of cells becomes dominant with respect to the other towards the end, meaning that part of the glucose is consumed (and thus part of the ethanol is produced) as a "required minimum" for the cells.

A nonlinear least-squares estimation technique as implemented in the MLAB computer program [9] (using the Marquardt-Levenberg method) was used to estimate the unknown parameters $k_{\alpha j}$ (j=1,2; α=p,a,i,d). The data sets of glucose and ethanol obtained in the two experiments were simultaneously used in the parameter estimation. The results of the fitting procedures follow the experimental data very closely. They are shown in Figure 17. Table 3 shows the optimal values of the estimated parameters and their standard deviations (SD). R^2 was 0.997.

The values of the parameters were determined utilizing only the glucose and ethanol data from experiments 1 and 2, since the true number of active cells and the real level of activation of the cells can not be measured. The model reflects the activity of the yeast which increases and then decreases when the effect of inhibition caused by ethanol becomes dominant.

The results reported in Table 3 are useful for interpreting the yeast biochemical process. It is important to note the null value of k_{2a}, meaning that no further cells are activated when the amount of glucose is close to zero and the minimum quantity of sugar available is used only for respiration. This is an important result of the model and it is not due to external constraints to the fitting procedure. Other parameters, e.g. k_{2d} and k_{2p} reported in Table 3, have precise biophysical meaning: the rate of consumption of glucose per cell and the rate of production of ethanol per cell respectively, when the amount of glucose available is close to zero. From Table 3 the requirement that $k_{jd} > (k_{ja} + k_{jp})$ is also verified.

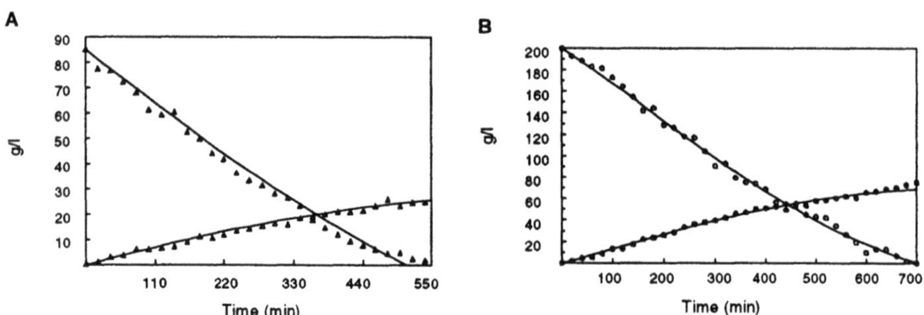

Fig. 17. Result of the simultaneous fit of data from experiment 1 (initial glucose concentration of 85 g/l)and 2 (initial glucose concentration of 200 g/l). Panel A; experiment 1: comparison of experimental data (ethanol=closed triangles; glucose=open triangles) and the results of fit (continuous lines). Panel B; experiment 2: comparison of experimental data (ethanol=closed circles; glucose=open circles) and the results of fit (continuos lines).

Table 3. Estimated values and coefficients of variation (CV) of the kinetics parameters

Kinetic Constants	Glucose 85-200 g/l	CV (%)
k_{1d}	3.992×10^{-4}	7
k_{2d}	4.253×10^{-2}	3
k_{1a}	1.284×10^{-5}	13
k_{2a}	0	-
K_i	1.471×10^{-5}	43
K_{1p}	2.058×10^{-4}	7
k_{2p}	9.019×10^{-3}	10

A close correlation between each component of the model and the biochemical function of the cells could be identified. In particular the constituents of box1 and box2 show the existence of "biological effectors", constituted by single proteins, enzyme systems, or bio-structures with a complex dynamics, to be identified.

Conclusions

The flexible structure of this modelling approach enables it to be applied in cases with different level of complexity. This may help to understand and identify functions, roles and mechanisms of the biological machinery. The constants of the model can constitute a useful tool for the comparison of microorganism performances, along the different pathways, thereby helping those scientists who try to improve the performances of selected strains of microorganisms.

Acknowledgments

This project was carried out with the support of National Research Council (CNR Contract n.98.00727PS13) and Consortium of Great Interphase Systems (CSGI).

References

1. Wüthrich K. (1989), *Science* 243, 45-47
2. Kleywegt G.J., Vuister G.W., Padilla A., Knegtel R.M.A., Boelens R. and Kaptein R. J. (1993), *Magn. Reson. Series B* 102, 166-176.
3. Ulianov N.B., Schmitz U., Kumar A. and James T.L. (1995), *Biophys. J.* 68, 13-21.

4. Bastianoni S., Gastaldelli A., Marchettini N., Renzoni D., Rossi C. and Ulgiati S. (1994), in Modelling and control in biomedical systems (B.W. Patterson ed.),IFAC, 356-357.
5. Magar M.E., (1972) Data Analysis in Biochemistry and Biophysics, Academic Press, New York.
6. Beechem J.M., (1992) in Methods in Enzimology, Vol. 210, (L. Brand and M.J. Johnson eds.), Academic Press, San Diego, 37-54.
7. Nielsen J. and Villadsen J. (1992), *Chem. Eng. Sci.*, 47, 4225-4270.
8. Shulman R.G., Brown T.R., Ugurbil K., Ogawa S., Cohen S.N. and den Hollander J.A. (1979), *Science* 205: 160-176.
9. Rossi C., Donati A., Medaglini D., Valassina M., Bastianoni S. and Cresta E. (1995), *Biomass and Bioenergy*, 8: 197-202.
10 Ørskov I., 1984. Genus V. Klebsiella Trevisan 1885. In: Bergey's Manual of Systematic Bacteriology (Krieg N.R. e Holt J.G. Eds.), Vol. 1, Williams and Wilkins, Baltimore, MD, pp.461-465.
11. Bunow B. and Knott G. (1992) MLAB, a mathematical modelling laboratory, Civilized Software Inc., Bethesda.
12. Jacquez J.A. (1996). Compartmental analysis in biology and medicine. BioMedware, Ann Arbor, Michigan, U.S.A
13. Odum H.T.(1972) in Systems analysis and simulation in ecology, vol.2. (B.C. Patten ed.) Academic Press, New York, 139-211.
14. Odum H.T. (1983) Systems Ecology. Wiley, New York.
15. Odum H.T., (1991) in Ecological Physical Chemistry, (C. Rossi and E. Tiezzi eds.) Elsevier Science Publisher, Amsterdam, 25-56.
16. Halvorson H.R. (1992) in Methods in Enzimology, (Brand L. and Johnson, M.J., Eds.), Vol. 210, 54-67, Academic Press, San Diego.
17. Bastianoni S., Gastaldelli A., Bonechi C., Mocenni C. and Rossi C. (1996), *Biochem. and Biophys. Res. Comm.*, 227, 41-46.
18. Rossi C., Porcelli M., Mocenni C., Marchettini N., Loiselle S. and Bastianoni S. (1998), *Ecological Modelling*, 113, 157-162.
19. den Hollander J.A., Brown T. R., Ugurbil K., and Shulman R. G., (1979) *Proc. Natl. Acad. Sci.*, 205, 6096-6100.
20. Shulman R. G., Brown T. R., Ugurbil K., Ogawa S., Cohen S. N., and den Hollander J. A., (1979) *Science*, 205, 160-176.
21. Bastianoni S., Donati A., Gastaldelli A., Marchettini N., Renzoni D., and Rossi C. (1996), *Biochem. and Biophys. Res. Comm.*, 227, 53-58.
22. Bastianoni S., Donati A., Gastaldelli, A., Marchettini N., Martini S., and Rossi C., (1999), *Chem. Phys. Lett.*, 310, 38-42.

Computer Simulation and Molecular Design of Model Liquid Crystals

Claudio Zannoni,

Dipartimento di Chimica Fisica ed Inorganica, Università di Bologna, Viale Risorgimento 4, I-40136 Bologna, Italy
vz3bod7a@sirio.cineca.it

Abstract. In this chapter we discuss some of the problems and peculiarities of the computer simulation of liquid crystals and we briefly summarize the state of the art in the field.

1. Models for mesogenic molecules

Liquid crystals are anisotropic fluids characterized by a long range orientational order and by a reduced (like in smectics) or altogether absent (like in nematics) positional order of the constituent molecules [1,2]. The properties of liquid crystals in certain thermodynamic conditions critically depend on their long-range molecular organisation and on the proximity of an order-disorder phase transition. Here we shall be concerned with molecular level models that can be used to simulate liquid crystals and their properties. The most natural model for molecules that yield liquid crystals (mesogenic molecules), like for any other molecule, is probably an atomistic one, where each atom or small group of atoms (e.g. a CH) is represented by a suitable attractive -repulsive centre. A suitable force field between bonded and non bonded centres is chosen [3] and the molecular representation obtained (cf. Fig.1) is very close to chemical intuition.

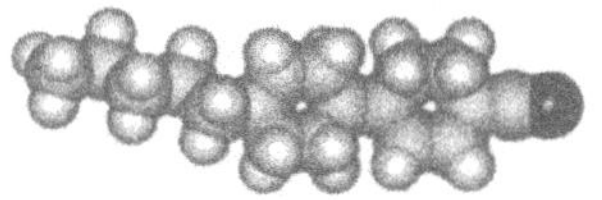

Fig. 1. An atomistic model of a typical mesogenic molecule: *trans*-4-(*trans*-4-n-pentylcyclohexyl) cyclohexylcarbonitrile (CCH5)

The simulation of the macroscopic properties and of the molecular organisation obtained for a system of N model molecules at a certain temperature and pressure

(T, P) typically proceeds through one of the two current mainstream methods of computational statistical mechanics: molecular dynamics (MD) or Monte Carlo (MC) [4,5]. MD sets up and solves step by step the equations of motions for all the particles in the system and calculates properties from the time trajectories obtained. MC calculates instead average properties for the system from equilibrium configurations generated with an algorithm designed to generate sets of positions and orientations of the N molecules with a frequency proportional to their Boltzmann factor.

Both methods, although quite different, proceed through repeated evaluations of the energy and thus of the intermolecular interactions in the sample (as well as of their derivatives to evaluate forces, at least in MD). Two examples of the equilibrium configurations obtained are shown in Fig.2, for a system of N=98 CCH5 molecules at conditions typical of the nematic and isotropic state [6].

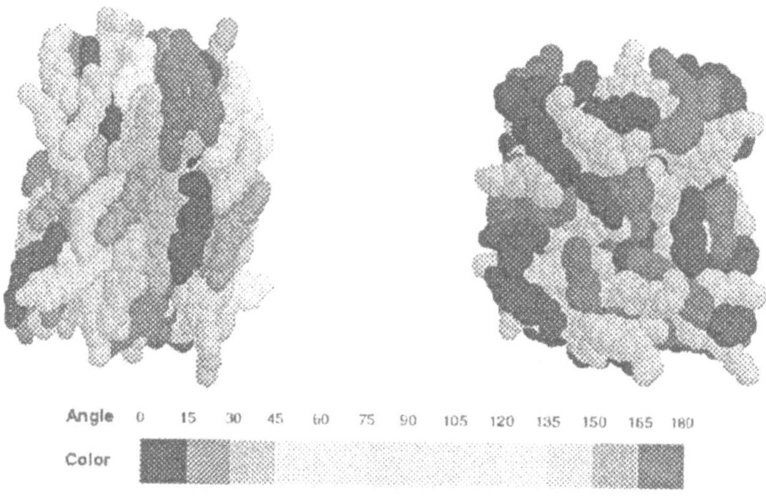

Fig. 2. A typical configuration of CCH5 molecules as obtained from molecular dynamics simulations in the isotropic phase at T= 380 K (right) and in the nematic at T=300 K (left). The grey tones indicate the molecular orientation according to the palette shown [6].

However, one peculiarity of nematic liquid crystals is that while their local structure is essentially that of an ordinary liquid, they also exhibit orientational transitions: from nematic to isotropic, where they lose their long-range orientational order or from nematic to smectic, where they form layered structures The observation of such phase transitions requires fairly large samples, of hundreds of preferably more than a thousand molecules.

The problem with the atomistic type models is thus simply with the number of interacting centres they contain that rapidly takes the simulation beyond the range of current capabilities (see, however for overviews of recent progresses the

contributions of Wilson, Glaser, Procacci in [5] and a review of Crain and Komolkin [7]).
One additional feature of liquid crystal simulations is that quite often we do not wish to investigate the detailed properties of a specific molecule, CCH5 say, that we already know to form a liquid crystal and whose structure we know. On the contrary we often have the problem of designing molecules that do not exist as yet! A typical problem could be the optimisation of the electric polarization properties of a liquid crystal formed of molecules containing a permanent dipole moment. The molecular design question could be posed as: what is the effect of changing the position, the orientation and the strength of the dipole on the structure of the liquid crystal phases obtained? In these cases it is essential to give up as many atomic details as possible and to consider lower resolution models where molecules are approximated with particles of simple shape. A simple choice often made is that of using purely repulsive models, e.g. hard spherocylinders or ellipsoids. This choice [8], which is consistent with the belief of many physicists [9] that it is only shape that determines the structure of a liquid and attractive for the relative simplicity of this type of models is more justified for ordered phases obtained with colloidal suspensions [10] then with the thermotropic materials we are interested with. Indeed in purely repulsive models [8] temperature plays no direct role, while the change from isotropic to nematic and then to smectic or crystal is temperature driven [1]. Moreover, although various ordered phases can be obtained, it is worth mentioning that no liquid - vapour transition is observed with purely repulsive models. We are thus particularly interested in considering simple molecular models, like the Gay-Berne potential [11] described in the following section, which are anisotropic but contain both an attractive and repulsive part.

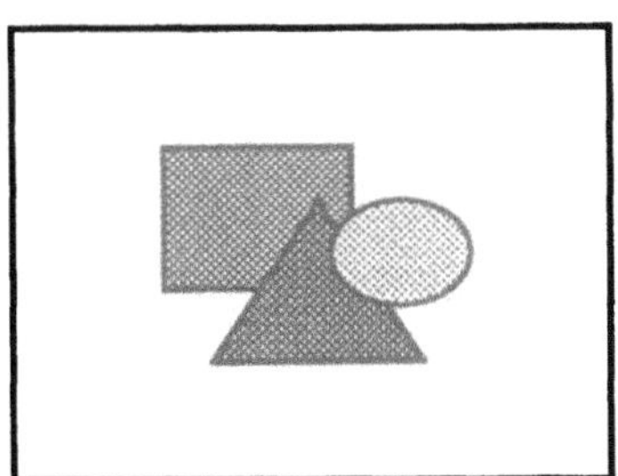

Fig. 3. A molecular level model of liquid crystals representing the molecules as ellipsoids

2. The Gay-Berne potential

The Gay-Berne (GB) potential in its uniaxial [11] and biaxial [12,13] version can be regarded as a generalized anisotropic and shifted version of the Lennard-Jones interaction commonly used for simple fluids [4], with attractive and repulsive contributions that decrease as inverse powers of distance:

$$U(r) = 4\varepsilon \left[\left(\frac{\sigma}{r}\right)^{12} - \left(\frac{\sigma}{r}\right)^{6} \right] \tag{1}$$

In the GB model the strength, ε, and the range, σ, parameters depend on the orientation vectors $\mathbf{u}_i$, $\mathbf{u}_j$ of the two particles and on their separation vector $\mathbf{r}$. Here we show the uniaxial version for simplicity:

$$U(\hat{\mathbf{u}}_i, \hat{\mathbf{u}}_j, \mathbf{r}) = 4\varepsilon_0 \varepsilon'^{\mu} \varepsilon^{\nu} \left[\left(\frac{\sigma_0}{r - \sigma(\hat{\mathbf{u}}_i, \hat{\mathbf{u}}_j, \hat{\mathbf{r}}) + \sigma_0} \right)^{12} - \left(\frac{\sigma_0}{r - \sigma(\hat{\mathbf{u}}_i, \hat{\mathbf{u}}_j, \hat{\mathbf{r}}) + \sigma_0} \right)^{6} \right] \tag{2}$$

where the cap indicates a unit vector, the anisotropic contact distance is

$$\sigma(\mathbf{u}_i, \mathbf{u}_j, \mathbf{r}) = \sigma_0 \left\{ 1 - \frac{\chi}{2} \left[\frac{(\mathbf{u}_i \cdot \mathbf{r} + \mathbf{u}_j \cdot \mathbf{r})^2}{1 + (\mathbf{u}_i \cdot \mathbf{u}_j)\chi} + \frac{(\mathbf{u}_i \cdot \mathbf{r} - \mathbf{u}_j \cdot \mathbf{r})^2}{1 - (\mathbf{u}_i \cdot \mathbf{u}_j)\chi} \right] \right\}^{-1/2} \tag{3}$$

and χ is an anisotropy parameter related to the length $\sigma_{\parallel}$ and the breadth $\sigma_{\perp}$ of the ellipsoid representing the molecule:

$$\chi = \frac{\sigma_{\parallel}^2 - \sigma_{\perp}^2}{\sigma_{\parallel}^2 + \sigma_{\perp}^2}. \tag{4}$$

Similarly the interaction anisotropy is the product of two terms:

$$\varepsilon'^{\mu}(\hat{\mathbf{u}}_i, \hat{\mathbf{u}}_j, \hat{\mathbf{r}}) = \left\{ 1 - \frac{\chi'}{2} \left[\frac{(\hat{\mathbf{u}}_i \cdot \hat{\mathbf{r}} + \hat{\mathbf{u}}_j \cdot \hat{\mathbf{r}})^2}{1 + (\hat{\mathbf{u}}_i \cdot \hat{\mathbf{u}}_j)\chi'} + \frac{(\hat{\mathbf{u}}_i \cdot \hat{\mathbf{r}} - \hat{\mathbf{u}}_j \cdot \hat{\mathbf{r}})^2}{1 - (\hat{\mathbf{u}}_i \cdot \hat{\mathbf{u}}_j)\chi'} \right] \right\}^{\mu} \tag{5}$$

$$\varepsilon^{\nu}(\hat{\mathbf{u}}_i, \hat{\mathbf{u}}_j) = \left[1 - (\hat{\mathbf{u}}_i \cdot \hat{\mathbf{u}}_j)^2 \chi^2 \right]^{-\nu/2} \tag{6}$$

where μ and ν, taken to be 2,1 in the original formulation [11,14-17], are parameters used to tune the shape of the potential and

$$\chi' = \frac{(\varepsilon_s / \varepsilon_e)^{1/\mu} - 1}{(\varepsilon_s / \varepsilon_e)^{1/\mu} + 1} \tag{7}$$

reflects the anisotropy in the potential well depths for the side-by-side and end- to-end configurations. Typical values used here for rod-like molecules (see Fig.1) are length to breadth $\sigma_\parallel / \sigma_\perp = 3$ and well depth anisotropy $\varepsilon_\perp / \varepsilon_\parallel = 5$. We also employ parameters $\mu=1$ and $\nu=3$ [18] that generate nematics with a wider temperature range than those in [11,13,14] and use σ_0, ε_0 as units of length and energy. Typical values for σ_0, ε_0 in real units could be $\sigma_0 = \sigma_\perp = 5$ Å, $\varepsilon_0/k = 100$K.

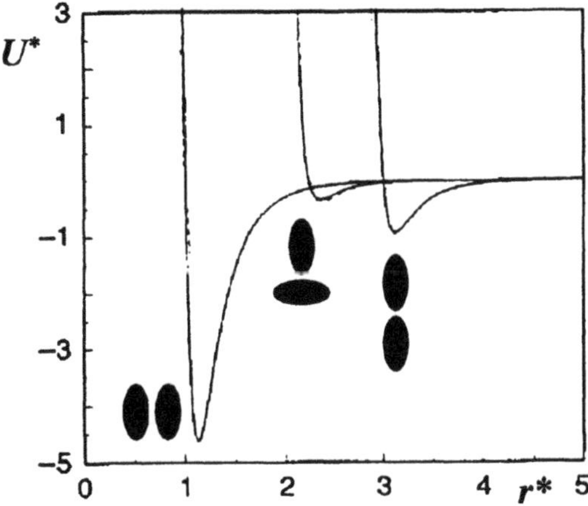

Fig. 4. Sections of the GB potential $U^* = U/\varepsilon_0$ with $\sigma_\parallel / \sigma_\perp = 3$, $\varepsilon_\perp / \varepsilon_\parallel = 5$ and energy parameters $\mu=1$ and $\nu=3$ [18] shown for side-by-side, tee and end-to-end orientations as a function of scaled distance $r^* = r/\sigma_0$.

The potential is strongly anisotropic and favours a side – side alignment, as we see from the few sections of the GB potential surface in Fig.4, but not to the point of necessarily giving a crystal when cooling from the isotropic phase. Although relatively simple to computer code the potential is already too complicated for a sufficiently accurate theoretical treatment and has to be studied using the numerical techniques of computer simulations. Indeed both Monte Carlo and Molecular Dynamics methods have been employed [5,13-18] to get the equilibrium phases generated by the GB potential under certain thermodynamic conditions and to construct at least in part its phase diagram.In Fig. 5 we show typical molecular organisations obtained for a system of $N=10^3$ GB particles using MC at a scaled density $\rho^*=0.3$ in the crystal, smectic B, nematic and isotropic phase [18]. We have grey scale coded the orientations of the molecules relative to the preferred direction of the liquid crystal (the director) as shown by the palette in Fig.5 to highlight orientational ordering.

We see that the GB phases have the expected characteristics: in the isotropic phase all orientations (and thus all shades of grey) are represented; the nematic shows a dominance of molecules parallel to the director (yellow here), but no positional order. Cooling further gives smectic layering and eventually crystal like three dimensional order.

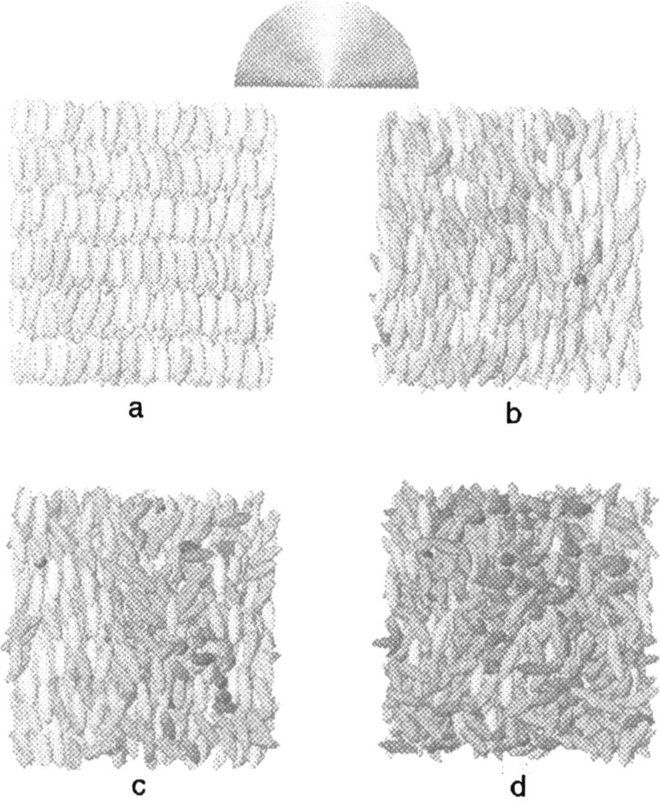

Fig. 5. Typical MC configurations of GB systems at four temperatures in the crystal (a), smectic B (b), nematic (c) and isotropic (d) phase. Details are given in [18].

Gay-Berne models can also be used to study the order of a liquid crystal close to a certain phase transition and to answer some basic questions, e.g. about the alignment at that interface.

In Fig.6 we show as an example the molecular organisation at the nematic-isotropic coexistence [19] that was obtained with a specially developed MD where the two halves of the cell containing N=12960 molecules are separately thermostated at temperatures slightly above and below the transition temperature. It is apparent that in this case molecules align parallel to the interface. Experimentally this is what happens, at least for some liquid crystals like n-(4 methoxy benzylidene)-4'-n butyl aniline (MBBA), although other types of alignment are also found for other materials.

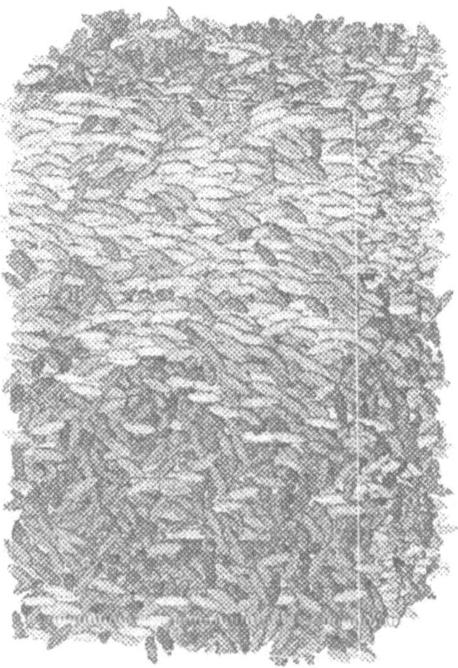

Fig. 6. A typical configuration for the GB model at the nematic-isotropic transition, showing that the molecules in the nematic phase coexisting with the isotropic one are aligned parallel to the interface[19].

The GB system has also been shown to give a nematic -vapour transition, at least when its parameters are changed to $\mu=1$ and $\nu=1$ [20]. Also in this case molecules at the phase interface tend to be aligned parallel to it. Experimentally various types of behaviour are found at a free interface: e.g. planar for 4,4'-dimethoxy azoxy benzene (PAA) and perpendicular in cyanobiphenyls (see refs. in [19, 20a,b]). A perpendicular alignment was also observed for a GB system with shorter particles with $\chi= 2$ and $\chi'=5$, $\mu=1$, $\nu=2$ [21]. Apart from the basic transition properties, a number of physical observables have been determined for GB systems, including translational and rotational correlations [16, 22], viscosity [23], elastic constants [24], thermal conductivity and diffusion coefficients [25], pretransitional properties [26]. The interaction of GB liquid crystals with surfaces to investigate the details of anchoring and structuring has been explored both for generic [27,28] and specific substrates like graphite [29]. Very large GB systems (over 80000 molecules) have also been recently studied to investigate some of the distinctive features of liquid crystals: topological defects [30,31], until now simulated only with lattice models [32].

In general the simple GB potential has proved able to yield the main liquid crystal phases and properties and thus to constitute an attractive reference potential for investigating trends of variation in the order and organisation of the nematic and smectic phases upon switching on of additional specific contributions. In the next section we shall see some examples of this approach.

Dipolar systems

We consider as an example the effect of changing molecular dipole position and orientation on the overall organisation of the dipoles in a liquid crystal phase [33-36]. This is not only of academic interest in view of the current efforts to obtain fluid ferroelectric liquid crystals, that would be of great technological importance and that, although not theoretically forbidden, have until now eluded the efforts of synthetic chemists [37]. Notice that a ferroelectric arrangement of dipoles would correspond to a somewhat non-intuitive overall polar phase, with dipoles pointing in the same direction.
The pair potential we consider is simply a sum of the previously seen Gay - Berne interaction and of a dipolar term:

$$U_{\mu\mu}(\hat{\mathbf{u}}_i,\hat{\mathbf{u}}_j,\mathbf{r}_d)=\frac{\mu_i^*\mu_j^*}{r_d^3}\left[\hat{\mathbf{u}}_i\cdot\hat{\mathbf{u}}_j-3(\hat{\mathbf{u}}_i\cdot\hat{\mathbf{r}}_d)(\hat{\mathbf{u}}_j\cdot\hat{\mathbf{r}}_d)\right] \tag{8}$$

where $\mathbf{r}_d$ is the vector joining the point dipoles μ_i and μ_j on the two molecules. In particular we consider [34] shifting an axial dipole from the centre to a position $d^*=d/\sigma=1$ towards the end of the molecule (see Fig.7). We consider a dipole strength $\mu^*=2$ in reduced units, which would correspond to about 2.4D in real units. While the preferred orientation of two dipoles at a certain distance and orientation can be easily guessed to be antiparallel, it is important to stress that the equilibrium organisation of a system of N polar molecules at a certain density and temperature cannot be reliably predicted without the use of computer simulations that can optimise the positions and orientations of all the N molecules at the same time.

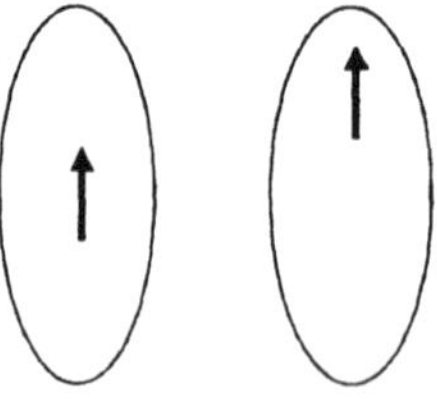

Fig. 7. A sketch of the central (left) and shifted (right) permanent dipole location in the two systems considered.

Thus if we start from the same density $\rho^{*}=0.30$ used in the previous section for the apolar GB system and we confine ourselves to the smectic phase, we find from MC simulations of N=1000 dipolar particles that the central dipole system behaves as we might have expected, with an essentially random distribution of up and down dipoles in each layer and with little interdigitation. On the other hand the simulation of the shifted dipole system gives the surprisingly very different dipole organisation shown in Fig. 8.

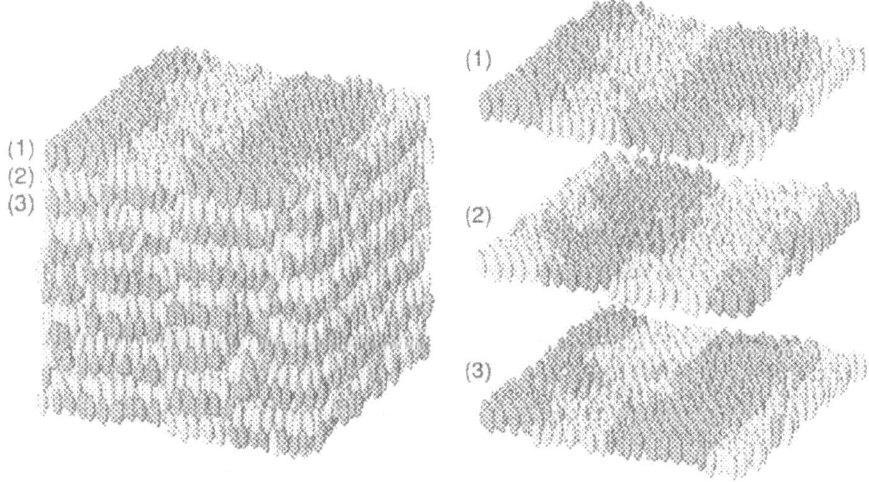

Fig. 8. The molecular organisation (left) for the N=8000 GB system with shifted axial dipoles, exhibiting the local polar domain pattern. The first three layers are exploded (right) to show how the domains are compensated anti - ferroelectrically by neighbouring layers. Dark and light grey indicate dipole up or down [34].

Here at short range the dipoles point in the same direction (same shade of grey here) and are compensated by the adjacent layers as shown in the shifted layers representation. This brings the dipoles of neighbouring layers in close contact, giving interdigitation and a large stabilising effect and the lowering of the energy. However, the organisation is not a fully bilayer one, but has a stripe domain structure. This self-organising ability is particularly striking and we have checked it in various ways and in particular simulating a much larger sample with N=8000 and verifying that the same type of structure (the one actually shown in Fig. 8) is obtained in both cases. It is worth noticing that even for these relatively simple systems the calculations are particularly demanding. Indeed, because of the long range nature of the dipolar interactions that are evaluated using the Ewald summation technique whose demand of computer time grows as $O(N^{3/2})$, we have found essential to use parallel computing techniques.

It is interesting to observe that these dipolar domain structures have been found experimentally by Levelut et al. [38] in rather complex liquid crystal mixtures.

The simple model above helps to single out a design feature that favours the domain formation. This is particularly interesting from the perspective of trying to optimise the position of the dipole towards the formation of a ferroelectric phase [37]. Changes of dipole strength also have a significant effect and in particular increasing the dipole strength gives rise to the strongly interdigitated partial bilayer phase [1] observed experimentally and called smectic A_d [39].
Although here we have only briefly recalled the case of axial dipoles, the effect of changing the orientation of the dipole from axial to transversal has also been studied and shown to produce interesting dipole chain structures in the plane perpendicular to the director [40], as also observed in hard spherocylinders with a central transverse dipole [41].

Discotic systems

The essential requirement for the formation of liquid crystal phases is that of having non spherical molecular shapes. Although historically anisotropic phases have been obtained from elongated molecules, in the last twenty years flat, discotic, molecules have been shown to yield interesting nematic and columnar organisations and these discotics [1] form one of the most rapidly growing and important family of liquid crystals [42]. The typical structure of a discotic is that of a flat aromatic core, e.g. triphenylenes, truxenes, superyines and many other moieties have been employed, with a certain number of chains attached [1,42]. Another interesting possibility is that offered by metallorganic compounds, where a suitable transition metal helps in organising a set of ligands with an appropriate nearly planar geometry [43,44].

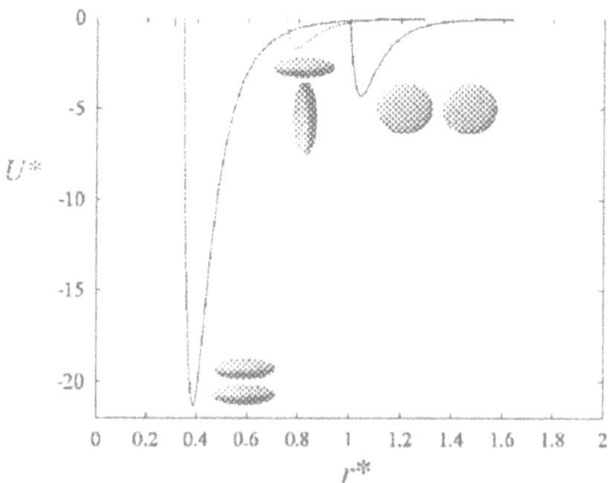

Fig. 9. Gay-Berne potential for discotic molecules with $\sigma_{\parallel}/\sigma_{\perp} = 0.345$, $\varepsilon_{\parallel}/\varepsilon_{\perp} = 5$ and energy parameters $\mu = 1$ and $\nu = 3$.

Here we consider a simple case of this kind, that of discs with an axial permanent dipole in the centre to examine their overall molecular and dipolar organisation. Structures resembling this exist for instance in polar discotic metallomesogens based on vanadyl 1,3-diketonate complexes [44] that have a large dipole moment normal to the plane of the molecule and have been found to give columnar phases. These systems are of particular interest as candidate for uniaxial ferroelectric phases, as an hexagonal structure of polar columns is expected to be globally ferroelectric, since the symmetry of the column lattice would not allow a cancellation of the dipoles of each stack of discs.

Despite the growing importance of discotic systems the number of computer simulations has been relatively small. Here we shall model the discotic mesogen using again a Gay Berne attractive-repulsive potential with an added dipole. We shall then employ Monte Carlo (MC) simulations to obtain the resulting equilibrium molecular organisation at a few selected temperatures.

Fig. 10. A GB discotic liquid crystal with axial dipole in the columnar phase (at $T^* = 2.0$). The light and dark grey spots are used to distinguish head from tail of the molecular dipoles[46].

It is worth pointing out that although the GB systems we have discussed until now were made of elongated particles, the GB model can actually be used equally

successfully for discotic particles, just utilizing oblate rather than prolate ellipsoid parameters [45]. The example considered here utilises a parameterisation related to that used by Emerson *et al.* in [45] and originally based on the dimensions of a triphenylene core, namely: shape and well depth anisotropy $\sigma_{\parallel}/\sigma_{\perp}= 0.345$ and $\varepsilon_{\parallel}/\varepsilon_{\perp} = 5$, but using instead energy parameters $\mu = 1$ and $\nu = 3$ [46]. We have studied a system of $N = 10^3$ discotic particles with central axial dipole using Monte Carlo computer simulations and we have shown that it gives isotropic, nematic and hexagonal columnar liquid crystal phases (see Fig.10), differently from the parameterisation [45] that yielded a tetragonal arrangement of columns. The system is then similar to the one studied experimentally, which was however not really ferroelectric. We have determined the overall polarization in our system and found that it is also not ferroelectric. Although this is somewhat disappointing we can however investigate why. We have thus also determined the molecular and dipolar distributions and the length of the polar domains in the columns [46]. At low temperature each column contains aligned dipolar domains but, as we can also see from Fig.10, we did not find fully polarized columns.
The typical length of each polar domain is less than ten molecules and the lack of consistent polarization of each column seems to be at the origin of the problem of lack of global phase polarization.

Conclusions

The molecular modelling problem that has often to be tackled in liquid crystals is that of designing molecules that have not yet been synthesised and that are able to yield mesophases with specific properties of interest for applications (such as ferroelectricity) rather than that of calculating in detail the properties of already known molecules. The simulation of model systems based on simple, molecular level, rather than atomistic, intermolecular potentials allows the identification of some of the physical features (e.g. molecular shape and attractive interaction anisotropy, biaxiality, electrostatic moments etc.) responsible of a certain collective behaviour, providing useful guidelines for the design of novel mesogenic molecules.
In this chapter we have presented a brief overview of recent results for systems of particles interacting with model potentials based on the Gay-Berne (GB) molecular level interaction that has shown to be a useful tool for studying nematic, smectic and columnar liquid crystals and we have presented and summarised the results of various examples of dipolar systems.
More complex molecular structures can be simulated by suitable combination of various ellipsoidal Gay-Berne and spherical Lennard-Jones particles e.g. to attempt modelling asymmetric molecules [47,48] or to include flexible chains [49]. Equally well other specific contributions, e.g. modelling quadrupolar [50] or hydrogen bond interactions could be added to the basic GB system [51]. With the continuous growth in computer resources it seems very likely that the level of detail and the

feasibility of large scale molecular modelling will correspondingly continue to increase and to provide a useful tool for understanding complex behaviour in liquid crystals in terms of molecular features.

Acknowledgements

I should like to thank University of Bologna, MURST (PRIN *Cristalli Liquidi*), CNR PF MSTA II and NEDO (Japan) for financial support to our work and R. Berardi, S. Boschi, A. Emerson, L. Muccioli, S. Orlandi, A. Porreca, P. Pasini for their contributions to the work described here.

References

1 S. Chandrasekhar, *Liquid Crystals*, 2nd ed., Cambridge U.P., **1992**.
2 G.R. Luckhurst, G.W. Gray eds., *The Molecular Physics of Liquid Crystals*, Academic Press, **1979**.
3 A.R. Leach, *Molecular Modelling Principles and Applications*, Longmans, **1996**.
4 M.P. Allen, D.J. Tildesley, *Computer Simulation of Liquids*, Clarendon Press, **1987**.
5 P. Pasini, C. Zannoni, eds., *Advances in the computer simulations of liquid crystals*, Kluwer, **2000**.
6 R. Berardi, L. Muccioli, C. Zannoni, *work in progress*, **1999**.
7 J. Crain, A.V. Komolkin, *Adv. Chem. Phys.*, **1999**, *109*, *39*.
8 M.P. Allen, G.T. Evans , D. Frenkel, B.M. Mulder, *Adv. Chem. Phys.*, **1993**, *86*, 1.
9 H.C. Andersen, D. Chandler, J.D. Weeks, , *Adv. Chem. Phys.*, **1976**, *34*, 105.
10 S. Fraden, in *Observation, prediction and simulation of phase transitions in complex fluids*, eds. M. Baus, L.F. Rull, J.P. Ryckaert, Kluwer, **1995**.
11 J.G. Gay, B.J. Berne, *J. Chem. Phys.*, **1981**, *74*, 3316.
12 R. Berardi, C. Fava, C. Zannoni, *Chem. Phys. Lett.*, **1995**, *236*, 462; *ibid*,**1998**, *297*, 8.
13 D.J. Cleaver, C.M.Care, M.P. Allen, M.P. Neal, *Phys. Rev. E*, **1996**, *53*, 1.
14 G.R. Luckhurst, R.A. Stephens, R.W. Phippen, *Liq. Cryst.*, **1990**, *8*, 451.
15 (a)M.K. Chalam, K.E. Gubbins, E. de Miguel, L.F. Rull, *Molec. Simul.*, **1991**, *7*, 357; (b) E. de Miguel, L. F. Rull, M. K. Chalam, K. E. Gubbins, *Mol. Phys.*, **1991**, *74*, 405.
16 L.F. Rull, *Physica A*, **1995**, 220, 113.
17 J.T. Brown, M.P. Allen, E. Martín del Río, E.de Miguel, *Phys. Rev. E*, **1998**, *57*, 6685.
18 R. Berardi, A.P.J. Emerson, C. Zannoni, *J. Chem. Soc. Faraday Trans.*, **1993**, *89*, 4069.
19 M. Bates, C. Zannoni, *Chem. Phys. Lett.*, **1977**, *280*, 40.
20 (a) A.P.J. Emerson, S. Faetti, C. Zannoni, *Chem. Phys. Lett.*, **1997**, *271*, 241; (b) E. de Miguel, E.M. del Rio, *Phys. Rev.E*, **1977**, *55*, 2916.
21 S.J. Mills, C.M. Care, M.P. Neal, D.J. Cleaver, *Phys. Rev. E*, **1998**,*58*,
22 E. de Miguel, L. F. Rull, and K. E. Gubbins, Phys. Rev. A 45, 3813 (1992)
23 A. M. Smondyrev, G. B. Loriot, R.A. Pelcovits, *Phys. Rev. Lett.*, **1995**, *75*, 2340.
24 J. Stelzer , L. Longa, H.-R. Trebin, *J. Chem. Phys.*, **1995**, 103, 3098; **1997**, *107*, 1295E.
25 S. Sarman, D. J. Evans, *J. Chem. Phys.*, **1993**,*99*, 620.
26 M. P. Allen, M.A. Warren, *Phys. Rev. Lett.*, **1997**, *78*, 1291.

27 J. Stelzer, P. Galatola, G. Barbero, L. Longa, *Phys. Rev. E*, **1997**, *55*, 477; J. Stelzer, L. Longa, H.-R. Trebin, *Phys. Rev. E*, **1997**, *55*, 7085 .
28 T. Gruhn, M. Schoen, *Mol. Phys.*, **1998**, *93*, 681.
29 V. Palermo, F. Biscarini, C. Zannoni, *Phys. Rev. E*, **1998**, *57*, 2519.
30 M.P. Allen, M.A. Warren, M.R. Wilson, Phys. Rev. E, **1998**, *57*, 5585.
31 J.L. Billeter, A.M. Smondyrev, G.B. Loriot, R.A. Pelcovits, *Phys. Rev. E*, **1999**,*60*, 6831
32 C. Chiccoli, O.D. Lavrentovich, P. Pasini, C. Zannoni, *Phys. Rev. Lett.*, **1997**,*79*, 4401.
33 K. Satoh, S. Mita, S. Kondo, *Chem. Phys. Lett.*, **1996**, *255*, 99.
34 R. Berardi, S. Orlandi, C. Zannoni, *Chem. Phys. Lett.*, **1996**, *261*, 357.
35 E. Gwozdz, A. Brodka, K. Pasterny, *Chem. Phys. Lett.*, **1997**, *267*, 557.
36 M.Houssa, A. Oualid, L.F. Rull, *Mol. Phys.*, **1998**, *94*, 439.
37 L.M. Blinov, *Liq. Cryst.*, **1998**, *24*, 143.
38 A.M. Levelut, R.J. Tarento, F. Hardouin, M.F. Achard, G. Sigaud, *Phys. Rev. A*, **1981**, *24*, 2180.
39 R. Berardi, S. Orlandi, C. Zannoni, *to be published.*
40 R. Berardi, S. Orlandi, C. Zannoni, *Int. J. Mod. Phys. C*, **1999**, *10* ,477.
41 A. Gil-Vilegas, S. McGrother, G. Jackson, *Chem. Phys. Lett.*, **1997**, *267*, 557.
42 D. Guillon, *Structure and Bonding*, **1999**, *95*, 41 .
43 S.A. Hudson , P.M. Maitlis , *Chem. Rev.*, **1993**, *93*, 861.
44 B. Xu, T.M. Swager, *J. Am. Chem. Soc.*, **1993**, *115*, 8879.
45 A. P. J. Emerson, G. R. Luckhurst, S. G. Whatling, *Mol. Phys.*, **1994**, *82*, 113.
46 R. Berardi, S. Orlandi, C. Zannoni, *J. Chem. Soc. Faraday Trans.*, **1997**, *93*, 1493.
47 J. Stelzer, R. Berardi, C. Zannoni, *Chem. Phys. Lett.*, **1999**, *299*, 9.
48 M.P. Neal, A.J. Parker, C.M. Care, *Mol. Phys.*, **1997**, *91*,603.
49 M. Wilson, *J. Chem. Phys.*, **1997**, *107*, 8654.
50 M.A. Bates, G.R. Luckhurst, *Liq. Cryst.*, **1998**, *24*, 229.
51 R. Berardi, M. Fehervari, C. Zannoni, *Mol. Phys*, **1999**,*97*, 1173.

Contributors

GPSR Compliance
The European Union's (EU) General Product Safety Regulation (GPSR) is a set of rules that requires consumer products to be safe and our obligations to ensure this.

If you have any concerns about our products, you can contact us on

ProductSafety@springernature.com

In case Publisher is established outside the EU, the EU authorized representative is:

Springer Nature Customer Service Center GmbH
Europaplatz 3
69115 Heidelberg, Germany

www.ingramcontent.com/pod-product-compliance
Ingram Content Group UK Ltd.
Pitfield, Milton Keynes, MK11 3LW, UK
UKHW021835190726
13853UKWH00003B/1303

* 9 7 8 3 6 6 2 0 4 1 5 5 0 *